Criminological Theory

A LIFE-COURSE APPROACH

EDITED BY:

Matt DeLisi, PhD
Iowa State University
Department of Sociology

Kevin M. Beaver, PhD
Florida State University
College of Criminology
and Criminal Justice

JONES & BARTLETT
LEARNING

World Headquarters
Jones & Bartlett Learning
5 Wall Street
Burlington, MA 01803
978-443-5000
info@jblearning.com
www.jblearning.com

Jones & Bartlett Learning books and products are available through most bookstores and online booksellers. To contact Jones & Bartlett Learning directly, call 800-832-0034, fax 978-443-8000, or visit our website, www.jblearning.com.

Substantial discounts on bulk quantities of Jones & Bartlett Learning publications are available to corporations, professional associations, and other qualified organizations. For details and specific discount information, contact the special sales department at Jones & Bartlett Learning via the above contact information or send an email to specialsales@jblearning.com.

Production Credits
Publisher: Cathleen Sether
Acquisitions Editor: Sean Connelly
Editorial Assistant: Caitlin Murphy
Editorial Assistant: Audrey Schwinn
Production Assistant: Leia Poritz
Marketing Manager: Lindsay White
Rights & Photo Research Associate: Lauren Miller
Manufacturing and Inventory Control Supervisor: Amy Bacus
Composition: Cenveo Publisher Services
Cover Design: Michael O'Donnell
Cover Image: © Leigh Prather/ShutterStock, Inc.
Printing and Binding: Edwards Brothers Malloy
Cover Printing: Edwards Brothers Malloy

Library of Congress Cataloging-in-Publication Data
Criminology theory : a life-course approach / [edited by] Matt DeLisi and Kevin M. Beaver.—2nd ed.
     p.   cm.
 Includes bibliographical references and index.
 ISBN 978-1-4496-8151-7 (pbk.)—ISBN 1-4496-8151-4 (pbk.)
 1. Criminology. 2. Crime. I. DeLisi, Matt. II. Beaver, Kevin M.
 HV6018.C732 2014
 364.01—dc23
                          2012035687

6048

Printed in the United States of America
17 16 15 14 13   10 9 8 7 6 5 4 3 2 1

Dedication

To James Q. Wilson: role model, gentleman, scholar.

—MD

To my children: Brooke, Jackson, and Belle; the center of my world and the inspiration for everything I do.

—KMB

Contents

PREFACE   xiii

LIST OF CONTRIBUTORS   xv

REVIEWERS   xvii

Part I	Aggression (Prenatal and Childhood)

CHAPTER 1   **Biosocial Bases of Antisocial Behavior**   3

*Yaling Yang, Yu Gao, Andrea Glenn, Melissa Peskin, Robert A. Schug,
and Adrian Raine*

Key Terms   3
Introduction   3
Empirical Findings on Antisocial Behavior   4
Biosocial Model of Antisocial Behavior   15
Conclusion   17
Glossary   18
Notes   18

CHAPTER 2   **Prenatal and Perinatal Predictors of Antisocial Behavior:
Review of Research and Interventions**   27

Stephen G. Tibbetts

Key Terms   27
Introduction   27
History of Research on Pre- and Perinatal Factors in
        Criminological Literature   27
Research on Pre- and Perinatal Biological Factors of Criminality   29
Research on Perinatal Sociological and Environmental
        Factors of Criminality   35
Intervention Programs and Policy   37
Conclusion   40
Glossary   40
Notes   40

CHAPTER 10   Media Violence and the Development of Aggressive Behavior   149
Edward L. Swing and Craig A. Anderson

Key Terms	149
Introduction	149
Aggression and Antisocial Behavior	150
Past Media Violence Effects Research	151
Aggression Theory and Media Violence	158
General Aggression Model	159
Conclusion	163
Glossary	163
Notes	164

CHAPTER 11   Substance Use Careers and Antisocial Behavior: A Biosocial Life-Course Perspective   167
Michael G. Vaughn and Brian E. Perron

Key Terms	167
Introduction	167
Substance Careers and Biosocial Life-Course Theory	169
Conclusion	173
Glossary	174
Notes	174

CHAPTER 12   Developmental Trajectories of Exposure to Violence   177
Daniel J. Flannery, Manfred H. M. van Dulmen, and Andrea D. Mata

Key Terms	177
Introduction	177
Correlates and Antecedents of Exposure to Violence	177
Group-Based Modeling of Middle Childhood Exposure to Violence: Previous Findings and Empirical Illustration	179
Discussion	183
Conclusion	184
Glossary	185
Notes	185

CHAPTER 13   A Partial Test of Social Structure Social Learning: Neighborhood Disadvantage, Differential Association with Delinquent Peers, and Delinquency   187
Chris L. Gibson, Traci B. Poles, and Ronald L. Akers

Key Terms	187
Introduction	187
Social Disorganization, Crime, and Delinquency	188
Neighborhoods and Child Development	190
Social Structure/Social Learning: An Expanded and Complementary Reason for the Link Between Neighborhood Structure and Delinquent Behavior	190

Current Focus   192
Methods   192
Results   195
Discussion   197
Conclusion   199
Glossary   199
Notes   200

CHAPTER 14   **Timing Is Everything: Gangs, Gang Violence, and the Life Course**   **201**
             Scott H. Decker and David Pyrooz

Key Terms   201
Introduction   201
Definition   202
Gangs, Violence, and the Life Course   202
Developmental and Life-Course Theory   206
Gang Desistance: Leaving the Gangs   207
Conclusion   210
Glossary   211
Notes   211

CHAPTER 15   **Gangs and Antisocial Behavior: A Critique and Reformulation**   **215**
             Matt DeLisi

Key Terms   215
Introduction   215
Critiques of the Gang Concept   217
Reformulation of the Gang Concept   219
Conclusion   225
Glossary   226
Notes   226

Part III   Crime (Adulthood)

CHAPTER 16   **Developmental and Life-Course Criminology: Theories
             and Policy Implications**   **233**
             David P. Farrington

Key Terms   233
Introduction   233
Two Important Issues   235
Three DLC Theories   237
ICAP Theory   240
Policy Implications   244
Conclusion   245
Glossary   246
Notes   246

CHAPTER 17 **Self-Control Theory and Antisocial Behavior**   249
George E. Higgins and Margaret Mahoney

Key Terms   249
Introduction   249
Self-Control Theory   250
Review of Self-Control Theory Literature   253
Conclusion   257
Glossary   257
Notes   258

CHAPTER 18 **Serial Crime: Psychology of Behavioral Consistency
and Applications to Linking**   261
C. Gabrielle Salfati

Key Terms   261
Introduction   261
Behavioral Consistency and Individual Differentiation   262
Contextual and Situational Influences on Behavior   263
Signatures Versus Psychological Themes   265
Legal Versus Psychological Definitions of Behaviors and Crimes   267
Conclusion   268
Glossary   268
Notes   268

CHAPTER 19 **Symbolic Interactionism and Crime in the Life Course**   271
Jeffery T. Ulmer

Key Terms   271
Introduction   271
What Is Symbolic Interactionism?   272
Symbolic Interactionism's Relationship to Criminological Theories   275
Labeling Theory   278
Interactionist Approaches to Criminal Careers   278
Conclusion   282
Glossary   283
Notes   283

CHAPTER 20 **A "Good Lives" Approach to Rehabilitation**   285
Edward Manier, Truce Ordoña, and C. Robert Cloninger

Key Terms   285
Introduction   285
Life-Course Development of Antisocial Potential   286
Psychological Effects of Incarceration   287

A Program for Decreasing Antisocial Potential and Increasing
   Capacity for Wellbeing Behind Bars   288
Personality: Temperament and Character   288
Applying Temperament and Character to Rehabilitation   291
Conclusion   295
Glossary   295
Notes   295

CHAPTER 21 **Never-Desisters: A Descriptive Study of the**
   Life-Course-Persistent Offender   **297**

Matt DeLisi, Anna E. Kosloski, Alan J. Drury, Michael G. Vaughn,
Kevin M. Beaver, Chad R. Trulson, and John Paul Wright

Key Terms   297
Introduction   297
Life-Course Desisters   298
Current Focus   299
Methodology   299
Results   300
Theoretical Discussion   307
Conclusion   308
Glossary   308
Notes   309

CHAPTER 22 **Evolutionary Psychological Perspectives on Men's**
   Partner-Directed Violence   **311**

Farnaz Kaighobadi and Todd K. Shackelford

Key Terms   311
Introduction   311
Paternity Uncertainty and Male Sexual Jealousy   312
Male Sexual Jealousy and Mate Retention Behaviors   312
Risk of Sperm Competition and Sexual Coercion   313
Intimate Partner Homicide   315
Conclusion   316
Glossary   316
Notes   317

INDEX   **321** scholar in the area of biosocial criminology. Part Two, "Delinquency," spans adolescence and early adulthood, covering an array of subject matter relating to delinquency and violence, as well as a diverse collection of theoretical perspectives that seek to explain these phenomena. This section also contains three new chapters written by leading scholars such as Robert Agnew, Stephen Baron, and Michael Leiber. Part Three, "Crime," includes theory, research, and policies that address continued offending among adults.

Overall, the second edition of *Criminological Theory: A Life-Course Approach* contains 22 chapters written by scholars whose training and expertise span the social, behavioral, and medical sciences. The contributors are eminent scholars, and we genuinely appreciate their exciting, interesting, and straightforward chapters that together provide a global look at criminological theory over the life course.

List of Contributors

Robert Agnew, PhD
Emory University

Ronald L. Akers, PhD
University of Florida

Craig A. Anderson, PhD
Iowa State University

J. C. Barnes, PhD
University of Texas, Dallas

Stephen W. Baron, PhD
Queen's University

Kevin M. Beaver, PhD
Florida State University

Brian B. Boutwell, PhD
Sam Houston State University

H. Harrington Cleveland, PhD
Pennsylvania State University

C. Robert Cloninger, MD
Washington University in Saint Louis

Scott H. Decker, PhD
Arizona State University

Matt DeLisi, PhD
Iowa State University

Alan J. Drury, PhD Candidate
Iowa State University

David P. Farrington, PhD
Cambridge University

Daniel J. Flannery, PhD
Kent State University

Yu Gao, PhD
University of Pennsylvania

Chris L. Gibson, PhD
University of Florida

Andrea Glenn, PhD
University of Pennsylvania

George E. Higgins, PhD
University of Louisville

Farnaz Kaighobadi, PhD
Florida Atlantic University

Anna E. Kosloski, PhD
University of Colorado, Colorado Springs

Michael J. Leiber, PhD
University of South Florida

Margaret Mahoney, PhD
University of Louisville

Edward Manier, PhD
University of Notre Dame

Andrea D. Mata, PhD
Kent State University

Truce Ordoña, MD
Private Medical Practice

Jennifer H. Peck, PhD Candidate
University of South Florida

Brian E. Perron, PhD
University of Michigan

Melissa Peskin, PhD Candidate
University of Pennsylvania

Traci B. Poles, BS
University of Florida

David Pyrooz, PhD
Sam Houston State University

Adrian Raine, PhD
University of Pennsylvania

C. Gabrielle Salfati, PhD
John Jay College of Criminal Justice

Robert A. Schug, PhD
University of Pennsylvania

Todd K. Shackelford, PhD
Florida Atlantic University

Edward L. Swing, PhD
Iowa State University

Stephen G. Tibbetts, PhD
California State University, San Bernardino

Chad R. Trulson, PhD
University of North Texas

Jeffery T. Ulmer, PhD
Pennsylvania State University

Manfred H. M. van Dulmen, PhD
Kent State University

Michael G. Vaughn, PhD
Saint Louis University

Anthony Walsh, PhD
Boise State University

Richard Wiebe, PhD
Fitchburg State University

John Paul Wright, PhD
University of Cincinnati

Yaling Yang, PhD
University of California, Los Angeles

Ilhong Yun, PhD
Boise State University

Reviewers

We would like to acknowledge the following individuals for reviewing the text:

Cliff Akiyama, *Philadelphia College of Osteopathic Medicine*
Geraldine M. Hendrix-Sloan, *Minnesota State University Moorhead*
Susan S. Hodge, *University of North Carolina at Charlotte*
Christine S. Janis, *Northern Illinois University*
Cheng-Hsien Lin, *Lamar University*
Hung-En Sung, *John Jay College of Criminal Justice*

CHAPTER ONE

Biosocial Bases of Antisocial Behavior

Yaling Yang, Yu Gao, Andrea Glenn, Melissa Peskin, Robert A. Schug, and Adrian Raine

KEY TERMS

Antisocial behavior
Biosocial interactions
Brain imaging
Executive functioning
Meta-analysis

INTRODUCTION

Antisocial behavior has long been a topic of interest among researchers in the fields of neuroscience, psychology, criminology, and sociology, who for decades have attempted to uncover the biological and social bases of this complex behavioral problem. Independently, several biological and social risk factors have been identified to predispose one to antisocial behavior. Biologically, factors such as autonomic underarousal, obstetrical factors, brain deficits, and neuropsychological impairments have been strongly linked to antisocial behavior. Numerous social and environmental factors have also been associated with antisocial behavior, including low socioeconomic status, peer influence, physical abuse, and parental rejection. Despite the establishment of these risk factors, the level of knowledge has been relatively limited to independent effects of either social or biological factors, with very little understanding of the interactions between these two factors. Indeed, across the literature it is noticeable that psychosocial researchers rarely use methods to measure the biological variables in their antisocial samples, whereas biological researchers often use social factors as covariate variables instead of moderators. However, due to the multidimensional nature of antisocial behavior, a multidisciplinary approach may not only help, but is also critically needed to further the understanding of the underlying mechanisms of antisocial behavior.

Nonetheless, many challenges make the approach to address both the independent and interaction effects of biological and social factors extremely difficult. One of the main challenges is that biological risk factors tend to correlate significantly with social factors, resulting in the lack of statistical power to detect the interaction effects. This often leads to the false-negative conclusions in studies. However, several recent innovative studies that are explored in this chapter include large twin and adoption samples, which managed to significantly increase their statistical power. Many have successfully detected the interaction effects of social and biological factors on the outcome of antisocial behavior. Findings from these studies are mainly in line with previous hypotheses or models of the biosocial bases of antisocial behavior and suggest that **biosocial interactions** may indeed contribute to the development of antisocial behavior.

In this chapter, the empirical findings on the known biological risk factors associated with antisocial behavior are reviewed first, with a focus placed in the research areas of psychophysiology, obstetrical factors, **brain imaging**, neuropsychology, neurology, hormones, neurotransmitters, and environmental toxins. Next, further discussions are conducted to review the evidence of biosocial interaction effects in relation to antisocial behavior in each of the key research areas. Finally, prior biosocial models of antisocial behavior are revisited and a model, extended from Raine's biosocial model of violence, is hypothesized for antisocial behavior.[1] It is worth mentioning that the term **antisocial behavior** is used throughout this chapter as an umbrella term for various behavioral problems, including violent, psychopathic, delinquent, and criminal behavior. We acknowledge there may be different underlying biosocial pathology among those behaviors; however, the approach of combining empirical data from studies on various antisocial-related behavior was used in the hope of providing a comprehensive review on the biosocial bases of antisocial behavior.

EMPIRICAL FINDINGS ON ANTISOCIAL BEHAVIOR

This section reviews major research findings on the physiological correlates of antisocial behavior.

Psychophysiological Impairments

Psychophysiological research has provided some of the most convincing evidence for a biological predisposition for antisocial behavior, including the repeatedly observed findings of reduced skin conductance and heart rate activity/reactivity, excessive slow-wave electroencephalogram (EEG), atypical EEG frontal asymmetry, and event-related potential (ERP) responses in antisocial children and adults.

Reduced Skin Conductance and Heart Rate Activity/Reactivity

Several studies and reviews have provided solid evidence for lower skin conductance levels in antisocial individuals. For example, Kruesi and colleagues showed that low skin conductance levels measured at 11 years of age predicted institutionalization at age 13.[2] Deficits in skin conductance responses to both neutral and emotional stimuli in antisocial individuals have also been reported. For example, Raine and Venables found antisocial adolescents with schizotypal features to show significantly lower skin conductance responsivity relative to that of their schizotypal-only counterparts.[3] Empirical studies also showed that antisocial populations exhibit poor skin conductance conditioning and that the association between poor skin conductance conditioning and antisocial behavior may be in place early on in childhood.[4]

One of the most replicable findings of psychophysiological impairments in antisocial individuals is that of low resting heart rate.[5] For children and adolescents, several studies reported that low resting heart rate is prospectively linked to antisocial behavior later in adulthood. For example, in a study involving more than 1800 children in the United Kingdom, Wadsworth reported that low resting heart rate at 11 years of age predicted delinquency at age 21. Furthermore, it was suggested that this relationship of lower heart rate and increased antisocial behavior is diagnostically specific: No other psychiatric condition (e.g., alcoholism, depression, schizophrenia, anxiety disorder) has been linked to low resting heart rate. In addition, some studies reported reduced heart rate reactivity to negative stimuli in antisocial individuals; however, the findings are somewhat mixed.[6]

Summarizing the main findings, it is evident that reduced skin conductance and heart rate activity/reactivity suggest a pathological pattern of underarousal and reduced responsivity in antisocial individuals. Underarousal in antisocial individuals indicated by lower skin conductance levels and lower resting heart rate may promote a need for thrill and sensation seeking in these individuals, whereas attenuated skin conductance responsivity and heart rate reactivity to aversive stimuli may suggest that antisocial individuals are less sensitive to the negative consequences of their behavior, which thus mitigates their moral development and their obedience of social rule. Specifically, both underarousal and reduced responsivity reflect low levels of anxiety or fear, which may predispose to antisocial and violent behavior, because such behavior in part requires a degree of fearlessness to execute. Lack of fear, especially in childhood, may explain poor socialization in antisocial individuals, because a low fear of punishment would reduce the effectiveness of parental socialization processes. Alternatively, individuals with lower activity/reactivity levels may attempt to maintain an optimal level of arousal by seeking out thrill and excitement, which in turn may lead to the development of antisocial or criminal behavior.

Excessive Slow-Wave EEG, Atypical EEG Frontal Asymmetry, and ERP Responses

Several extensive reviews were conducted that suggested fairly consistent findings of higher rates of EEG abnormalities in antisocial individuals relative to control subjects.[7] In general, these reviews indicate that higher levels of slow-wave activity (particularly delta wave, frequency < 4 Hz) have been repeatedly found in aggressive, antisocial individuals. For example, one quantitative EEG analysis conducted by Convit, Czobor, and Volavka[8] found significant correlations between EEG delta activity and the number of violent incidents among psychiatrically hospitalized patients. In addition, a few prospective longitudinal studies demonstrated that excessive slow-wave EEG (i.e., delta wave) precedes the onset of significant criminal behavior in both high-risk (i.e., parents with schizophrenia or personality disorders) and community samples. For example, Raine, Venables, and Williams reported that slower frequency EEG activity at age 15 predicted antisocial behavior at age 24. It is generally considered that the enhanced, slow-wave EEG activity is an indication of cortical immaturity among those with antisocial behavior.[9]

Another line of research has revealed that atypical frontal EEG asymmetry (right > left) is associated with antisocial/externalizing behavior problems in children and adult populations.[10] In general, relatively greater left frontal activity (i.e., relatively reduced left alpha wave activity) is associated with positive affect and approach behavior, whereas relatively greater right frontal activity (i.e., relatively reduced right alpha wave activity) is related to negative affect and withdrawal behavior.[11] It is therefore considered that atypical EEG frontal asymmetry is an indication of aberrant emotional regulation among those with antisocial behavior.

Regarding the ERP, one postulated neurobiological marker for antisocial behavior is reduced amplitude of the P300 component of the ERP in "oddball" tasks.[12] The P300 has been viewed as an orienting response ("what is it?"), and the reduction of its amplitude has been considered to indicate inefficient deployment of neural resources to process cognitive task-relevant information.[13] A more recent **meta-analysis** confirmed the presence of reduced P300 amplitudes in antisocial individuals.[14] Because P300 is generated maximally over the parietal lobe (a region important for working memory),[15] it is suggested that a reduction in P300 amplitude may imply a higher level cognitive impairment in antisocial, violent individuals, such as poor decision making and behavioral disinhibition.

Biosocial Interactions

Increasing evidence shows that social factors may interact with psychophysiological predispositions in the development of antisocial and violent behavior. For example, Farrington reported that boys with low resting heart rate are more likely to become adult violent criminals if they also have a poor relationship with their parents and come from a large family. Similarly, boys with low resting heart rates are more likely to be rated as aggressive by their teachers if their mother was pregnant as a teenager, if they were from a family of low socioeconomic status (SES), or if they were separated from a parent before age 10.[16] Alternatively, a number of studies found that biological impairments, including skin conductance and heart rate, show stronger relationships to antisocial behavior in those from benign social backgrounds that lack the classic psychosocial risk factors for crime. For example, Hemming observed poor skin conductance conditioning among criminals from relatively stable social backgrounds.[17] Similarly, Raine and Venables found poor skin conductance conditioning specifically in antisocial children of higher SES but not in those of lower SES.[18] In a prospective longitudinal study, Raine et al. found that low heart rate at 3 years of age predicted aggression at age 11 in children from high but not low social classes. These findings, as argued by the "social push" hypothesis, suggest that psychophysiological impairments may assume greater importance when social predispositions to crime are minimized.[19] In contrast, social causes may be more important explanations of antisocial behavior in those exposed to adverse early home conditions.

Obstetrical Factors

Of all the areas of biological research on antisocial behavior, studies on obstetrical factors have provided the most compelling evidence for biosocial interaction effects on antisocial behavior. Several prenatal and perinatal factors, including minor physical anomalies, prenatal nicotine or alcohol exposure, and birth complications, have been most closely linked to antisocial behavior and thus are the focus of this section.

Minor Physical Anomalies

Minor physical anomalies (e.g., low-seated ears, adherent ear lobes) have been associated with pregnancy disorders and are considered to be indicators of fetal neural maldevelopment near the end of the first or the beginning of the second trimester of pregnancy.[20] Because the integument and the central nervous system have shared embryological origins, minor physical anomalies are seen as indirect markers of atypical central nervous system and brain development. Several studies found a relationship between more minor physical anomalies and increased antisocial behavior in children, adolescents, and adults. In particular, minor physical anomalies have been linked to violent as opposed to nonviolent offending. For instance, Arseneault and colleagues showed that minor physical anomalies measured at 14 years of age in 170 males predicted violent but not nonviolent delinquency at age 17. The authors reported that these effects were independent of childhood physical aggression or family adversity.[21] In another study, Kandel, Brennan, Mednick, and Michelson assessed minor physical anomalies in 265 12-year-old Danish children and found that recidivistic violent offenders had a greater number of minor physical anomalies compared with subjects with one or no violent offenses at an average age of 21 years.[22] These studies suggest that prenatal insults toward the end of the first 3 months of pregnancy may increase risk for violent behavior as a result of abnormal brain development.

Prenatal Nicotine and Alcohol Exposure

Extensive evidence on prenatal nicotine exposure has established beyond a reasonable doubt that children who are exposed to maternal smoking during pregnancy are at increased risk for later antisocial behavior that extends over the life course.[23] Specifically, prenatal exposure to nicotine has been linked to childhood externalizing behavior, conduct disorder, delinquency, and adult criminal and violent offending.[24] Several studies also reported a dose–response relationship between the extent of maternal smoking during pregnancy and the severity of later antisocial behavior in offspring.[25]

In addition to nicotine exposure, it has also been established that fetal alcohol exposure significantly increases risk for antisocial behavior in children, adolescents, and adults.[26] Heavy alcohol consumption during pregnancy can result in fetal alcohol syndrome, which is characterized by a host of cognitive, behavioral, social, and physical deficits. However, deficits are observed even in those who have been prenatally exposed to alcohol but do not meet diagnostic criteria for fetal alcohol syndrome.[27] For example, research found high rates of delinquency in children and adolescents with heavy fetal alcohol exposure, even if they do not have fetal alcohol syndrome.[28] In addition, studies showed that adolescents who were prenatally exposed to alcohol are overrepresented in the juvenile justice system. For example, Fast and colleagues[29] found that 3 percent of adolescents in a juvenile inpatient forensic psychiatry unit were diagnosed with fetal alcohol syndrome, and 22 percent were diagnosed with fetal alcohol effects. Another study reported that 61 percent of adolescents, 58 percent of adults, and 14 percent of children between the ages of 6 and 11 years with fetal alcohol exposure had a history of trouble with the law.

Birth Complications

Complications at birth (e.g., pre-eclampsia, deprivation of oxygen) have been demonstrated in several studies to predict future engagement of antisocial behavior. The first study was conducted by Pasamanick and colleagues in which a significant link between birth complications and behavior disorders in children was established.[30] The findings were replicated by several studies.[31] However, other studies also questioned the independent effects of birth complications on the development of antisocial behavior by showing that only when combined with social risk factors (e.g., maternal rejection, disadvantageous family background) did birth complications associate with violent behavior.[32]

Biosocial Interactions

Multiple studies have shown a significant interaction effect between obstetrical and social risk factors in predisposing one to antisocial behavior. Regarding minor physical anomalies, for example, one study by Mednick and Kandel measured minor physical anomalies in 129 boys during visits to a pediatrician at age 12. The authors found that minor physical anomalies were associated with violent behavior when the subjects were 21 years old. Interestingly, however, when the authors divided subjects into those from unstable, nonintact families and those from stable families, they found that minor physical anomalies only predicted later criminal involvement for those reared in unstable, nonintact homes.[33] A similar finding was reported by Brennan, Mednick, and Raine, who evaluated adult violent offenses in a sample of 72 male offspring of parents with psychiatric diagnoses. The authors found particularly high rates of violent offending in individuals who had both family adversity and minor physical anomalies compared with those who had only one of these risk factors.[34] In another study, Pine et al. investigated the interaction of minor physical anomalies and social risk factors, such as low socioeconomic status, spousal conflict, and marital disruption, in predicting later disruptive behavior disorders. The authors found individuals who had both increased minor physical anomalies and social risk at age 7 showed greater antisocial behavior at age 17.[35]

In terms of prenatal nicotine/alcohol exposure, several studies also documented interactions between maternal prenatal smoking and social risks in the prediction of later violence. These studies are notable for their large sample sizes, assessment of long-term outcomes, prospective data collection, and control for potential confounders such as drug use and socioeconomic status difference. One striking study conducted by Rasanan et al. found that the offspring of women who smoked during pregnancy had a twofold increase in violent crime at age 26, and when combined with being raised in a single-parent family, recidivistic violent offending increased 11.9 times. Moreover, prenatal nicotine exposure led to a 14.2 times increase in recidivistic violence when combined with a number of other social risk factors (i.e., teenage pregnancy, single-parent family, unwanted pregnancy, and developmental motor delays).[36] It is worth mentioning that in this study the biosocial effect was more prominent for persistent violent offending rather than violence in general or property crime.

A number of well-designed studies also demonstrated that birth complications interact with social risk factors in predicting antisocial behavior in adulthood. For example, Werner found that birth complications combined with a disruptive family environment (i.e., maternal separation, illegitimacy, marital discord, parental mental health problems, paternal absence) predisposed to delinquency over and above either biological or psychosocial risk factors independently.[37] Two prospective longitudinal studies by Raine, Brennan, and Mednick also provide evidence for the presence of biosocial interactions in antisocial behavior.[38] In brief, Raine et al. evaluated whether the early experience of extreme maternal rejection (e.g., unwanted pregnancy, attempts to abort the fetus, institutional care of the infant during the first year of life) interacted with birth complications to predispose to antisocial behavior in a sample of 4269 males born in Copenhagen, Denmark between 1959 and 1961. The authors found that birth complications significantly interacted with maternal rejection in predisposing to antisocial behavior at 18 years of age. The importance of this finding is highlighted by the fact that whereas only 4 percent of the sample experienced both birth complications and maternal rejection, this group was responsible for 18 percent of the violent offenses perpetrated by the whole sample. In a follow-up study reassessing this sample at age 34, the authors replicated the biosocial interaction and found the effect to be specifically strong for serious and early-onset antisocial behavior. Similar biosocial interactions between birth complications and various social risks have been reported in studies using large samples from around the world.[39]

Overall, findings from these studies provide robust evidence suggesting that obstetrical factors, particularly minor physical anomalies, prenatal nicotine/alcohol exposure, and birth complications, may significantly increase the likelihood of antisocial behavior when combined with social risk factors. Also, the presence of both biological and social risk factors may be needed to predispose one to severe antisocial behavior.

Brain Deficits

A significant body of evidence has accumulated suggesting that brain deficits—regional structural/functional abnormalities or brain damage—may act as precursors to the development of antisocial behavior.

Structural and Functional Abnormalities

Brain imaging has become increasingly popular among researchers in recent years. As a relatively objective method for assessing the structure and function of the brain, several studies have used brain imaging to reveal the neuropathological impairments in antisocial children and adults. Most of these studies used positron emission tomography (PET; measures glucose metabolism), single-photon emission computed tomography (SPECT; assesses blood flow), and functional magnetic resonance imaging (fMRI; measures blood oxygen level changes) to evaluate brain function, and anatomical magnetic resonance imaging (aMRI) to assess global and regional brain structural alterations in individuals with antisocial behavior.

Using PET, several studies found metabolic abnormalities during resting states in antisocial individuals. For example, Volkow et al. observed significantly reduced glucose metabolism in both prefrontal and medial temporal regions in violent adult patients with antisocial personality disorder (APD) compared with normal control subjects.[40] Similar results have been reported for children with antisocial behavior. For example, in a PET study on aggressive children with epilepsy, Juhasz et al. found a significant correlation between higher severity of aggression and lower metabolism in the bilateral medial prefrontal and left temporal cortex in this sample of antisocial children. On the other hand, instead of using a resting state, some studies examined the metabolic response in antisocial individuals during a challenge task (e.g., a continuous performance task).[41] For example, using an auditory continuous performance task, Goyer et al. showed that the number of impulsive-aggressive acts in patients with APD and other personality disorders was negatively associated with glucose metabolism in the orbitofrontal, left anterior frontal, and anterior medial frontal cortices.[42] Reduced glucose metabolism has also been found in the anterior medial prefrontal, orbitofrontal, and superior frontal cortex in murderers compared with normal control subjects during a continuous performance task. Similar indications of abnormal functioning in the frontal and temporal regions have been reported in studies using single photon emission computed tomography. For example, Amen et al. found decreased regional cerebral blood flow activity in the prefrontal cortex, and increased activity in the anterior medial frontal and left temporal cortex in aggressive psychiatric patients.[43] Using 21 individuals convicted of impulsive violent offenses, Hirono et al. also found that violent patients with dementia showed reduced regional cerebral blood flow in the left anterior temporal, bilateral dorsofrontal, and right parietal cortex compared with nonviolent dementia patients.[44]

The use of fMRI has furthered knowledge by revealing both cognitive and emotional impairments in antisocial children and adults. For example, during the viewing of negative affective pictures, decreased activation in the amygdala–hippocampal complex and increased activation in the frontotemporal region were observed in criminal psychopaths.[45] Sterzer et al. also found reduced activation in the amygdala and hippocampus in aggressive children with conduct disorders while viewing negative emotional pictures. In addition, abnormal conditioning response was found in antisocial individuals.[46] By using an aversive conditioning task, Schneider et al. found an increase in activation in the dorsolateral prefrontal cortex and amygdala during the acquisition phase of aversive conditioning in individuals with APD.[47] To assess cognitive functioning in antisocial individuals, Raine et al. used a working memory task and revealed reduced activation in the right temporal cortex in violent offenders with a history of abuse compared with control subjects.[48] Similar findings have been reported by Kumari et al., showing activation deficits in the left frontal gyrus and anterior cingulate cortex in violent patients compared with normal control subjects during a working memory task.[49]

In terms of structural abnormalities, aMRI has become the most common method to be used in examining brain structure in recent years. Several studies to date have found volumetric abnormalities in the prefrontal and temporal regions in antisocial individuals. For example, Raine and colleagues found that individuals with APD show a significant gray matter volume reduction in the prefrontal cortex compared with control subjects.[50] Woermann et al. also found reduced left prefrontal gray volumes in aggressive epileptic patients compared with nonaggressive epileptic patients.[51] Similarly, Laakso et al. reported reduced gray matter volume in the dorsolateral prefrontal, medial frontal, and orbitofrontal cortex in alcoholics with APD compared with control subjects.[52] Another study conducted by Dolan et al. found a volume reduction in the temporal lobe in violent APD patients, but no such reduction was found in the prefrontal cortex.[53] More recently, Yang et al. found a volume reduction in prefrontal gray matter in psychopaths with prior convictions (i.e., unsuccessful psychopaths) compared with both those without convictions (i.e., successful psychopaths) and nonpsychopathic control subjects. These findings further suggest that the relatively intact prefrontal cortex may act as a protective factor in preventing successful psychopaths from getting convicted.[54] In addition to frontal and temporal regions, structural abnormalities have also been found for the hippocampus and corpus callosum in antisocial individuals.[55] However, more replications are needed to confirm the effects in these regions.

Overall, functional abnormalities and volumetric reductions in the frontal and temporal cortex have been repeatedly linked to antisocial behavior. Due to the crucial roles of both frontal and temporal regions in the process of decision making, emotional regulation, and moral judgment, deficits to these two areas may therefore predispose one to antisocial violent behavior.

Brain Lesions

The examination of patients with brain lesions has proven to be promising in establishing a causal link between brain damage and antisocial behavior. In fact, some of the most striking evidence for the role of certain brain regions in antisocial behavior comes from descriptions of patients with acquired brain damage who have subsequently developed antisocial and psychopathic-like behavior. A number of case studies point to the role of the frontal lobe in antisocial behavior. One of the earliest cases is that of Phineas Gage, a railway foreman who had a tamping iron blown through his frontal lobe in an accident involving explosives. He survived the injury and recovered his physical and intellectual abilities, but his personality changed dramatically and he became callous, irritable, obnoxious, and irresponsible.[56] Similar personality changes have been observed in various case studies of frontal lobe damage.[57] Common features after prefrontal damage include lack of empathy, difficulties with emotion regulation, impulsivity, disinhibited behavior, poor planning, and blunted emotions. In essence, these individuals develop a psychopathic-like personality or what has been referred to as "acquired sociopathy."[58] Antisocial or psychopathic characteristics seem to develop particularly when damage occurs to the orbital or ventromedial regions of the frontal lobe. For example, Grafman et al. examined a large group of Vietnam War veterans and found that aggressive and violent attitudes were heightened in veterans who had suffered lesions to the ventromedial region of the frontal lobe when compared with control subjects and individuals with damage to other brain regions.[59]

Developmentally, it appears that when brain damage occurs very early in life, the effects on antisocial behavior can be even more pronounced. Anderson et al. found that patients who incurred brain damage before the age of 16 months developed irresponsible and criminal behavior, abusive behavior toward others, and a lack of empathy or remorse.[60] These antisocial characteristics and behaviors were more severe than those observed in patients who suffer ventromedial prefrontal damage in adulthood. It has been suggested that intact functioning of the ventromedial prefrontal cortex is important for moral development. When this region is damaged very early in life, the process of moral socialization may be disrupted. Brain damage during childhood has also been found to lead to clinical diagnoses of antisocial behavior. Pennington and Bennetto found that seven of nine patients who had incurred damage to the frontal lobes before the age of 10 years developed conduct disorder.[61]

Together, these studies demonstrate that brain impairments, particularly in the orbitofrontal/ventromedial region of the prefrontal cortex, may lead to antisocial behavior. However, not all patients with brain damage become antisocial, suggesting that other factors may influence whether an individual with brain impairment becomes antisocial.

Biosocial Interactions

Very few brain-imaging studies have been conducted to date that evaluated interactions between social influences and brain deficits in antisocial individuals. The first was conducted in 1998 by Raine, Stoddard, Bihrle, and Buchsbaum using PET to address the issue of how social deficits moderate the relationship between brain function and antisocial behavior. In brief, the authors divided a sample of murderers into those with and those without psychosocial deprivation. In addition, ratings of psychosocial deprivation took into account early physical and sexual abuse, neglect, extreme poverty, foster home placement, having a criminal parent, severe family conflict, and a broken home. Compared with normal control subjects, murderers with psychosocial deprivation showed relatively good prefrontal functioning, whereas nondeprived murderers showed significantly reduced prefrontal functioning. Specifically, a 14.2 percent reduction in functioning in the right orbitofrontal cortex was found in murderers from good homes. These results suggest that the association between biological impairment and antisocial behavior is more prominent in those lacking social risk factors for antisocial behavior.[62] By using fMRI, Raine and colleagues found a similar effect of biosocial interactions by comparing violent individuals with and without a child abuse history. More specifically, they found that violent offenders who had suffered severe child abuse show reduced right hemispheric functioning, particularly in the right temporal cortex. Further analyses revealed that abused individuals who had refrained from serious violence showed relatively lower left, but higher right, activation of the temporal lobe. The results further suggest that a higher functioning right temporal region may act as a protective factor in preventing one with social risk factors to develop antisocial behavior.[63]

The interaction effect found in brain imaging studies on antisocial individuals is consistent with evidence in patients with brain damage, suggesting that social factors, in combination with the biological risk factor of brain impairment, can influence whether patients develop antisocial behavior. For example, studies by Lewis et al. of young children, adolescents, and adults found that across the lifespan, exposure to violence and abuse in the family is the strongest factor that leads to violence in individuals with neurological impairment. Alternatively, there is also evidence that social factors can act as protective factors to prevent individuals with brain damage from becoming antisocial.[64] Mataró et al. describe a patient in Spain who suffered a similar accident to Phineas Gage when his frontal lobe was impaled by the spike of an iron gate in 1937; however, he showed no signs of hostility, outbursts, or irritability. Such a different outcome may have been because his childhood sweetheart stood by him and married him after the accident, and his family was highly protective and caring and gave him a job in his father's factory where he could be supervised.[65] This finding suggests that social factors such as a nurturing family environment may be able to lower the risk that an individual with brain impairments will become antisocial.

Overall, these findings suggest that structural or functional brain deficits when combined with a social risk factor can predispose one to antisocial behavior. They also suggest that biological or social factors may protect against the outcome of antisocial behavior.

Neuropsychological Impairments

Neuropsychological research has contributed significantly to our understanding of the pathological bases of antisocial behavior. Neuropsychological investigations of violence, crime, and aggression have generally focused on different domains of cognitive functioning in an attempt to understand these phenomena by identifying associated behavioral expressions of brain dysfunction. As a result, several neuropsychological impairments have been identified, including lower intelligence and impaired executive functioning in antisocial individuals, suggesting a dysfunction in the brain, particularly the prefrontal cortex.

Lower Intelligence

Literature available to date suggests that lower intelligence (i.e., IQ or Full Scale IQ) is one of the best replicated cognitive correlates of antisocial behavior among non–mentally ill individuals.[66] In addition, both Verbal and Spatial/Performance IQ have been examined in adults and children with antisocial behavior. Regarding antisocial children and adolescents, lowered Verbal IQ appears to be a crucial characteristic of this population. For example, in a study of New Zealand birth cohort children, Moffitt, Lynam, and Silva reported that verbal deficits at 13 years of age predicted delinquency with persistent, high-level offenses at age 18.[67] The findings are consistent with prior arguments that verbal deficits may affect the development of self-control,[68] leading ultimately to socialization failure and antisocial behavior.[69] Findings for Spatial/Performance IQ have been inconclusive; however, a study by Raine et al. using a community sample of 325 adolescents linked both Verbal and Spatial IQ deficits to antisocial behavior. Furthermore, one prospective longitudinal study showed that low Spatial (but not Verbal) IQ at 3 years of age predisposed to life-course-persistent offending, whereas several other studies of childhood antisocial behavior have also observed spatial ability impairment.[70] These authors proposed that early visuospatial deficits may potentially interfere with mother–infant bonding and may reflect right hemisphere dysfunction that disrupts the processing and regulation of emotions, in turn contributing to life-course antisocial behavior. Although similar deficits in Spatial/Performance or Verbal IQ have not been reported in adult antisocial individuals, reduced Verbal as opposed to Performance IQ has been reported in adult antisocial populations.[71]

Executive Dysfunction

Executive functioning (EF) refers to the cognitive processes that allow for goal-oriented, contextually appropriate behavior and effective self-serving conduct.[72] Executive dysfunction is thought to represent frontal lobe impairment and is indicated by poor performance on neuropsychological measures of strategy formation, cognitive flexibility, or impulsivity. Neuropsychological investigations of EF deficits in antisocial behavioral research have typically focused on diagnostic categories (i.e., APD, conduct disorder, psychopathy) and legal/judicial concepts (i.e., criminality and delinquency). A prominent meta-analysis found overall EF deficits in antisocial individuals compared with control subjects, with the strongest effects found for the Porteus Maze Test and criminal behavior. More recently, EF deficits have been found in a variety of adult antisocial populations, including male and female violent and nonviolent criminals and individuals with APD.[73] However, results for executive dysfunction in children and adolescents with antisocial behavior are less than conclusive. Earlier studies of EF in children reported mixed evidence for a link between delinquency and EF deficits, although this may be attributable to methodological weaknesses, inconsistent definitions of EF, or both.[74] More recent findings are also mixed, with EF deficits characterizing some antisocial youths and not others.[75]

Biosocial Interactions

Neuropsychological studies have suggested a possible interaction of lower intelligence/executive dysfunction and adverse social influences with significantly increased levels of antisocial behavior later on in life. For example, Lewis et al. conducted a study on juvenile delinquents at age 15 and found that a combination of neuropsychological impairments and child abuse was associated with a significant increase in violent offenses in adulthood compared with those with either neuropsychological impairments alone or child abuse alone. These findings are supported by several recent longitudinal studies. For example, in a group of 435 children, Moffitt reported that those with both low neuropsychological performance and family adversity had aggressive scores four times higher than those with either neuropsychological deficits or family adversity only. Using a high-risk sample of 370 Australian adolescents, Brennan et al. also found that although the independent presence of biological risk factors including neuropsychological impairments (e.g., low age 5 vocabulary ability, poor age 15 Verbal IQ and EF) or social risk factors (e.g., lack of paternal control or acceptance, poor educational background of the mother, poverty, harsh discipline style) predicted later antisocial behavior, an interaction of early social risks with later biological risks predicted persistent antisocial violent behavior.[76] Furthermore, a lifetime cumulative interaction of these risks is a stronger predictor of persistent antisocial behavior than when the risks were presented only in childhood or adolescence. Results from these studies suggest that a combined focus on neuropsychological impairments and social risk factors may be influential in the developmental patterns of antisocial behavior.

Abnormal Hormones, Neurotransmitters, and Toxins

Although very few studies have been conducted, several additional biological risk factors have been suspected to predispose one to antisocial behavior. These risk factors are abnormal levels of hormones, neurotransmitter dysfunctions, and the presence of high-level environmental toxins.

Hormonal Imbalances

Common hormones associated with antisocial behavior are cortisol and testosterone. There is considerable evidence suggesting that cortisol levels are reduced in antisocial children, adolescents, and adults. In children, low cortisol levels have been associated with aggression, externalizing behavior and low anxiety, and symptoms of conduct disorder. Low cortisol has been observed in adolescents with conduct disorder, callous and unemotional traits, and conduct problems.[77] In a 5-year longitudinal study, Shoal et al. found that low cortisol in preadolescent boys (ages 10–12 years) was associated with low harm avoidance, low self-control, and more aggressive behavior during adolescence (ages 15–17 years). Finally, low cortisol levels have been found in violent adults and psychopathic offenders.[78] Lower levels of cortisol may indicate that individuals are less responsive to stressors and may be less fearful of negative consequences such as potential punishment.

Testosterone has also been associated with aggressive behavior. Males have several times the amount of testosterone as females. Because there are large gender differences in antisocial behavior, with a male-to-female ratio of about 4:1 for APD and as large as 10:1 for violent crimes, it has been hypothesized that testosterone may be involved in aggressive behavior.[79] Elevated testosterone levels have been linked to antisocial behavior and violent crime in adults, yet studies of aggressive children and adolescents have yielded mixed results.[80] Nevertheless, in a meta-analysis by Brook, Starzyk, and Quinsey, a modest but robust association between testosterone levels and antisocial behavior was confirmed.[81] It has been argued that testosterone may not be linked to aggression per se but to social dominance, whether it be within healthy or antisocial groups, which may account for some of the mixed results.[82]

Neurotransmitter Dysfunction

One of the neurotransmitters found to be particularly linked to antisocial behavior is serotonin and related molecules, including the enzyme that metabolizes serotonin (i.e., monoamine oxidase A [MAO-A]) and the serotonin metabolite 5-hydroxyindoleacetic acid. In a review of the literature, Berman and Coccaro concluded that reduced serotonin activity is related to aggressive behavior, particularly in those who commit or attempt to commit crimes with significant potential for harming others, such as arson and homicide.[83] Lower concentrations of the serotonin metabolite 5-hydroxyindoleacetic acid in the cerebrospinal fluid have been found among antisocial populations.[84] In a meta-analysis by Moore, Scarpa, and Raine, the authors concluded that there is a significant association between 5-hydroxyindoleacetic acid levels and antisocial behavior.[85] Levels of MAO-A activity, although an indirect measure of metabolization of serotonin and dopamine, have been associated with violent, antisocial behavior. In addition to these studies, several studies also found that antidepressants, which increase serotonin, are linked to reductions in aggressive behavior.[86]

Environmental Toxins

The toxins discussed here are metals found in the environment that are known to be harmful to humans when exposure is high. Research has demonstrated a possible link between metal toxicity and criminal behavior. For example, studies of prison inmates found that hair levels of manganese, lead, and cadmium were significantly higher in violent offenders than in nonviolent offenders or control subjects.[87] Environmental lead exposure has been associated with aggressive behavior in children, adolescents, and adults.[88] For example, Burns et al. found that school children from a lead smelting community had increased externalizing behavior problems, when controlling for other variables. It has been suggested that metal toxins may be related to antisocial behavior because they can affect neurotransmitter levels.[89]

Biosocial Interactions

Although the interaction between the biological factors of abnormal levels of hormones, neurotransmitters, and toxins and social risk factors has rarely been studied, results from a few studies suggest such interactions may indeed influence the outcome of antisocial behavior. For example, Dabbs and Morris found that high testosterone was associated with higher levels of childhood and adult delinquency in subjects with low socioeconomic status but not in those with high socioeconomic status. Another study conducted by Mazur found that biological factors of abnormal levels of hormones, including cortisol, testosterone, and thyroxin, in combination with social factors, including age, education, and income, were better at predicting delinquent behavior than either factor alone.[90]

In terms of neurotransmitters, Moffitt et al. found that violent offenders with high blood serotonin levels and a conflicted family background were more than three times as likely to become violent by age 21 than men with only the biological or the social risk factor.[91] Caspi and his colleagues, using a large sample of 1037 children, found maltreated children with a genotype conferring low levels of MAO-A expression were more likely to exhibit antisocial behavior. High levels of MAO-A expression appeared to act as a biological protective factor in maltreated individuals, as the rates of antisocial behavior were lower.[92] A similar pattern was found in anther study of 514 twins conducted by Foley et al., showing that family adversity (e.g., parental neglect, exposure to interparental conflict, and inconsistent parental discipline) was more significantly associated with conduct disorders in those with low MAO-A activity than in those with high MAO-A activity.[93]

Overall, it appears that abnormal levels of hormones, neurotransmitters, and toxins, when combined with social risk factors, may increase the likelihood of developing antisocial behavior. However, future studies are needed to further elucidate moderator effects of social factors on the relationship between hormones, neurotransmitters, and toxins and antisocial behavior.

BIOSOCIAL MODEL OF ANTISOCIAL BEHAVIOR

Several researchers proposed a biosocial model for antisocial behavior. The first model, proposed by Eysenck, suggests that certain biological factors increase the risk for antisocial outcomes, particularly when a certain social upbringing is present.[94] This model suggests that antisocial behavior is intrinsically rewarding, and that family and social environment inhibit antisocial behavior by repeatedly pairing antisocial behavior with punishment (i.e., classical conditioning). This process should effectively reduce antisocial behavior if the child has a nervous system that responds normally to conditioning. This model, although intriguing and supported in some studies, has not been rigorously tested.

Later models of antisocial behavior focused instead on a "dual hazards" interaction of antisocial behavior, predicting that a negative social environment in combination with biological deficits increases the likelihood of predisposing one to antisocial behavior. For example, Mednick proposed that children with deficits in the autonomic nervous system (i.e., the biological factor) who are also raised in inadequate family environments (i.e., the social factor) are at the highest risk for developing poor avoidance conditioning and an inability to learn law-abiding behavior. It was argued that passive avoidance (i.e., avoiding committing an act that has previously been punished) occurs because of the child's fear of punishment, a necessary "civilization" process in normal development. The model suggests that a social environment, when presented with consistent and adequate punishment for antisocial acts (to induce the fear) or a well-functioning autonomic nervous system (to quickly dissipate the fear), will result in fast learning of passive avoidance and a successful inhibition of the nonpreferable (i.e., antisocial) behavior. If both biological and social components are absent, the child is more likely to display antisocial behavior.[95] This model receives some support from the empirical studies, particularly the biological component of poor conditioning in antisocial individuals. However, this model does not fully explain some results of biosocial interaction effects, such as why antisocial individuals from benign families show greater biological impairments.

A "social push" model proposed by Raine incorporated empirical findings and highlighted the key influences of genetic and environmental processes in giving rise to social and biological risk factors that both individually and interactively predispose one to antisocial behavior. Striking evidence suggested a strong interaction effect between genetic and environmental factors in predisposing one to antisocial behavior. One of the earliest studies by Cloninger et al. showed that 40 percent of adoptees with both genetics (i.e., biological parents were criminals) and environmental risk factors (i.e., negative parenting in adoptive parents) were criminals compared with 12.1 percent of those with only genetic factors present, 6.7 percent of those with only a bad family environment, and 2.9 percent for those with neither risk factor.[96] Results show that genetic and environmental factors indeed interact and that the interaction results in a nonadditive increase in antisocial behavior in individuals. The interaction effect was consistent with findings of several other adoption studies confirming the strong influence of these two basic variables in the development of antisocial behavior.[97] This model also incorporated several social and biological protective factors, many of which were mentioned earlier in the review of the empirical findings. The involvement of additional protective biological (e.g., increased prefrontal volume, higher EF) or social factors (e.g., higher SES, supporting family) further explains why some individuals with biosocial precursors did not exhibit antisocial behavior.

In this chapter, a biosocial model extending Raine's model is proposed, suggesting that three key factors—social/environmental, genetics/biological, and protective factors—independently and interactively influence the outcome of antisocial behavior (**Figure 1-1**). This biosocial model hypothesizes that genetic predispositions could lead to a variety of biological risk factors, including abnormal hormones/neurotransmitters, brain abnormalities, and psychophysiological/

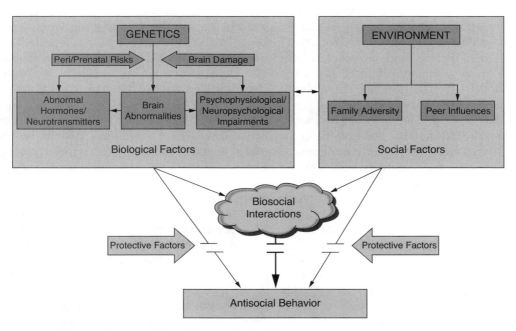

FIGURE 1-1. A Biosocial Model for Antisocial Behavior

neuropsychological impairments, which ultimately trigger antisocial behavior. The involvement of peri- and prenatal risks and brain damage could further contribute to the development of those biological deficits in this biosocial mechanism of antisocial behavior. It is proposed that the presence of both social and biological factors greatly increases the likelihood of developing a more severe form of antisocial behavior (i.e., repetitive violent offending). In addition, protective factors (e.g., intact prefrontal cortex, benign family) are also included in this model because they may interfere with the outcome of antisocial behavior.

Nonetheless, this model is unlikely to be applicable to different groups of antisocial individuals with a wide variety of symptom manifestations. Specifically, it has been argued that theoretical models involving biological vulnerabilities and maladaptive early home environments are better served in explaining the pathology underlying life-course-persistent offenders (i.e., individuals with stable, continuous, lifelong antisocial behavior that begins in early childhood) but not their adolescent-limited counterparts (i.e., individuals with late-onset antisocial behavior who recover by early adulthood).[98] On the other hand, some unique subgroups of antisocial individuals, such as psychopaths, have also been found to show a different pathological pattern, such as the lack of executive dysfunction or prefrontal deficits, and thus may not share the same underlying biosocial mechanisms with other antisocial populations. Despite its limitations, this biosocial model provides several testable hypotheses that could guide future research in examining the epidemiology of antisocial behavior.

CONCLUSION

In general, studies on antisocial individuals have presented convincing evidence for a biological contribution to antisocial behavior. Psychophysiological studies have shown that lower skin conductance and heart rate activity/reactivity, excessive slow-wave EEG, atypical EEG frontal asymmetry, and ERP response are among the most robust findings on antisocial individuals. Prenatal and perinatal studies have linked increased numbers of minor physical anomalies, prenatal nicotine and alcohol exposure, and birth complications to antisocial behavior, particularly with respect to repetitive violent behavior. Imaging studies on antisocial individuals found promising results suggesting that abnormal brain structure and function, particularly in the prefrontal and temporal cortex, may predispose to antisocial behavior. These findings are consistent with lesion studies showing that damage to the frontal and temporal regions is followed by an increase in antisocial behavior. Neuropsychological findings of lower general intelligence as well as poorer performance on EF in antisocial individuals have also been observed. Other biological factors including hormones, neurotransmitters, and toxins have also been examined, and antisocial individuals have been found to show lower cortisol, higher testosterone, reduced serotonin activity, and a high level of metal toxicity. Despite some null findings, these biological risks were found to be associated with antisocial behavior in both children and adults.

The findings from studies examining biological and social interactions suggest that such interactions predispose to antisocial behavior. The significant number of biosocial studies on obstetrical factors provides the strongest evidence for such interaction effects, although supporting evidence was also accumulated in other areas of biological research in antisocial behavior. Overall, the findings suggest that biological risk factors, particularly low heart rate, obstetrical risks, and abnormal levels of hormones and neurotransmitters, when combined with social risk factors (i.e., low socioeconomic status, family adversity) significantly increase the likelihood of an antisocial outcome.

Using a different approach, a number of studies also found that antisocial individuals from relatively benign home backgrounds are more likely to exhibit higher biological risk factors compared with their counterparts from bad homes. Essentially, a greater degree of biological deficits may be needed to predispose individuals to antisocial behavior when their social backgrounds are otherwise relatively normal. Nonetheless, the most consistent findings in studies on biosocial interactions have demonstrated that individuals are most likely to engage in antisocial behavior when both the social and biological risk factors were present. In fact, these individuals are more likely to show persistent, violent antisocial behavior than those with only one risk factor. By reviewing the empirical findings and proposing a biosocial model for antisocial behavior, it is hoped that this chapter will encourage researchers to consider both biological and social factors in their research work, as such practice, if it becomes a standard, will generate a new body of knowledge for antisocial research and may lead to new insights into developmental pathways to antisocial behavior.

GLOSSARY

Antisocial behavior—an umbrella term for various behavioral problems, including violent, psychopathic, delinquent, and criminal behavior

Biosocial interactions—a process whereby the presence of a biological risk factor and a social risk factor increases the odds of antisocial behavior above and beyond the individual effects of either biological or social factors alone

Brain imaging—techniques used to examine the structure and functioning of the brain

Executive functioning—the cognitive processes that allow for goal-oriented, contextually appropriate behavior and effective self-serving conduct

Meta-analysis—a methodological technique used to quantitatively summarize all research conducted on a particular topic

NOTES

1. Raine, A. (2002). Biosocial studies of antisocial and violent behavior in children and adults: A review. *Journal of Abnormal Child Psychology, 30*, 311–326.
2. Kruesi, M. J., Casanova, M. F., Mannheim, G., & Johnson-Bilder, A. (2004). Reduced temporal lobe volume in early onset conduct disorder. *Psychiatry Research, 132*, 1–11.
3. Raine, A., & Venables, P. H. (1984). Electrodermal nonresponding, antisocial behavior, and schizoid tendencies in adolescents. *Psychophysiology, 21*, 424–433.
4. Birbaumer, N., Veit, R., Lotze, M., Erb, M., Hermann, C., Grodd, W., et al. (2005). Deficient fear conditioning in psychopathy. *Archives of General Psychiatry, 62*, 799–805; Fairchild, G., van Goozen, S. H., Stollery, S. J., & Goodyer, I. M. (2008). Fear conditioning and affective modulation of the startle reflex in male adolescents in early-onset of adolescence-onset conduct disorder and healthy control subjects. *Biological Psychiatry, 63*, 279–285; Gao, Y., Raine, A., Venables, P. H., Dawson, M. E., & Mednick, S. A. (2010). Reduced electrodermal fear conditioning from ages 3 to 8 years is associated with aggressive behavior at age 8 years. *Journal of Child Psychology and Psychiatry, 51*(5), 550–558; Hare, R. D. (1978). Electrodermal and cardiovascular correlates of psychopathy. In R. D. Hare & D. Schalling (Eds.), *Psychopathic behavior: Approaches to research* (pp. 107–144). New York: Wiley; Raine, A. (1993). *The psychopathology of crime: Criminal behavior as a clinical disorder.* San Diego: Academic Press.

5. Ortiz, J., & Raine, A. (2004). Heart rate level and antisocial behavior in children and adolescents: A meta-analysis. *Journal of Academy of Child and Adolescent Psychiatry, 43,* 154–162; Wadsworth, M. E. J. (1976). Delinquency pulse rate and early emotional deprivation. *British Journal of Criminology, 16,* 245–256.

6. Pehlham, W. E., Milich, R., Cummings, E. M., Murphy, D. A., Schaughency, E. A., & Greiner, A. R. (1991). Effects of background anger, provocation, and methylphenidate on emotional arousal and aggressive responding in attention-deficit hyperactivity disorders boys with and without concurrent aggressiveness. *Journal of Abnormal Child Psychology, 19,* 407–426.

7. Milstein, V. (1988). EEG topography in patients with aggressive violent behavior. In T. E. Moffitt & S. A. Mednick (Eds.), *Biological contributions to crime causation.* Dordrecht, North Holland: Martinus Nijhoff; Volavka, J. (1987). Electroencephalogram among criminals. In S. A. Mednick, T. E. Moffitt, & S. Stack, (Eds.), *The causes of crime: New biological approaches* (pp. 137–145). Cambridge, UK: Cambridge University Press; Yaralian, P. S., & Raine, A. (2000). Biological approaches to crime: Psychophysiology and brain dysfunction. In R. Paternoster & R. Bachman (Eds.), *Explaining criminals and crime: Essays in contemporary criminological theory* (pp. 57–72). Los Angeles: Roxbury; Mednick, S. A., Volavka, J., Gabrielli, W. F., & Itil, T. M. (1981). EEG as a predictor of antisocial behavior. *Criminology, 19,* 219–229.

8. Convit, A., Czobor, P., & Volavka, J. (1991). Lateralized abnormality in the EEG of persistently violent psychiatric inpatients. *Biological Psychiatry, 30,* 363–370.

9. Raine, A., Venables, P. H., & Williams, M. (1990). Autonomic orienting responses in 15-year-old male subjects and criminal behavior at age 24. *American Journal of Psychiatry, 147,* 933–937.

10. Baving, L., Laucht, M., & Schmidt, M. H. (2000). Oppositional children differ from healthy children in frontal brain activation. *Journal of Abnormal Child Psychology, 28,* 267–275; Deckel, A. W., Hesselbrock, V., & Bauer, L. (1996). Antisocial personality disorder, delinquency, and frontal brain dysfunctioning: EEG and neuropsychological findings. *Journal of Clinical Psychology, 52,* 639–650; Peterson, C. K., Shackman, A. J., & Harmon-Jones, E. (2008). The role of asymmetrical frontal cortical activity in aggression. *Psychophysiology, 45,* 86–92; Santesso, L. D., Dana, R., Schmidt, L. A., & Segalowitz, S. J. (2006). Frontal electroencephalogram activation asymmetry, emotional intelligence, and externalizing behaviors in 10-year-old children. *Child Psychiatry and Human Development, 36,* 311–328.

11. Harmon-Jones, E., & Allen, J. J. B. (1997). Behavioral activation sensitivity and resting frontal EEG asymmetry: Covariation of putative indicators related to risk for mood disorders. *Journal of Abnormal Psychology, 106,* 159–163; Sutton, S. K., & Davidson, R. J. (1997). Prefrontal brain asymmetry: A biological substrate of the behavioral approach and inhibition systems. *Psychological Science, 8,* 204–210.

12. Barratt, E. S., Stanford, M. S., Kent, T. A., & Felthous, A. R. (1997). Neuropsychological and cognitive psychophysiological substrates of impulsive aggression. *Biological Psychiatry, 41,* 1045–1061; Bauer, L. O., & Hesselbrock, V. M. (1999). P300 decrements in teenagers with conduct problems: Implications for substance abuse risk and brain development. *Biological Psychiatry, 46,* 263–272; Branchey, M. H., Buydens-Branchey, L., & Lieber, C. S. (1988). P3 in alcoholics with disordered regulation of aggression. *Psychiatry Research, 25,* 49–58; Gerstle, J. E., Mathias, C. W., & Stanford, M. S. (1998). Auditory P300 and self-reported impulsive aggression. *Progress in Neuropsychopharmacology & Biological Psychiatry, 22,* 575–583; Hermon-Jones, E., Barratt, E. S., & Wigg, C. (1997). Impulsiveness, aggression, reading and the P300 component of the event-related potential. *Personality and Individual Differences, 22,* 439–445; Iacono, W. G., Carlson, S. R., Taylor, J., Elkins, I. J., & McGue, M. (1999). Behavioral disinhibition and the development of substance use disorders: Findings from the Minnesota Twin Family Study. *Development and Psychopathology, 11,* 869–900; Mathias, C. W., & Stanford, M. S. (1999). P300 under standard and surprise conditions in self-reported impulsive aggression. *Progress in Neuropsychopharmacological and Biological Psychiatry, 23,* 1037–1051.

13. Polich, J. (2003). Overview of P3a and P3b. In J. Polich (Ed.), *Detection of change: Event-related potential and fMRI findings* (pp. 83–98). Boston: Kluwer Academic Press.

14. Gao, Y., & Raine, A. (2009). P3 event-related potential impairments in antisocial and psychopathic individuals: A meta-analysis. *Biological Psychology, 82,* 199–210.

15. Donchin, E., & Coles, M. G. H. (1998). Is the P300 a manifestation of context updating? *Behavioral and Brain Sciences, 11,* 355–372.

16. Farrington, D. P. (1997). The relationship between low resting heart rate and violence. In A. Raine, P. A. Brennan, D. P. Farrington, & S. A. Mednick (Eds.), *Biosocial bases of violence* (pp. 89–106). New York: Plenum Press.

17. Hemming, J. H. (1981). Electrodermal indices in a selected prison sample and students. *Personality and Individual Differences, 2,* 37–46.

18. Raine, A., & Venables, P. H. (1981). Classical conditioning and socialization—A biosocial interaction? *Personality and Individual Differences, 2,* 273–283.

19. Raine, A., Venables, P. H., & Mednick, S. A. (1997). Low resting heart rate at age 3 years predisposes to aggression at age 11 years: Evidence from the Mauritius Child Health Project. *Journal of Academy of Child and Adolescent Psychiatry, 36,* 1457–1464.

20. Firestone, P., & Peters, S. (1983). Minor physical anomalies and behavior in children: A review. *Journal of Autism and Developmental Disorders, 13,* 411–425.

21. Arseneault, L., Tremblay, R. E., Boulerice, B., Seguin, J. R., & Saucier, J. F. (2000). Minor physical anomalies and family adversity as risk factors for violent delinquency in adolescence. *American Journal of Psychiatry, 157,* 917–923.

22. Kandel, E., Brennan, P. A., Mednick, S. A., & Michelson, N. M. (1989). Minor physical anomalies and recidivistic adult criminal behavior. *Acta Psychiatrica Scandinavica, 79,* 103–107.

23. Wakschlag, L. S., Pickett, K. E., Cook, E. C., Benowitz, N. L., & Leventhal, B. L. (2002). Maternal smoking during pregnancy and severe antisocial behavior in offspring: A review. *American Journal of Public Health, 92,* 966–974.

24. Brennan, P. A., Grekin, E. R., & Mednick, S. A. (1999). Maternal smoking during pregnancy and adult male criminal outcomes. *Archives of General Psychiatry, 56,* 215–219; Brennan, P. A., Grekin, E. R., Mortensen, E. L., & Mednick, S. A. (2002). Relationship of maternal smoking during pregnancy with criminal arrest and hospitalization for substance abuse in male and female adult offspring. *American Journal of Psychiatry, 159,* 48–54; Fergusson, D. M., Horwood, L. J., & Lynskey, M. T. (1993). Maternal smoking before and after pregnancy. *Pediatrics, 92,* 815–822; Fergusson, D. M., Woodward, L. J., & Horwood, J. (1998). Maternal smoking during pregnancy and psychiatric adjustment in late adolescence. *Archives of General Psychiatry, 55,* 721–727; Orlebeke, J. F., Knol, D. L., & Verhulst, F. C. (1997). Increase in child behavior problems resulting from maternal smoking during pregnancy. *Archives of Environmental Health, 52,* 317–321; Rantakallio, P., Laara, E., Isohanni, M., & Moilanen, I. (1992). Maternal smoking during pregnancy and delinquency of the offspring: An association without causation? *International Journal of Epidemiology, 21,* 1106–1113; Wakschlag, L. S., Lahey, B. B., Loeber, R., Green, S. M., Gordon, R., & Leventhal, B. L. (1997). Maternal smoking during pregnancy and the risk of conduct disorder in boys. *Archives of General Psychiatry, 54,* 670–676; Weissman, M. M., Warner, V., Wickramaratne, P. J., & Kandel, D. B. (1999). Maternal smoking during pregnancy and psychopathology in offspring followed to adulthood. *Journal of the American Academy of Child and Adolescent Psychiatry, 38,* 892–899.

25. Maughan, B., Taylor, C., Taylor, A., Butler, N., & Bynner, J. (2001). Pregnancy smoking and childhood conduct problems: A causal association? *Journal of Child Psychology and Psychiatry, 42,* 1021–1028; Maughan, B., Taylor, A., Caspi, A., & Moffitt, T. E. (2004). Prenatal smoking and early childhood conduct problems. *Archives of General Psychiatry, 61,* 836–843.

26. Fast, D. K., Conry, J., & Loock, C. A. (1999). Identifying fetal alcohol syndrome among youth in the criminal justice system. *Journal of Developmental and Behavioral Pediatrics, 20,* 370–372; Olson, H. C., Streissguth, A. P., Sampson, P. D., Barr, H. M., Bookstein, F. L., & Thiede, K. (1997). Association of prenatal alcohol exposure with behavioral and learning problems in early adolescence. *Journal of the American Academy of Child and Adolescent Psychiatry, 36,* 1187–1194; Streissguth, A. P., Barr, H. M., Kogan, J., & Bookstein, F. L. (1996). *Understanding the occurrence of secondary disabilities in clients with fetal alcohol syndrome (FAS) and fetal alcohol effects (FAE).* Washington, DC: Centers for Disease Control and Prevention.

27. Schonfeld, A. M., Mattson, S. N., & Riley, E. P. (2005). Moral maturity and delinquency after prenatal alcohol exposure. *Journal of Studies on Alcohol, 66,* 545–554.

28. Mattson, S. N., & Riley, E. P. (2000). Parent ratings of behavior in children with heavy prenatal alcohol exposure and IQ-matched controls. *Alcoholism: Clinical and Experimental Research, 24,* 226–231;

Roebuck, T. M., Mattson, S. N., & Riley, E. P. (1999). Behavioral and psychosocial profiles of alcohol-exposed children. *Alcoholism: Clinical and Experimental Research, 23,* 1070–1076.

29. Fast et al. (1999), Olson et al. (1997), Streissguth et al. (1996), see Note 26.

30. Pasamanick, B., Rodgers, M. E., & Lilienfield, A. M. (1956). Pregnancy experience and development of behavior disorders in children. *American Journal of Psychiatry, 112,* 613–618.

31. Cocchi, R., Felici, M., Tonni, L., & Venanzi, G. (1984). Behavior troubles in nursery school children and their possible relationship to pregnancy or delivery difficulties. *Acta Psychiatrica Belgica, 84,* 173–179.

32. Piquero, A., & Tibbetts, S. (1999). The impact of pre/perinatal disturbances and disadvantaged familial environment in predicting criminal offending. *Studies on Crime and Crime Prevention, 8,* 52–70; Raine, A., Brennan, P., & Mednick, S. A. (1994). Birth complications combined with early maternal rejection at age 1 year predispose to violent crime at age 18 years. *Archives of General Psychiatry, 51,* 984–988.

33. Mednick, S. A., & Kandel, E. S. (1988). Congenital determinants of violence. *Bulletin of the American Academy of Psychiatry and the Law, 16,* 101–109.

34. Brennan, P. A., Mednick, S. A., & Raine, A. (1997). Biosocial interactions and violence: A focus on perinatal factors. In A. Raine, P. A. Brennan, D. Farrington, & S. A. Mednick (Eds.), *Biosocial bases of violence* (pp. 163–174). New York: Plenum.

35. Pine, D. S., Shaffer, D., Schonfeld, I. S., & Davies, M. (1997). Minor physical anomalies: Modifiers of environmental risks for psychiatric impairment? *Journal of the American Academy of Child and Adolescent Psychiatry, 36,* 395–403.

36. Rasanan, P., Hakko, H., Isohanni, M., Hodgins, S., Jarvelin, M. R., & Tiihonen, J. (1999). Maternal smoking during pregnancy and risk of criminal behavior among adult male offspring in the northern Finland 1996 birth cohort. *American Journal of Psychiatry, 156,* 857–862.

37. Werner, E. E. (1987). Vulnerability and resiliency in children at risk for delinquency: A longitudinal study from birth to young adulthood. In J. D. Burchard & S. N. Burchard (Eds.), *Primary prevention of psychopathology* (pp. 16–43). Newbury Park, CA: Sage.

38. Raine, A., Brennan, P., & Mednick, S. A. (1997). Interaction between birth complications and early maternal rejection in predisposing individuals to adult violence: Specificity to serious, early-onset violence. *American Journal of Psychiatry, 154,* 1265–1271.

39. Arseneault, L., Tremblay, R. E., Boulerice, B., & Saucier, J. F. (2002). Obstetrical complications and violent delinquency: Testing two developmental pathways. *Child Development, 73,* 496–508; Brennan, P. A., Mednick, B. R., & Mednick, S. A. (1993). Parental psychopathology, congenital factors, and violence. In S. Hodgins (Ed.), *Mental disorder and crime* (pp. 244–261). Thousand Oaks, CA: Sage; Hodgins, S., Kratzer, L., & McNeil, T. F. (2001). Obstetric complications, parenting, and risk of criminal behavior. *Archives of General Psychiatry, 58,* 746–752.

40. Volkow, N. D., Tancredi, L. R., Grant, C., Gillespie, H., Valentine, A., Nullani, N., et al. (1995). Brain glucose metabolism in violent psychiatric patients: A preliminary study. *Psychiatry Research: Neuroimaging, 61,* 243–253.

41. Juhasz, C., Behen, M. E., Muzik, O., Chugani, D. C., & Chugani, H. T. (2001). Bilateral medial prefrontal and temporal neocortical hypometabolism in children with epilepsy and aggression. *Epilepsia, 42,* 991–1001.

42. Goyer, P. F., Andreason, P. J., Semple, W. E., & Clayton, A. H. (1994). Positron-emission tomography and personality disorders. *Neuropsychopharmacology, 10,* 21–28.

43. Amen, D. G., Stubblefield, M., Carmicheal, B., & Thisted, R. (1996). Brain SPECT findings and aggressiveness. *Annuals of Clinical Psychiatry, 8,* 129–137.

44. Hirono, N., Mega, M. S., Dinov, I. D., Mishkin, F., & Cummings, J. L. (2000). Left fronto-temporal hypoperfusion in associated with aggression in patients with dementia. *Archives of Neurology, 57,* 861–866.

45. Kiehl, K. A., Smith, A. M., Hare, R. D., Mendrek, A., Forster, B. B., & Brink, J. (2001). Limbic abnormalities in affective processing by criminal psychopaths as revealed by functional magnetic resonance imaging. *Biological Psychiatry, 50,* 677–684.

46. Sterzer, P., Stadler, C., Krebs, A., Kleinschmidt, A., & Poustka, F. (2005). Abnormal neural responses to emotional visual stimuli in adolescents with conduct disorder. *Biological Psychiatry, 57,* 7–15.

47. Schneider, F., Habel, U., Kessler, C., Posse, S., Grodd, W., & Muller-Gartner, H. W. (2000). Functional imaging of conditioned aversive emotional responses in antisocial personality disorder. *Neuropsychobiology, 42,* 192–201.

48. Raine, A., Park, S., Lencz, T., Bihrle, S., LaCasse, L., Widom, C. S., et al. (2001). Reduced right hemisphere activation in severely abused violent offenders during a working memory task: An fMRI study. *Aggressive Behavior, 27,* 111–129.

49. Kumari, V., Das, M., Hodgins, S., Zachariah, E., Barkataki, I., Howlett, M., & Sharma, T. (2005). Association between violent behavior and impaired prepulse inhibition of the startle response in antisocial personality disorder and schizophrenia. *Behavioral and Brain Research, 158,* 159–166.

50. Raine, A., Lencz, T., Bihrle, S., LaCasse, L., & Colletti, P. (2000). Reduced prefrontal gray matter volume and reduced autonomic activity in antisocial personality disorder. *Archives of General Psychiatry, 57,* 119–127.

51. Woermann, F. G., Van Elst, L. T., Koepp, M. J., Free, S. L., Thompson, P. J., Trimble, M. R., et al. (2000). Reduction of frontal neocortical grey matter associated with affective aggression in patients with temporal lobe epilepsy: An objective voxel by voxel analysis of automatically segmented MRI. *Journal of Neurology, Neurosurgery, & Psychiatry, 68,* 162–169.

52. Laakso, M. P., Gunning-Dixon, F., Vaurio, O., Repo-Tiihonen, E., Soininen, H., & Tiihonen, J. (2002). Prefrontal volume in habitually violent subjects with antisocial personality disorder and type 2 alcoholism. *Psychiatry Research, 114,* 95–102.

53. Dolan, M. C., Deakin, J. F., Roberts, N., & Anderson, I. M. (2002). Quantitative frontal and temporal structural MRI studies in personality-disordered offenders and control subjects. *Psychiatry Research, 116,* 133–149.

54. Yang, Y., Raine, A., Lencz, T., Bihrle, S., LaCasse, L., & Colletti, P. (2005). Volume reduction in prefrontal gray matter in unsuccessful criminal psychopaths. *Biological Psychiatry, 57,* 1130–1138.

55. Laakso, M. P., Vaurio, O., Koivisto, E., Savolainen, L., Eronen, M., & Aronen, H. J. (2001). Psychopathy and the posterior hippocampus. *Behavioural Brain Research, 118,* 187–193.

56. Harlow, J. M. (1848). Passage of an iron bar through the head. *Boston Medical Surgery Journal, 13,* 389–393.

57. Damasio, A. (1994). *Descartes' error: Error, reason, and the human brain.* New York: Grosset/Putnam; Damasio, A. R., Tranel, D., & Damasio, H. (1990). Individuals with sociopathic behavior caused by frontal damage fail to respond autonomically to social stimuli. *Behavioural Brain Research, 41,* 81–94.

58. Eslinger, P. J., & Damasio, A. R. (1985). Severe disturbance of higher cognition after bilateral frontal lobe ablation: Patient EVR. *Neurology, 35,* 1731–1741.

59. Grafman, J., Schwab, K., Warden, D., Pridgen, A., Brown, H. R., & Salazar, A. M. (1996). Frontal lobe injuries, violence, and aggression: A report of the Vietnam Head Injury Study. *Neurology, 46,* 1231–1238.

60. Anderson, S. W., Bechara, A., Damasio, H., Tranel, D., & Damasio, A. R. (1999). Impairment of social and moral behavior related to early damage in human prefrontal cortex. *Nature Neuroscience, 2,* 1032–1037.

61. Pennington, B. F., & Bennetto, H. (1993). Main effects or transactions in the neuropsychology of conduct disorder: Commentary on the neuropsychology of conduct disorder. *Development and Psychopathology, 5,* 153–164.

62. Raine, A., Stoddard, J., Bihrle, S., & Buchsbaum, M. (1998). Prefrontal glucose deficits in murderers lacking psychosocial deprivation. *Neuropsychiatry, Neuropsychology, and Behavioral Neurology, 11,* 1–17.

63. Raine et al. (2001), see Note 48.

64. Lewis, D. O. (1990). Neuropsychiatric and experiential correlates of violent juvenile delinquency. *Neuropsychology Review, 1,* 125–136; Lewis, D. O., Yeager, C. A., Blake, P., Bard, B., & Strenziok, M. (2004). Ethics questions raised by the neuropsychiatric, neuropsychological, educational, developmental, and family characteristics of 18 juveniles awaiting execution in Texas. *Journal of the American Academy of Psychiatry and the Law, 32,* 408–429.

65. Mataró, M., Jurado, M. A., García-Sánchez, C., Barraquer, L., Costa-Jussá, F. R., & Junqué, C. (2001). Long-term effects of bilateral frontal brain lesion: 60 years after injury with an iron bar. *Archives of Neurology, 58,* 1139–1142.

66. Wilson, J. Q., & Herrnstein, R. J. (1985). *Crime and human nature.* New York: Simon & Schuster.

67. Moffitt, T. E., Lynam, D. R., & Silva, P. A. (1994). Neuropsychological tests predicting persistent male delinquency. *Criminology, 32,* 277–300.

68. Luria, A. (1980). *Higher cortical functions in man* (2nd ed.). New York: Basic Books.
69. Eriksson, K., Hodgins, S., & Tengström, A. (2005). Verbal intelligence and criminal offending among men with schizophrenia. *International Journal of Forensic Mental Health, 4,* 191–200.
70. Raine, A., Moffitt, T. E., Caspi, A., Loeber, R., Stouthamer-Loeber, M., & Lynam, D. (2005). Neurocognitive impairments in boys on the life-course persistent antisocial path. *Journal of Abnormal Psychology, 114,* 38–49.
71. Barkataki, I., Kumari, V., Das, M., Taylor, P., & Sharma, T. (2006). Volumetric structural brain abnormalities in men with schizophrenia or antisocial personality disorder. *Behavioural Brain Research, 169,* 239–247; Kosson, D. H., Miller, S. K., Byrnes, K. A., & Leveroni, C. L. (2007). Testing neuropsychological hypotheses for cognitive deficits in psychopathic criminals: A study of global-local processing. *Journal of the International Neuropsychological Society, 13,* 267–276.
72. Lezak, M. D., Howieson, D. B., Loring, D. W., Hannay, H. J., & Fischer, J. S. (2004). *Neuropsychological assessment* (4th ed.). New York: Oxford University Press; Morgan, A. B., & Lilienfeld, S. O. (2000). A meta-analytic review of the relationship between antisocial behavior and neuropsychological measures of executive function. *Clinical Psychology Review, 20*(1), 113–136; Spreen, O., & Strauss, E. (1998). *A compendium of neuropsychological tests* (2nd ed.). New York: Oxford University Press.
73. Broomhall, L. (2005). Acquired sociopathy: A neuropsychological study of executive dysfunction in violent offenders. *Psychiatry, Psychology, and Law, 12,* 367–387; Dolan, M., & Park, I. (2002). The neuropsychology of antisocial personality disorder. *Psychological Medicine, 32,* 417–427; Moffitt, T. E. (1988). Neuropsychology and self-reported early delinquency in an unselected birth cohort. In T. E. Moffitt & S. A. Mednick (Eds.), *Biological contributions to crime causation* (pp. 93–120). New York: Martinus Nijhoff; Stanford, M. S., Conklin, S. M., Helfritz, L. E., & Kockler, T. R. (2007). P3 amplitude reduction and executive function deficits in men convicted of spousal/partner abuse. *Personality and Individual Differences, 43,* 365–375; Teichner, G., & Golden, C. J. (2000). The relationship of neuropsychological impairment to conduct disorder in adolescence: A conceptual review. *Aggression and Violent Behavior, 5,* 509–528; Teichner, G., Golden, C. J., Van Hasselt, V. B., & Peterson, A. (2001). Assessment of cognitive functioning in men who batter. *International Journal of Neuroscience, 111,* 241–253.
74. Moffitt, T. E., & Henry, B. (1989). Neuropsychological assessment of executive functions in self-reported delinquents. *Development and Psychopathology, 1,* 105–118.
75. Nigg, J. T., Glass, J. M., Wong, M. M., Poon, E., Jester, J., Fitzgerald, H. E., et al. (2004). Neuropsychological executive functioning in children at elevated risk for alcoholism: Findings in early adolescence. *Journal of Abnormal Psychology, 113,* 302–314; White, J. L., Moffitt, T. E., Caspi, A., Jeglum, D., Needles, D. J., & Stouthamer-Loeber, M. (1994). Measuring impulsivity and examining its relationship to delinquency. *Journal of Abnormal Psychology, 103,* 192–205.
76. Brennan, P. A., Hall, J., Bor, W., Najman, J. M., & Williams, G. (2003). Integrating biological and social processes in relation to early-onset persistent aggression in boys and girls. *Developmental Psychology, 39,* 309–323; Lewis et al. (2004), see Note 64.
77. Loney, B. R., Butler, M. A., Lima, E. N., Counts, C. A., & Eckel, L. A. (2006). The relation between salivary cortisol, callous-unemotional traits, and conduct problems in an adolescent non-referred sample. *Journal of Child Psychology and Psychiatry, 47,* 30–36; McBurnett, K., Lahey, B. B., Rathouz, P. J., & Loeber, R. (2000). Low salivary cortisol and persistent aggression in boys referred for disruptive behavior. *Archives of General Psychiatry, 57,* 38–43; Oosterlaan, J., Geurts, H. M., & Sergeant, J. A. (2005). Low basal salivary cortisol is associated with teacher-reported symptoms of conduct disorder. *Psychiatry Research, 134,* 1–10; Pajer, K., Gardner, W., Rubin, R. T., Perel, J., & Neal, S. (2001). Decreased cortisol levels in adolescent girls with conduct disorder. *Archives of General Psychiatry, 58,* 297–302; van Goozen, S. H. M., Matthys, W., Cohen-Hettenis, P. T., Wied, C. G., Wiegant, V. M., & van Engeland, H. (1998). Salivary cortisol and cardiovascular activity during stress in oppositional defiant disorder boys and normal controls. *Biological Psychiatry, 43,* 531–539.
78. Holi, M., Auvinen-Lintunen, L., Lindberg, N., Tani, P., & Virkkunen, M. (2006). Inverse correlation between severity of psychopathic traits and serum cortisol levels in young adult violent male offenders. *Psychopathology, 39,* 102–104; Virkkunen, M. (1985). Urinary free cortisol secretion in habitually violent offenders. *Acta Psychiatrica Scandinavica, 72,* 40–44; Cima, M., Smeets, T., & Jelicic, M. (2008). Self-reported trauma, cortisol

levels, and aggression in psychopathic and non-psychopathic prison inmates. *Biological Psychiatry*, *78*, 75–86; Shoal, G., Giancola, P. R., & Kirillova, G. (2003). Salivary cortisol, personality, and aggressive behavior in adolescent boys: A 5-year longitudinal study. *Journal of the American Academy of Child & Adolescent Psychiatry*, *42*, 1101–1107.

79. van Honk, J., & Schutter, D. J. L. G. (2007). Testosterone reduces conscious detection of signals serving social correction: Implications for antisocial behavior. *Psychological Science*, *18*, 663–667.

80. Banks, T., & Dabbs, J. M. (1996). Salivary testosterone and cortisol in a delinquent and violent urban subculture. *Journal of Social Psychology*, *136*, 49–56; Dabbs, J. M., Frady, R. L., & Carr, T. S. (1987). Saliva testosterone and criminal violence in young adult prison inmates. *Psychosomatic Medicine*, *49*, 174–182; Dabbs, J. M., & Morris, R. (1990). Testosterone, social class, and antisocial behavior in a sample of 4,462 men. *Psychological Science*, *1*, 209–211; Maras, A., Laucht, M., Gerdes, D., Wilhelm, C., Lewicka, S., Haack, D., et al. (2003). Association of testosterone and dihydrotestosterone with externalizing behavior in adolescent boys and girls. *Psychoneuroendocrinology*, *28*, 932–940.

81. Brook, A. S., Starzyk, K. B., & Quinsey, V. L. (2001). The relationship between testosterone and aggression: A meta-analysis. *Aggression and Violent Behavior*, *6*, 579–599.

82. Archer, J. (2006). Testosterone and human aggression: An evaluation of the challenge hypothesis. *Neuroscience and Biobehavioral Reviews*, *30*, 319–345.

83. Berman, M. E., & Coccaro, E. F. (1998). Neurobiologic correlates of violence: Relevance to criminal responsibility. *Behavioral Sciences and the Law*, *16*, 303–318.

84. Virkkunen, M., Eggert, M., Rawlings, R., & Linnoila, M. (1996). A prospective follow-up study of alcoholic violent offenders and fire setters. *Archives of General Psychiatry*, *53*, 523–529.

85. Moore, T. M., Scarpa, A., & Raine, A. (2002). A meta-analysis of serotonin metabolite 5-HIAA and antisocial behavior. *Aggressive Behavior*, *28*, 299–316.

86. Alm, P. O., Alm, M., Humble, K., Leppeter, J., Sörensen, S., Lidberg, L., & Orland, L. (1994). Criminality and platelet monoamine oxidase activity in former juvenile delinquents as adults. *Acta Psychiatrica Scandinavica*, *89*, 41–45; Belfrage, H., Lidberg, L., & Oreland, L. (1992). Platelet monoamine oxidase activity in mentally disordered violent offenders. *Acta Psychiatrica Scandinavica*, *85*, 218–221; Coccaro, E. F., Kavoussi, R. J., & Hauger, R. L. (1997). Serotonin function and anti-aggressive response to fluoxetine a pilot study. *Biological Psychiatry*, *42*, 546–552.

87. Masters, R. D., Hone, B., & Doshi, A. (1998). Environmental pollution, neurotoxicity, and criminal violence. In J. Rose (Ed.), *Environmental toxicology* (pp. 13–48). London: Gordon and Breach.

88. Thomson, G. O., Raab, G. M., Hepburn, W. S., Hunter, R., Fulton, M., & Laxen, D. P. (1989). Blood-lead levels and children's behavior-results from the Edinburgh Lead Study. *Journal of Child Psychology and Psychiatry*, *30*, 515–528; Needleman, H. L., Riess, J. A., Tobin, M. J., Biesecker, G. E., & Greenhouse, J. B. (1996). Bone lead levels and delinquent behavior. *Journal of the American Medical Association*, *275*, 363–369.

89. Burns, J. M., Baghurst, P. A., Sawyer, M. G., McMichael, A. J., & Tong, S. L. (1999). Lifetime low-level exposure to environmental lead and children's emotional and behavioral development at ages 11–13 years: The Port Pirie Cohort Study. *American Journal of Epidemiology*, *149*, 740–749; Liu, J., & Wuerker, A. (2005). Biosocial bases of aggressive and violent behavior—Implications for nursing studies. *International Journal of Nursing Studies*, *42*, 229–241.

90. Dabbs, & Morris (1990), see Note 80; Mazur, A. (1995). Biosocial models of deviant behavior among male army veterans. *Biological Psychology*, *41*, 271–293.

91. Moffitt, T. E., Caspi, A., Fawcett, P., Brammer, G. L., Raleigh, M., Yuwiler, A., & Silva, P. (1997). Whole blood serotonin and family background relate to male violence. In A. Raine, P. A. Brennan, D. P. Farrington, & S. A. Mednick (Eds.), *Biosocial bases of violence* (pp. 231–249). New York: Plenum.

92. Caspi, A., McClay, J., Moffitt, T. E., Mill, J., Martin, J., Craig, I. W., et al. (2002). Role of genotype in the cycle of violence in maltreated children. *Science*, *297*, 851–854; Kim-Cohen, J., Caspi, A., Taylor, A., Williams, B., Newcombe, R., Craig, I. W., & Moffitt, T. E. (2006). MAOA, maltreatment, and gene-environment interaction predicting children's mental health: New evidence and a meta-analysis. *Molecular Psychiatry*, *11*, 903–913.

93. Foley, D. L., Eaves, L. J., Wormley, B., Silberg, J. L., Maes, H. H., Kuhn, J., & Riley, B. (2004). Childhood adversity, monoamine oxidase a genotype, and risk for conduct disorder. *Archives of General Psychiatry, 61,* 738–744.

94. Eysenck, H. J. (1977). *Crime and personality* (3rd ed.). St. Albans, England: Paladin.

95. Mednick, S. A. (1977). A biosocial theory of the learning of law-abiding behavior. In S. A. Mednick & K. O. Christiansen (Eds.), *Biosocial bases of criminal behavior* (pp. 1–8). New York: Gardner.

96. Raine (2002), see Note 1; Cloninger, R., Sigvardsson, S., Bohman, M., & von Knorring, A. (1982). Predisposition to petty criminality in Swedish adoptees. *Archives of General Psychiatry, 39,* 1242–1247.

97. Cadoret, R. J., Cain, C. A., & Crowe, R. R. (1983). Evidence for gene-environment interaction in the development of adolescent antisocial behavior. *Behavior Genetics, 13,* 301–310; Cadoret, R. J., Yates, W. R., Troughton, E., Woodworth, G., & Stewart, M. A. (1995). Genetic-environmental interaction in the genesis of aggressivity and conduct disorder. *Archives of General Psychiatry, 52,* 916–924; Crowe, R. R. (1974). An adoption study of antisocial personality. *Archive of General Psychiatry, 31,* 785–791.

98. Moffitt, T. E. (1993). Adolescence-limited and life-course-persistent antisocial behavior: A developmental taxonomy. *Psychological Review, 100,* 674–701.

Prenatal and Perinatal Predictors of Antisocial Behavior: Review of Research and Interventions

Stephen G. Tibbetts

KEY TERMS

Amygdala
Behavioral genetics
Hypoglycemia
Interventions
Neurotransmitters
Polymorphism
Testosterone

INTRODUCTION

In this chapter we first review the history of research regarding pre- and perinatal factors in the criminological literature. Recent research has given far more attention to this area since 2000, and the role of these pre- and perinatal factors is now considered one of the most accepted theories of serious, chronic offending. We then review the current state of research regarding physiological predictors of future criminality, with an emphasis on the sociological and environmental factors that tend to interact with and exaggerate the physiological factors in the perinatal stages. Finally, we examine some of the intervention programs and policy implications that have resulted from the findings of this research.

HISTORY OF RESEARCH ON PRE- AND PERINATAL FACTORS IN CRIMINOLOGICAL LITERATURE

Although prenatal and perinatal factors have become an important topic in modern criminological literature, this attention was relatively limited until recent decades. Historically, notable attention (albeit relatively primitive) in the late 19th and early 20th centuries was given to early physiological factors present at birth or during early development. Researchers such as Lombroso, Sheldon, and

Glueck and Glueck performed research on physiological dispositions toward crime.[1] However, such investigations took an extreme downturn after the mid-point of the 20th century, especially regarding the very early perinatal factors that may influence future criminality in individuals; rather, sociological factors became the primary target for research in criminological literature.

This downturn in attention given to such perinatal factors, or any physiological factors for that matter, was largely due to a very important scientific debate that occurred between Eleanor and Sheldon Glueck—the champions of the multifactor approach—and Edwin Sutherland—the champion of the sociological perspective who rejected all other causal explanations of crime other than social factors. The winner of this scientific/literary debate was clearly Sutherland, which is manifested in the criminological literature in the sense that the discipline became dominated by only social factors, to the exclusion of biological factors, in the middle to late portions of the 20th century. This resulted in what one former president of the American Society of Criminology referred to as "systemic reductionism" due to the primary emphasis of research in the criminological discipline being focused on only sociological factors while neglecting factors from other fields, such as biology, psychology, and medicine.[2]

Beginning in the early 1990s, however, pre- and perinatal factors began to receive more attention. Specifically, between 1990 and 1995, there were 20 published works cited by *Criminal Justice Abstracts (CJA)* on pre- or perinatal factors, which was the same amount in this 5-year period as that published during the 2 previous decades. This pace remained relatively constant through the 1990s and then dramatically increased in the 21st century, with the publications cited in *CJA* showing 72 publications between 2000 and 2007. Thus, it appears from examining the entries in the *CJA* database that the attention given to pre- and perinatal factors regarding crime is growing at an exponential rate.

A similar increase in studies on pre- and perinatal factors in the development of criminality or aggression can be seen in related disciplines, specifically in psychology and sociology. Regarding research in psychology, a search of the entries in the PsycINFO database for "aggression" and "prenatal" or "perinatal" key words revealed 142 entries. It should be noted that this total was significantly more than the total number of entries in *CJA*, which is likely due to two factors. The first is that the PsycINFO database goes back farther in years than *CJA*, with the first relevant entry in PsycINFO appearing in 1964. Second, it is readily acknowledged by criminologists that researchers in psychology were more advanced in their acknowledgment of pre- and perinatal factors in causing aggression or criminality than were the researchers in the criminal justice field in the latter portion of the 20th century.

An increase in the study of pre- and perinatal factors in developing criminal propensity can also be seen when surveying the entries in the primary sociological database, Sociological Abstracts. A search of this database (at the time this text was written) for "criminal" or "deviant" or "illegal" and "prenatal" or "perinatal" as key words anywhere in Sociological Abstracts found 85 entries, with the first published in 1977. Additionally, the mid-point of the entries relevant to this search was in 2004, which is notably later than that of *CJA*.

Not surprisingly, the criminological literature shows more attention to pre- and perinatal factors than does the sociological literature. Thus, although criminologists appear to be behind the learning curve on perinatal factors as compared with psychologists, criminologists seem to be ahead of sociologists when it comes to examining early biosocial predictors of criminality. However, it is important to note that even in sociology, studies on perinatal factors are experiencing a high rate of growth, with more than half of the entries in Sociological Abstracts occurring in the last handful of years. Ultimately, when examining the primary databases of criminology, psychology, and sociology, it is quite clear that research on pre- and perinatal factors in the development of criminality and aggression is very much on the rise in all fields, with psychology leading the way in this trend.

In addition to database surveys of the growing importance of pre- and perinatal factors in predicting criminality, this growing attention to such factors was also supported by a recent survey of criminologists that was conducted by Ellis and colleagues. In this study, Ellis et al. surveyed 387 criminologists to discover which theories each of them considered the most viable explanations for serious/persistent criminal behavior, and findings showed the second (after social learning theory) most chosen theoretical framework for explaining serious offending was the life-course/developmental theory, which places an emphasis on following individuals from the perinatal stage to old age. Perhaps more importantly for the purposes of this chapter, the theory that ranked sixth on this scale was biosocial theory, which was ranked much higher than in similar previous studies.[3] Furthermore, biosocial theory was ranked higher than traditionally dominant theories, such as differential association, strain and general strain theory, routine activities, rational choice, conflict/critical theory, and labeling theory. Therefore, it is clear that both the life-course/developmental and biosocial perspectives are currently given much more acceptance and respect from most criminological researchers than in previous decades.

At this point, with perinatal factors receiving due attention in the literature, it is good to see that many criminologists, as scientists, are open to altering their opinions and beliefs when provided with strong empirical evidence for a given perspective. This alone should be seen as a positive aspect that has occurred recently in the criminological discipline, and the importance of this transformation cannot be overstated.

RESEARCH ON PRE- AND PERINATAL BIOLOGICAL FACTORS OF CRIMINALITY

In this section, we review the extant research on the pre- or perinatal factors that predict chronic offending, with an emphasis on the most contemporary research in this area. We concentrate on the genetic and physiological factors that tend to predispose individuals toward a propensity to serious, violent offending. However, it should be noted that most of these factors are likely to be a significant predictor of chronic offending when coupled with sociological or environmental factors, which are discussed in the next section.

Genetic and Cytogenetic Factors

One of the primary areas of research examining the importance of genetic or biological contributions to persistent criminality is that of **behavioral genetics**, which typically compares statistically the relative effects of genetic and environmental effects among samples of twins. Such studies have found that variations in hyperactivity and conduct problems are almost entirely attributed to genetic factors.[4] Because of the consistency of the scientific findings, the combination of these studies provides convincing conclusions regarding the vital importance of genetic factors in predicting criminality, although many of these factors were even further enhanced by environmental variables. Such studies reveal that there is no longer any doubt, at least regarding scientific findings, that genetic factors contribute a significant amount to the future criminality of individuals.

One of the most predictive genetic variations linked to early, persistent offending is the monoamine oxidase A (MAO-A) **polymorphism**.[5] MAO is an enzyme that helps to break down certain

neurotransmitters, such as serotonin; studies indicate that an MAO-A polymorphism is linked to lower levels of MAO activity among chronic offenders.[6] Specifically, not only is the MAO-A polymorphism linked to low serotonin levels, which are consistently linked to habitual offending (and will be discussed later), but it also has been linked to higher levels of **testosterone**, which also have been consistently linked to chronic offending (discussed later). The finding of a link between MAO-A and criminality was further supported by a meta-analysis that showed a significant effect of this interaction between low MAO-A and child maltreatment across various studies.[7] However, by itself MAO-A has been found to be a relatively inconsistent predictor of criminality.[8]

It is believed that key causal mechanisms in which genes are most implicated in predicting criminality are those that deal with regulating the levels of a certain neurotransmitter, specifically dopamine production. Dopamine, discussed later, is a neurotransmitter that aids in many functions, particularly pleasure and general wellbeing. Extremely low or high levels of dopamine have been found to be linked with future criminality, as well as depression and substance abuse. So it is not surprising that studies have implicated the interaction of two dopamine receptor polymorphisms (i.e., DRD2 and DRD4) in the development of antisocial behavior or criminal offending. Perhaps the only study to estimate the interactions between these two polymorphisms revealed that neither polymorphism had significant independent effects on antisocial behavior, but that DRD2 and DRD4 interacted to predict both adolescent and adult antisocial behavior in males.[9]

Furthermore, one study implicated the short version of the 5-HTTLPR gene-linked polymorphism as being significantly associated with impulsivity, childhood psychopathology, and personality disorders linked to aggression.[10] This genetic disposition is linked to lower production levels of serotonin, a primary neurotransmitter that is key in information processing. This study is consistent with other studies that have shown similar results regarding this type of polymorphism.[11] Such complications with regulation of levels of serotonin caused by this genetic disposition are consistently associated with future criminality.

There are likely numerous other genetic polymorphisms or mutations that predispose an individual to criminality, but such links must be examined by future research. However, what is known at this point is that certain genetics do indeed contribute to a higher likelihood that an individual will be predisposed to criminality due to genetic polymorphisms. At this time most polymorphisms in genetic makeup linked to criminality have been those that are vital in regulating individuals' levels of neurotransmitters. However, other genetic types that predict (or cause) criminality may not be responsible for neurotransmitter levels but rather hormones or some other form of brain function. Future research will explore these various aspects, and it remains to be seen what other polymorphisms or other types of predictors it may uncover.

Regarding cytogenetic factors, or mutations that are not hereditary but rather happen by chance, studies show that the mutations, such as a chromosome makeup of XYY, that produce higher levels of androgens—male hormones, such as testosterone—result in variant genotypes and higher dispositions toward criminal offending and deviance. A good test of this proposition was offered by Anthony Walsh, who estimated the extent of androgens on the recorded deviance of subjects on a continuum from those of genetic compositions that produce less androgens versus those that produce more androgens (e.g., XYY). The conclusion from his findings are that the higher levels of androgens in a given individual are likely to predispose that individual to criminal offending throughout the lifespan.[12]

Whereas a normal female chromosomal makeup is XX and a normal male is XY, sometimes mutations occur which result in a variety of chromosomal abnormalities. However, as an individual's genetics include more Y-chromosomal makeup, with XYY being the extreme end of the continuum, the more likely that person will be disposed toward committing criminal and deviant offenses. In other words, the more "male" the mutation is, the higher the likelihood that the individual will become involved in criminal offending; on the other hand, the more "female" the mutation is (e.g., Klinefelter's syndrome of XXY), the less likely that the individual will be involved in criminality. Thus, the findings of the cytogenetic studies lead directly to the studies on hormones (discussed later); specifically, the more androgens an individual has, the more likely that individual is to become a chronic offender.

Head Trauma and Central Nervous System Factors

One of the first cited studies in *CJA* that dealt with pre- and perinatal factors involved a study of prenatal histories of labor and delivery problems. These studies revealed that certain complications of pregnancy and delivery tend to occur in the history of most children who show evidence of brain injury and various learning or behavioral disorders. This early work also discussed the link between violent behavior and structural brain damage caused by oxygen deprivation (i.e., anoxia), especially when combined with drug usage and environmental factors. However, it was also noted that this condition could be improved with enough notice to provide intervention.[13]

Additionally, another early study cited by *CJA* compared the medical histories of incarcerated and nonincarcerated delinquent children. This study found that the incarcerated youth were significantly more likely than the nonincarcerated delinquents to have sustained severe head or facial injury. Perhaps most striking is the finding that such differences in injury were evident by age 2. Furthermore, in this study perinatal difficulties and psychological impairments were not only more common in incarcerated youth, but the violent youth had an even higher rate of perinatal problems, as well as accidents and bodily injuries. Lewis and colleagues concluded that a combination of early central nervous system trauma along with parental problems and/or social deprivation is primarily responsible for serious, violent delinquency.[14]

A review regarding head trauma concluded that "brain damage . . . is a robust finding in the field of the biosocial bases to antisocial and violent behavior . . . evidence for this position converges together from studies of head injury, neuropsychological testing, and brain imaging."[15] Most studies done on head injuries and brain functioning implicate trauma to and/or reduced glucose metabolism of the frontal lobe, especially the prefrontal cortex, which is the region of the brain most responsible for higher level executive functions, such as problem solving, decision making, and inhibiting the impulsive drives of the lower brain/limbic system. Another brain region that head trauma/imaging studies find is consistently linked to future chronic offending is that of the temporal lobe region, which is located just above the ear. Other brain regions implicated by studies of head trauma/functioning in predicting chronic offending are the corpus callosum, which is responsible for communication between the two hemispheres, and the left angular gyrus, which is located at the junction of the temporal, parietal, and occipital lobes of the brain and plays a key role in integrating information from various lobes of the brain. Further studies have linked structural and/or activity abnormalities of several limbic structures, such as the **amygdala** (which is responsible for emotional responses) and hippocampus (which is responsible for memory), to violent and persistent offending.[16]

Antisocial tendencies are found in individuals with structurally lower amounts of gray matter (as a ratio to white matter) in their prefrontal region. Although this finding was made regarding adults, it likely applies to young offenders due to the emphasis on early brain structure and growth. This finding makes a lot of sense because gray matter is more of the substantive, thinking portion of the brain, whereas the white matter is more of the communication of one region to another. Thus, it is likely that the less there is of the substantive portion of the prefrontal cortex, which is the region that inhibits the emotional responses of the subcortical structures, the less likely impulsive behaviors will be inhibited or thought through in terms of consequences. Furthermore, it is likely that this relatively lower volume of gray matter may help explain why studies using electroencephalographic data consistently show that chronic offenders/psychopaths have significantly slower brain-wave patterns than nonoffenders.

Furthermore, the earlier the age that this head trauma occurs, the worse it is for the child in predicting future criminality.[17] Likewise, as discussed later in this chapter, the greatest impact in terms of intervention can be made at the earliest point of development, namely during pregnancy or in infancy or toddler stages. The basic rule of thumb at this point is the earlier the intervention, the better. Perhaps this is true because early trauma or problematic factors in development pose the greatest risk for criminality in the long term. Numerous studies have consistently shown that trauma to the skull/brain has an extremely strong effect on future criminality, and the general consensus is that most of this impact is due to the influence that such trauma has on the cognitive ability of such youths.

Cognitive ability has been one of the key areas of research, especially that of scores on various IQ or aptitude tests given at early ages, which represents one of the primary manifestations of difficulties in central nervous system processing. Such difficulties are often the result of early abuse and head trauma, as well as parental neglect. One study in this area found that higher scores on such exams were related to significantly less criminality, and the authors concluded that "higher scores on tests of cognitive abilities protected against life-course-persistent patterns of offending."[18] Consistently, another study from that same year showed that cognitive ability maintained a robust inverse relationship with the likelihood of onset of delinquency, early onset of delinquency, and the persistence of criminality. This finding is also consistent with traditional studies that have shown a protective element in having a high level of cognitive aptitude among juvenile nondelinquents across various races/ethnicities and social classes. Lower scores on such aptitude measures, especially verbal intelligence, are highly predictive of criminality, especially when combined with social disadvantages.[19]

Further, an additional study examined the prenatal trauma and neuropsychiatric impairment of a sample of six male adoptees who later committed murder. The authors of this study found that "in all six cases, central nervous system development was compromised in utero or perinatally."[20] Therefore, it appears that early trauma was evident in all these cases of killing. Despite the small sample, the finding of all youths becoming killers seems a rather robust finding, especially in light of the prior studies regarding the profound effects of head trauma.

Other Early Physiological Factors and Biosocial Effects

Regarding biosocial interactions, one study found that both low birth weight and low socioeconomic status had significant effects on early offending, but that the combined influence of risk factors was more likely to result in early offending than for those youths who only had one of these factors. Another study in this area examined an index of numerous pre- and perinatal risk factors, such as low birth weight, maternal venereal/infectious diseases during pregnancy, presence of neurological/psychological conditions in the mother, complications with pregnancy, maternal cigarette smoking during pregnancy, prolapsed cord during delivery, and irregular heartbeat in the infant before/during delivery to determine whether these factors predicted violent criminality; however, only when combined with a weak or broken family structure did the index have a predictive effect on violent offending in a cohort study. Interestingly, in this study the interaction between early perinatal complications and weak family structure only predicted violent offending, but did not predict other forms of offending.[21]

One of the more recent studies found that early perinatal neuropsychological deficits interact with disadvantaged familial environments in predicting life-course-persistent offending. However, this study found evidence of this effect only among non-Caucasians in disadvantaged neighborhoods, and the macro-level structural factors appeared to moderate the influence of individual and familial risks. Thus, Turner and colleagues concluded that the neighborhoods that poor non-Caucasians live in are ecologically distinct from those that poor Caucasians tend to live in and that this trend "exacerbates the criminogenic effects of individual-level deficits and family disadvantage."[22]

Also, it should be noted that infant maltreatment is far more likely when the child is born at a low birth weight, which compounds the difficulties in the development of the child. For example, a study that examined 15 perinatal and sociodemographic variables related to infant maltreatment found that one of the primary predictors of such infant maltreatment was infants born at low birth weight.[23] Unfortunately, the empirical research seems to suggest that the very populations of children who are most at risk of perinatal factors, such as low birth weight, are the same who suffer from infant/child maltreatment. Such an interaction is highly detrimental and strongly predicts future criminality among such children, especially for violence.

The nutrition of pregnant mothers is also implicated in the development of persistent criminality. Specifically, studies have shown that malnutrition among pregnant mothers can have a significant impact on the future antisocial tendencies of their offspring, such as in a population of pregnant women in Holland, who were essentially starved during World War II.[24] This study showed that the pregnant women who were starved during a blockade by the Germans had offspring who had 2.5 times the rate of antisocial disorders by adulthood compared with a control group. Additional studies of nutrition in early development found that infants/children who have dietary deficiencies of protein, zinc, iron, omega-3 fatty acids, and other micronutrients consistently showed future aggression and antisocial behavior.

Other early physiological factors that predispose an infant/child to criminality are hypoglycemia and epilepsy. **Hypoglycemia** is a condition in which a person has a physiological detriment in the production of glucose—a sugar that is the main fuel used for brain functioning. Studies consistently show that individuals who are hypoglycemic are significantly predisposed to criminality, especially violence. Regarding epilepsy, individuals who experience seizures, which studies show are largely hereditary but are often set off by environmental conditions, tend to be predisposed to criminality. It is quite likely that the seizures of epileptics are a manifestation of problems in the functioning of the brain (especially the limbic structures and motor regions), which is consistent with the findings from studies of head trauma/central nervous system malfunctioning discussed earlier.[25]

Hormones and Neurotransmitters

Comprehensive scientific reviews of the extant empirical findings in this area concluded that variations in certain hormones and neurotransmitters have a profound effect on criminal behavior. Specifically, higher testosterone levels in youth are strongly linked with higher levels of criminality. Thus, it is inherent that males are far more likely to engage in violence, even at an early age, than are females. This largely stems from the perinatal formation of testes, which produce high levels of testosterone in males, and which occurs in the prenatal stages of life and continues throughout development.

Most experts now recognize that this high level of male androgens is a significant cause for the worldwide, consistent findings of more aggression among males than females; those who do not accept this conclusion have not examined the extant literature on the subject regarding both humans and animals. After all, if there is anything close to being a "law" in criminology, it is that in all societies, over time and place, males always commit more violent, chronic offending than do females. The most likely, and by far most probable, explanation of this universally consistent observation (even in the animal kingdom) is because males have higher levels of testosterone. Furthermore, the link between high testosterone levels and criminality may go a long way toward explaining why mesomorphs, or youths with an athletic/muscular build, seem to be predisposed toward criminal offending, as compared with obese or thin peers.

It should be noted that such effects also impact female offending; studies show that during the premenstrual cycle, in which females are far more likely to commit violent offenses, females experience a hormonal makeup that most relatively mimics male hormonal makeup (due to varying levels of estrogen and progesterone).[26] This observation is consistent with findings regarding the importance of testosterone/androgens, albeit indirectly.

Another hormone found to be linked to criminality is cortisol, which is typically released due to stress. Most studies examining this link found that offenders have lower levels of cortisol, which is consistent with other studies that show psychopaths/habitual offenders have far lower functioning of the autonomic nervous system (i.e., anxiety levels) than nonoffenders. A manifestation of this lower level of anxiety is consistently found by studies that examine heart rates of chronic offenders; the heart rate of offenders is typically found to be approximately 10 heartbeats per minute slower on average (at rest), and their heartbeat rates do not increase as much as nonoffenders when presented with a stressful situation. Although most studies regarding cortisol levels or heart rate have older samples, it is quite likely that such findings can be applied to perinatal stages.

These findings are related to studies that found a consistent relationship between galvanic skin response and skin conductivity, which, like heart rate, is an indicator of anxiety. Specifically, studies examining this measure of sweat, which produces sodium (which provides a conductor for electricity) that can be measured by electrodes, consistently found that chronic offenders have lower skin conductivity than nonoffenders. Thus, it appears that studies regarding a variety of indicators of stress, anxiety, or arousal are quite consistent in showing that individuals who exhibit low levels of arousal or anxiety are predisposed to becoming chronic offenders. Although most of these studies have not involved infants or young children, it is assumed that such low levels of arousal were present at younger ages.

Neurotransmitters are chemicals that transport electric signals throughout our nervous system. Several types of neurotransmitters have been found to have a significant impact on criminality, and the default levels of such chemicals are largely determined by an individual's genetics/heredity. Specifically, low levels of serotonin, which is required for information processing and adequate feelings of esteem, have consistently been linked to chronic offending. Furthermore, very high and low levels of both dopamine and norepinephrine (adrenaline) have been found to be related to criminality; thus, there appears to be a curvilinear effect regarding these chemicals. However, these are just a few of the many neurotransmitters, which are fundamental chemicals in the body. Therefore, far more research is required in this area to form more concrete conclusions regarding these substances, as well as the many other neurotransmitters that have not been thoroughly examined to date. Many neurotransmitters beyond the ones discussed here have not been carefully researched in terms of their influence on criminality, and resources should be given to future research that can fully examine all neurotransmitters and their link (or not) to criminality.

International Studies

One of the earliest international studies on perinatal factors cited by *CJA* followed a Danish cohort of 31,436 males born in 1944–1947 concluded that there was "evidence demonstrating that aggressive behavior is a consistent pattern in some males from early childhood . . . these results suggest some factors operating early in life, e.g., heritability and perinatal health, may contribute to the emergence of persistent violent behavior."[27] Thus, even some of the earliest international studies in the criminological literature implicate perinatal factors in the development of chronic offending.

Since that time, many of the key studies of pre- and perinatal factors and criminality have come from abroad. For example, key studies done in Finland, Denmark, Mexico, Canada, and England[28] found solid evidence that such perinatal factors are significantly important to future criminality. Furthermore, a conference called the NATO Advanced Study Institute on the Biosocial Bases of Violence, held in Greece in 1996, included 29 papers on this topic. Such international attention and research is extremely beneficial for building a more robust database regarding the influence of perinatal factors on criminality, and the findings are highly consistent with those found in the United States.[29] Given the consistency of the evidence regarding the importance of perinatal factors in predicting future criminality, there is little doubt that such conferences and symposiums will likely become more prevalent, even if they are not hosted by criminologists in the United States.

RESEARCH ON PERINATAL SOCIOLOGICAL AND ENVIRONMENTAL FACTORS OF CRIMINALITY

In this section, we review the extant research on the perinatal sociological and environmental factors that predict chronic offending, with an emphasis on recent research on environmental factors that appear to condition or interact with such biological and physiological factors in producing criminality. We concentrate on the familial and other sociological factors that tend to predispose individuals toward a propensity to serious, violent offending, whether independently or when coupled with biological factors.

Socioeconomic Status and Poverty

Relatively recent studies show definitively that pre- and perinatal problems combined with socioeconomic status are together far more likely to be risk factors for future violent offending than either of these two risk factors alone. For example, Piquero and Tibbetts found that although low birth weight and socioeconomic status were both independently predictive of an early onset of criminal offending, these two factors combined had far more predictive strength than the other two factors considered separately. Furthermore, they examined 15 perinatal and sociodemographic variables related to infant maltreatment and found that one of the primary predictors of such infant maltreatment was infants born to mothers/families that were Medicaid beneficiaries. This study found that a combination of maltreatment and being born into a poor household was one of the most significant predictors in an infant being physically maltreated, which unfortunately combines to create a high likelihood of future criminal behavior among such children. Sometimes the effects of poverty are quite extreme. For example, one study showed that neonaticide, the killing of young infants, is far more likely among poor women. This same study showed that such mothers also tend to be young and unmarried, with virtually no perinatal care; however, these factors are strongly related to (or causes of) their being poor and lacking the resources to gain the needed care that may have prevented such killings of their children. For children who survive such conditions, the likelihood of future criminal behavior is high, especially due to the perinatal factors and familial environments that are typical in such households.[30] Many other studies have shown that infants face many more predictive risk factors when they come from mothers/households that are poor or from lower socioeconomic classes.[31]

Abuse of Mothers and Infants

Some of the recent research in the criminological literature regarding perinatal trauma/ interventions has focused on intimate partner violence (IPV) against pregnant mothers and mothers with infants. One study of pregnant mothers found that such abuse is significantly linked with poor physical health, increased likelihood of sexually transmitted diseases, preterm labor, and low-birth-weight infants.[32] Another study examined whether or not women who have experienced IPV during pregnancy have a higher child abuse potential than women who have not experienced such abuse. This study showed a significant association between IPV and child abuse potential, even after controlling for sociodemographics. In fact, the odds of having a high level of child abuse potential was three times greater for women who had been victims of IPV, as compared with nonvictims of IPV.[33]

Such IPV during pregnancy has been linked to certain groups, especially African American, Native American, and Hispanic women, which unfortunately also tend to be the very groups that are more likely to be poor and disadvantaged in terms of resources. Lipsky et al. also found that such abused women were more likely to be younger, to have inadequate prenatal care, and to have histories of induced abortion; alcohol abuse was also found to be a factor in IPV. Thus, like many culminations of factors in criminology, it appears that the very groups most vulnerable to risk factors appear to have the most of them.[34]

McCartan and Gunnison examined the relationship between prenatal injury and future development of low self-control, which is one of the most important predictors of habitual offending. They concluded that "the results indicate that prenatal injury encapsulated within a poor rearing environment is predictive of low self-control. Both parenting factors and low self-control are also predictive of late adolescent delinquency."[35]

Furthermore, research shows that women who experienced physical assault or sexual coercion by their intimate partners during pregnancy have significantly higher levels of depressive symptoms, and that these depressive symptoms are linked to higher levels of aggression among such women.[36] Such high aggression propensities are likely to translate to the way they treat their infants and children.

At the furthest extreme, Chang and colleagues examined risk factors associated with pregnancy-associated homicide or women who died as a result of homicide (during or within 1 year of pregnancy). The researchers found that the most significant risk factors were mothers' younger than 20 years of age, African American ethnicity, and, perhaps most notably, having late or no prenatal care.[37] Jasinski found that consequences of pregnancy-related violence were associated with late entry into prenatal care, low birth weight, premature labor, fetal trauma, and unhealthy behaviors by mothers. Jasinski concluded that a minority of healthcare providers are adequately trained to screen or deal with pregnant women who have been exposed to violence.[38] A study of pregnant substance abusers found that the prevalence of abuse against the mother was even higher than expected; specifically, four times as many perinatal substance abusers reported physical abuse, as compared with average obstetrical patients.[39]

INTERVENTION PROGRAMS AND POLICY

Before we begin our review of various programs and policy implications shown to be effective in the dealing with such pre- or perinatal complications, it is important to note that the most important factor is when such **interventions** occur. Specifically, the earlier the interventions take place, the more effect they can have on the child's future criminality. The greatest impact in terms of intervention can be made at the earliest point of development, namely during pregnancy and the infancy and toddler stages. Of course, this starts before a mother becomes pregnant and during her pregnancy; such mothers-to-be should be taking supplements, such as folic acid and other prepregnancy vitamin supplements. Medical studies clearly show that when mothers avoid substances and/or take supplements before becoming pregnant, the embryo/fetus/infant has a far better chance for a healthy development. Studies have further shown that simply advising pregnant women regarding health, nutrition, and child rearing led to a reduction in delinquency among their offspring at 15 years of age. Therefore, investing in the resources to provide such counseling and/or provide for such supplements would go a long way toward reducing crime in the long term.

Programs for Youths of Incarcerated Parents

One of the earliest recorded programs for children of incarcerated mothers that was cited in *CJA* was the Pleasanton Children's Center, which was administered in a federal prison in Pleasanton, California. This program aimed to strengthen the bonds between mothers and their children by providing training and education on early childhood, providing services, building connections to the outside world, and creating a relaxed environment for forming these bonds. This program, evaluated by researchers from the National Council on Crime and Delinquency Research Center West, found that participating inmates were highly enthusiastic about the program and that prenatal services provided for pregnant inmates were beneficial for both the mothers and their children.[40]

Parents and Children Together (PACT), a more recent program studied in a federal prison for expectant mothers in Lexington, Kentucky, emphasized time spent in productive job searches, parenting skills, communication techniques, and taking educational courses, and was found to show some effectiveness. Another study found that a live-in nursery program for female inmates showed success; specifically, participants felt stronger bonds with their children, and this program was highly praised by media, inmates, and others related to the program. Furthermore, the initial results of the program showed a decrease in misconduct reports and reduced recidivism among members who completed the program, which was also found to be a cost-saving program.[41]

Home Visitation Programs

Another category of programs for perinatal children and mothers is that of home visitation programs. An effective program in England involved advice and support provided to the mothers in an attempt to improve child-rearing methods and parental knowledge about infant health and development. Olds and colleagues evaluated a home visitation program of high-risk families and youth that included numerous visits both before birth and up to the second year after birth, with a 15-year follow-up and control group. This evaluation found that adolescents "born to women who received nurse visits during pregnancy and post-natally . . . reported fewer instances of running away, fewer arrests, fewer convictions and violations of probation, fewer lifetime sex partners, and fewer days of having consumed alcohol."[42] These researchers also concluded that this program can reduce reports of serious antisocial behavior and use of substances. More recently, Olds concluded that the program had consistent positive effects on prenatal health behaviors, parental care of the child, child abuse/neglect, child health/development, and, perhaps most relevantly, a lower criminal involvement of the mothers and children.[43]

Thus, it appears that such home visitation programs show a very high level of promise for dealing with the perinatal factors addressed in this chapter. Adding a home visitation component to clinic-based programs is perhaps the most cost-effective way to prevent the maltreatment of children and to deal with various perinatal complications that may predispose a child to criminality.

Nutrition and Dietary Interventions

Perhaps one of the most cost-effective and logistically easy forms of intervention includes simply providing high-risk infants and children (as well as pregnant mothers) with vitamin and mineral supplements. As discussed earlier, deficiencies in basic vitamins and minerals (zinc, iron, omega-3 fatty acids, protein, etc.) have been consistently linked to criminality. A randomized trial study demonstrated that when experimental groups of institutionalized youth were given supplements to make up for such deficiencies, they had approximately 40 percent less violent and antisocial behaviors than the placebo control group. Others consistently found similar results.[44] Such nutritional and dietary supplementation, as simple as it may seem, can go a long way toward preventing future criminal behavior by such high-risk youth and, even more importantly, at-risk pregnant mothers.

Risk Assessment Instruments

There are now numerous scientific scales and instruments that can be used to determine risk factors regarding which children are at risk for criminality. Some of these measures are far more useful than the *Diagnostic and Statistical Manual of Mental Disorders, Fourth Edition Text Revision (DSM-IV-TR)*, published by the American Psychiatric Association. This is primarily because the *DSM-IV-TR* is only used to diagnose a clinical condition, not to alert and inform before the development of such a condition. Most of the measures we discuss here are primarily used to measure risk factors for future development of problems, and thus can be seen as preventive measures if they are used for early intervention. Unfortunately, these have not been recognized or used by mainstream service providers in the United States.

The first of these risk assessment scales is the Mednick Perinatal Complications Scale. The scores from this scale are not only more prospective than the *DSM-IV-TR*, but they are categorically different from those of the criteria contained in the *DSM-IV-TR*. Also available is the Delivery Complications Scale, which, combined with the previous scale, can offer a very robust measure of risk. One study that included both measures found that delivery complications were significantly associated with adolescent and adult violent offending.[45]

Another type of measurement that can be used to assess risk at the pre- or perinatal stage is that of instruments used to assess risk of violence against women (e.g., when they are pregnant). For example, the Severity of Violence Against Women Scale has been used and validated by empirical studies in the criminological literature.[46] Additional scales that also measure violence against pregnant women are the Conflict Tactics Scale 2 and the Women's Experience with Battering Scale, both of which showed predictive validity in an empirical study. This study showed that women who had been in violent relationships for longer periods were the most likely to express perceptions of vulnerability, which in turn was highly likely to affect the environment of their children.[47]

Another instrument that can be highly useful in addressing perinatal complications is the Danger Assessment Instrument.[48] Surprisingly, the study that used this measure showed that women who stayed with their abusers actually experienced less violent risk factors during pregnancy as compared with the previous year. Such findings contradict the evidence of other studies that found that pregnant women who had been in violent relationships for longer periods were most vulnerable, as were their offspring.

Measures have also been developed to help prenatal care providers identify adolescents who are at risk for mistreating their children. In a study examining the Family Stress Checklist from a sample of 262 adolescent, predominantly single, and poor mothers who fit the criteria of the study, Stevens et al. found that, after controlling for preexisting sociodemographic variables, the scores on the Family Stress Checklist significantly predicted which children were more likely to be mistreated. Furthermore, the researchers concluded that "prenatal care providers can use the Family Stress Checklist to systematically identify a subgroup of adolescent mothers . . . with a propensity for mistreating their children which suggests the need for additional support services."[49]

The Addiction Severity Index scale measures prenatal substance usage.[50] The index measures not only the alcohol and drug usage of women, but also measures medical, familial, and social difficulties involved. Another study also used this measure, showing that high scores predicted victimization during pregnancy. Such a measure is likely to be useful for screening pregnant mothers before they give birth to flag cases that require intervention for the child's benefit and to predict pre- or perinatal abuse for both the mother and child.

Legal Issues

Perhaps the most prominent legal issues when it comes to pre- and perinatal factors is whether or not to legally prosecute, or otherwise legally intervene, in cases that involve pregnant mothers using substances (mostly illegal) while they are pregnant. There have been many studies regarding this issue, with most of them concluding that prosecution and/or incarceration is not the recommended policy. Rather, most of these studies advise working with the mother to try to provide rehabilitation and other services. However, other experts disagree, claiming a more effective and reasonable third option: legally removing the children from the home of addicted parents.[51]

Unfortunately, outside of pregnant women taking illegal substances, there does not seem to be much attention in the legal literature to other major factors that lead to pre- and perinatal factors. For example, few criminological studies emphasized the need for universal health care for pregnant mothers or infants. After all, obtaining adequate health care is perhaps the most important (and effective) way to reduce future criminality and violence by the offspring, and it is rather surprising that there has been virtually no discussion in the recent criminological literature regarding such early medical care.

A study examined the extrinsic barriers to substance abuse treatment among pregnant women and reported a variety of difficulties in helping such women, such as time, distance, cost, availability, organization of services, discrimination, and several other barriers. The authors concluded that the "elimination of extrinsic barriers to treatment would provide access to numerous beneficial options for both drug dependent women and their children."[52]

CONCLUSION

One of the earliest works cited by *CJA* regarding perinatal factors was a synopsis of a colloquium in 1978 that dealt with the correlates and determinants of criminal behavior. The primary conclusions of this colloquium were "to develop a multidisciplinary center for prospective and longitudinal studies to foster data collection on family interaction, perinatal experiences, motor measures, biologic markers, cognitive functions, and maturational changes."[53] Thus, it is difficult to understand why relatively few attempts have been made to do so in the more than 3 decades that have passed since these conclusions were made. It is time to take this advice and put it into practice. Granted, we would also want to measure all other major influences and factors, especially those regarding families and neighborhoods. However, there is no longer any excuse for not placing the highest priority on gathering information on the pre- and perinatal factors for all youth, especially those who are at high risk of becoming chronic, violent offenders. Furthermore, there is no excuse for not making early intervention in development among pregnant mothers and infants a primary policy issue.

GLOSSARY

Amygdala—an almond-shaped structure in the brain that is part of the limbic system and is responsible for creating feelings, including anger and fear

Behavioral genetics—a line of research that examines the genetic and environmental effects on behaviors, personality traits, and other measurable human characteristics

Hypoglycemia—a condition in which a person has a physiological detriment in the production of glucose, which is a sugar and the main fuel used for brain functioning

Interventions—programs designed to eliminate or reduce criminogenic risk factors and/or antisocial behaviors

Neurotransmitters—chemical messengers in the brain that are responsible for transporting information from neuron to neuron

Polymorphisms—genes that vary from person to person

Testosterone—a male sex hormone produced by the testes in males and by the ovaries in females and that has been found to be related to antisocial behaviors

NOTES

1. Lombroso, C. (1876). *The criminal man (L'uomo Delinquente)*. Milan: Hoepli; Glueck, S., & Glueck, E. (1950). *Unraveling juvenile delinquency*. Cambridge, MA: Harvard University Press; Sheldon, W. (1949). *Varieties of delinquent youth*. New York: Harper and Row.
2. Wellford, C. (1989). Towards an integrated theory of criminal behavior. In S. F. Messner, M. D. Krohn, & A. E. Liska, (Eds.), *Theoretical integration in the study of deviance and crime* (pp. 119–128). Albany, NY: State University of New York Press.
3. Ellis, L., Cooper, J. A., & Walsh, A. (2008). Criminologists' opinions about causes and theories of crime and delinquency: A follow-up. *The Criminologist, 33*, 23–26.
4. DeLisi, M., Beaver, K. M., Wright, J. P., & Vaughn, M. G. (2008). The etiology of criminal onset: The enduring salience of nature and nurture. *Journal of Criminal Justice, 36*, 217–223.

5. Caspi, A., McClay, J., Moffitt, T. E., Mill, J., Martin, J., & Craig, D. (2002). Role of genotype in the cycle of violence in maltreated children. *Science, 297,* 851–854.

6. Ellis, L. (2005). A theory explaining biological correlates of criminality. *European Journal of Criminology, 2,* 287–315.

7. Kim-Cohen, J., Caspi, A., Taylor, A., Williams, B., Newcombe, R., Craig, W., & Moffitt, T. E. (2006). MAOA, maltreatment, and gene-environment interaction predicting children's mental health. *Molecular Psychiatry, 11,* 903–913.

8. Beaver, K., Wright, J. P., DeLisi, M., Walsh, A., Vaughn, M., Boisvert, D., & Vaske, J. (2007). A gene × gene interaction between DRD2 and DRD4 is associated with conduct disorder and antisocial behavior in males. *Behavioral and Brain Functions, 3,* 30–42.

9. Rowe, D. C. (2002). *Biology and crime.* Los Angeles: Roxbury.

10. Retz, W., Retz-Junginger, P., Supprian, T., Thome, J., & Rosler, M. (2004). Association of serotonin transporter promoter gene polymorphism with violence: Relation with personality disorders, impulsivity, and childhood ADHD psychopathology. *Behavioral Sciences and the Law, 22,* 415–425.

11. Volavka, J. (1999). The neurobiology of violence: An update. *Journal of Neuropsychiatry and Clinical Neurosciences, 11,* 307–314.

12. Walsh, A. (1995). Genetic and cytogenetic intersex anomalies: Can they help us to understand gender differences in deviant behavior? *International Journal of Offender Therapy and Comparative Criminology, 39,* 151–166.

13. Hippchen, L. J. (1978). *Ecological-biochemical approaches to treatment of delinquents and criminals.* New York: Van Nostrand Reinhold.

14. Lewis, D. O., Shanok, S. S., & Balla, D. A. (1979). Perinatal difficulties, head and face trauma, and child abuse in the medical histories of seriously delinquent children. *American Journal of Psychiatry, 136,* 419–423.

15. Liu, J., & Wuerker, A. (2005). Biosocial bases of aggressive and violent behavior—implications for nursing studies. *International Journal of Nursing Studies, 42,* 229–241, p. 233.

16. Fishbein, D. (2001). *Biobehavioral perspectives in criminology.* Belmont, CA: Wadsworth; Raine, A. (1993). *The psychopathology of crime: Criminal behavior as a clinical disorder.* San Diego: Academic Press; Wright, J. P., Tibbetts, S. G., & Daigle, L. (2008). *Criminals in the making: Criminality across the life course.* Thousand Oaks, CA: Sage.

17. Bufkin, J., & Luttrell, V. (2005). Neuroimaging studies of aggressive and violent behavior: Current findings and implications for criminology and criminal justice. *Trauma, Violence, & Abuse, 6,* 176–191.

18. Piquero, A. R., & White, N. A. (2003). On the relationship between cognitive abilities and life-course-persistent offending among a sample of African Americans: A longitudinal test of Moffitt's hypothesis. *Journal of Criminal Justice, 31,* 399–409, p. 399.

19. McGloin, J. M., & Pratt, T. C. (2003). Cognitive ability and delinquent behavior among inner-city youth: A life-course analysis of main, mediating, and interaction effects. *International Journal of Offender Therapy and Comparative Criminology, 47,* 253–271; Gibson, C. L., Piquero, A. R., & Tibbetts, S. G. (2001). The contribution of family adversity and verbal IQ to criminal behavior. *International Journal of Offender Therapy and Comparative Criminology, 45,* 574–592.

20. Lewis, D. O., Yeager, C. A., Gidlow, B., & Lewis, M. (2001). Six adoptees who murdered: Neuropsychiatric vulnerabilities and characteristics of biological and adoptive parents. *The Journal of the American Academy of Psychiatry and the Law, 29,* 390–397, p. 390.

21. Piquero A. R., & Tibbetts, S. G. (1999). The impact of pre/perinatal disturbances and disadvantaged familial environment in predicting criminal offending. *Studies on Crime and Crime Prevention, 8,* 52–70; Tibbetts, S. G., & Piquero, A. R. (1999). The influence of gender, low birth weight, and disadvantaged environment in predicting early onset of offending: A test of Moffitt's interactional hypothesis. *Criminology, 37,* 843–878.

22. Turner, M. G., Hartman, J. L., & Bishop, D. M. (2007). The effects of prenatal problems, family functioning, and neighborhood disadvantage in predicting life-course-persistent offending. *Criminal Justice and Behavior, 34,* 1241–1261, p. 1241.

23. Wu, S. S., Ma, C., & Carter, R. L. (2004). Risk factors for infant maltreatment: A population-based study. *Child Abuse & Neglect, 28,* 1253–1264.

24. Neugebauer, R., Hoek, H., & Susser, E. (1999). Prenatal exposure to wartime famine and developmental of antisocial personality disorder in early adulthood. *Journal of the American Medical Association, 282,* 455–462.
25. Liu, J., & Raine, A. (2006). The effect of childhood malnutrition on externalizing behavior. *Current Opinion in Pediatrics, 18,* 565–570.
26. Carlson, J. R. (2001). Prison nursery 2000: A five-year review of the prison nursery at the Nebraska correctional center for women. *Journal of Offender Rehabilitation, 33,* 75–97.
27. Moffitt, T. E., Mednick, S. A., & Gabrielli, W. F. (1989). Predicting careers of criminal violence: Descriptive data and predispositional factors. In D. A. Brizer & M. L. Crowner (Eds.), *Current approaches to the prediction of violence* (pp. 13–34). Washington, DC: American Psychiatric Press, p. 13.
28. Castro, R., Peek-Asa, C., & Ruiz, A. (2003). Violence against women in Mexico: A study of abuse before and during pregnancy. *American Journal of Public Health, 93,* 1110–1116; Farrington, D. P. (1994). Early developmental prevention of juvenile delinquency. *Criminal Behaviour and Mental Health, 4,* 209–227; Kandel, E. (1989). Genetic and perinatal factors in antisocial personality in a birth cohort. *Journal of Crime and Justice, 12,* 61–77; Kandel, E., & Mednick, S. A. (1991). Perinatal complications predict violent offending. *Criminology, 29,* 519–529; Kendall, K., Glenn, A., & Pease, K. (1992). Health histories of juvenile offenders and a matched control group in Saskatchewan, Canada. *Criminal Behaviour and Mental Health, 2,* 269–286; Koskinen, O., Sauvola, A., Valonen, P., Hakko, H., Järvelin, M. R., & Räsänen, P. (2001). Increased risk of violent recidivism among adult males is related to single-parent family during childhood: The northern Finland 1966 birth. *Journal of Forensic Psychiatry, 12,* 539–548; Pagani, L., Tremblay, R. E., & Vitaro, F. (1998). Does preschool help prevent delinquency in boys with a history of perinatal complications? *Criminology, 36,* 245–267.
29. Raine, A., Brennan, P., & Mednick, S. A. (1997). Interaction between birth complications and early maternal rejection in predisposing individuals to adult violence: Specificity to serious, early-onset violence. *American Journal of Psychiatry, 154,* 1265–1271.
30. Piquero, & Tibbetts (1999), Tibbetts, & Piquero (1999), see Note 21; Hatters Friedman, S., McCue Horwitz, S., & Resnick, P. J. (2005). Child murder by mothers: A critical analysis of the current state of knowledge and a research agenda. *American Journal of Psychiatry, 162,* 1578–1587.
31. Stevens, S. C., Nelligan, D., & Kelly, L. (2001). Adolescents at risk for mistreating their children, part I: Prenatal identification. *Child Abuse and Neglect, 25,* 737–751.
32. Sharps, P. W., Laughon, K., & Giangrande, S. K. (2007). Intimate partner violence and the childbearing year: Maternal and infant health consequences. *Trauma, 8,* 105–116.
33. Casanueva, C. E., & Martin, S. L. (2007). Intimate partner violence during pregnancy and mothers' child abuse potential. *Journal of Interpersonal Violence, 22,* 60–72.
34. Lipsky, S., Holt, V. L., & Easterling, T. R. (2005). Police-reported intimate partner violence during pregnancy: Who is at risk? *Violence and Victims, 20,* 69–86.
35. McCartan, L. M., & Gunnison, E. (2007). Examining the origins and influence of low self-control. *Journal of Crime & Justice, 30,* 35–62, p. 35.
36. Martin, S. L., Li, Y., & Casanueva, C. (2006). Intimate partner violence and women's depression before and during pregnancy. *Violence Against Women, 12,* 221–239.
37. Chang, J., Berg, C. J., & Saltzman, L. E. (2005). Homicide: A leading cause of injury deaths among pregnant and postpartum women in the United States, 1991–1999. *American Journal of Public Health, 95,* 471–477.
38. Jasinski, J. L. (2004). Pregnancy and domestic violence: A review of the literature. *Trauma, 5,* 47–64.
39. Haller, D. L., & Miles, D. R. (2004). Personality disturbances in drug dependent women: Relationship to childhood abuse. *American Journal of Drug and Alcohol Abuse, 30,* 269–286.
40. National Council on Crime and Delinquency Research Center West. (1980). *The Pleasanton Children's Center Program: Second-year report and evaluation, 1979–1980.* San Francisco: National Council on Crime and Delinquency Research Center West.
41. Carmouche, J., & Jones, J. (1989). Her children, their future. *Federal Prisons Journal, 1,* 23, 26–27.
42. Olds D., Henderson C. R., & Robert, C. (1998). Long-term effects of nurse home visitation on children's criminal and antisocial behavior. *Journal of the American Medical Association, 280,* 1238–1244, p. 1238.

43. Olds, D. L. (2007). Preventing crime with prenatal and infancy support of parents: The nurse-family partnership. *Victims & Offenders, 2*, 205–225.
44. Schoenthaler, S., & Bier, I. (2000). The effect of vitamin-mineral supplementation on juvenile delinquency among American schoolchildren: A randomized, double-blind placebo-controlled trial. *Journal of Alternative and Complementary Medicine, 6*, 7–17.
45. Kandel, & Mednick (1991), see Note 28.
46. Wiist, W. H., & McFarlane, J. (1998). Utilization of police by abused pregnant Hispanic women. *Violence Against Women, 4*, 677–693.
47. Goldstein, K. M., & Martin, S. L. (2004). Intimate partner physical assault before and during pregnancy: How does it relate to women's psychological vulnerability? *Violence and Victims, 19*, 387–398.
48. Decker, M. R., Martin, S. L., & Moracco, K. E. (2004). Homicide risk factors among pregnant women abused by their partners: Who leaves the perpetrator and who stays? *Violence Against Women, 10*, 498–513.
49. Stevens et al. (2001), p. 737, see Note 31.
50. Roberts, A. C., Nishimoto, R. H., & Kirk, R. S. (2003). Cocaine abusing women who report sexual abuse: Implications for treatment. *Journal of Social Work Practice in the Addictions, 3*, 5–24.
51. Compare Gustavsson, N. S., & MacEachron, A. E. (1997). Criminalizing women's behavior. *Journal of Drug Issues, 27*, 673–687; Dinsmore, J. (1992). *Pregnant drug users: The debate over prosecution.* Alexandria, VA: National Center for Prosecution of Child Abuse; Huffaker, M. L. (2001). A parent's addiction: The judicial disposition of children to drug abusing parents. *Law and Psychology Review, 25*, 145–160; Huth-Bocks, A. C., Levendosky, A., & Bogat, G. A. (2002). The effects of domestic violence during pregnancy on maternal and infant health. *Violence and Victims, 17*, 169–185; Maschke, K. J. (1995). Prosecutors as crime creators: The case of prenatal drug use. *Criminal Justice Review, 20*, 21–33; Smith, G. B., & Dabiri, G. M. (1991). Prenatal drug exposure: The constitutional implications of three governmental approaches. *Seton Hall Law School Constitutional Law Journal, 2*, 53–126; Terres, J. L. (1990). Prenatal cocaine exposure: How should the government intervene? *American Journal of Criminal Law, 18*, 61–86.
52. Jessup, M. A., Humphreys, J. C., & Brindis, C. D. (2003). Extrinsic barriers to substance abuse treatment among pregnant drug dependent women. *Journal of Drug Issues, 33*, 285–304, p. 285.
53. MITRE Corporation. (1978). *Colloquium on the correlates of crime and the determinants of criminal behavior.* McLean, VA: MITRE Corporation, p. 1.

CHAPTER THREE

Prenatal Insults and the Development of Persistent Criminal Behavior

John Paul Wright

KEY TERMS

DNA
Fetal alcohol syndrome
Fetal development
Gene × environment interactions
Magnetic resonance imaging
Teratogens

INTRODUCTION

Since the 1960s, criminal behavior theories have focused on "environmental," "social," and "exogenous" variables. An examination of any introductory criminology textbook or the perusal of the academic journals reveals the pervasiveness of environmentally based explanations for criminal behavior. This fact has been extensively documented with calls critical of the dominance of environmental explanations left largely unheard.[1] Focus on environmental explanations of crime has also been consequential. The U.S. federal government, states, local communities, and virtually any organization focused on human interventions (e.g., housing, schools, and police departments) are informed by these environmental theories. Social disorganization theory, for example, connects conditions in neighborhoods to aberrant behavior. Perhaps unintentionally, this theory has become the intellectual bedrock that has justified the expenditure of trillions of tax dollars on community organizing, community investment, and the many other efforts directed at changing the environment in which people develop.

At first glance, the focus on environment may not seem especially problematic. After all, some environmental stimuli, such as delinquent peers, are clearly associated with a host of negative behavioral outcomes and appear differentially distributed across society. And, arguably, some of the expenditures of tax money may have ameliorated some criminogenic forces in some places during some time periods. It is difficult to know, however, how much crime has been reduced, where, and at what cost by general welfare spending. First glances, of course, can be deceiving. On closer inspection, two problems immediately appear. First, theorists rarely define and operationalize the term *environment*. Indeed, although the term is bantered about in academic and popular discourse, there exists no agreed-upon definition of what constitutes an environment. Working without a precise definition of what constitutes social or environmental characteristics is also consequential. Virtually any measured characteristic can be defined as social or as environmental. Such a broad definition allows for, if not encourages, gross oversimplification in thinking and in logic and may lead to policies that have few desirable effects and some that have not so desirable effects.

Research on communities serves as a perfect illustration. Communities are thought to influence criminal behavior through a variety of mechanisms. Some "communities" have more informal control than others; other communities are said to suffer from poverty, a lack of good jobs, and high rates of racial or class discrimination.[2] Of course, what constitutes a community is typically left unaddressed because of the problems inherent in its definition. Communities thus become mere mental abstractions that offer much in the way of popular conversation but little scientific specificity. Just as communities can be defined as nearly anything at all, so too can the terms *environment* and *social*. They are defined so broadly, with such a lack of precision, that they are for all intents and purposes undefined.

Problems with definitional ambiguity have long been recognized and systematically ignored.[3] Why? Clearly, an overarching possibility has to do with the sociological training, and hence the focus of most criminologists. The central dogma of most sociologically trained criminologists is that "social facts require social explanations." To explain social facts, criminologists typically use vague concepts, such as anomie or informal social control, in an effort to create a body of knowledge that prioritizes perceived social factors over others. Consequently, use of these "fuzzy" concepts, along with their equally fuzzy theories, usually results in the active exclusion of other bodies of knowledge. Nowhere is this more obvious than in the continued exclusion of human biology from criminology.

Much has been written on the exclusion of biology from serious criminological consideration, and it is not the purpose of this chapter to repeat this history.[4] However, a large number of studies now exists that readily document that genetic and biological forces affect the development of antisocial behavior, its maintenance over time, and even desistance from a lifetime of criminal behavior.[5] With this noted, a fear over anything biological and an isolating focus on everything environmental has blinded criminologists to the very real conditions under which "environmental" factors can and do influence human development.

Studies on the nexus between prenatal exposure to neurotoxins and their resulting impact on the human nervous system highlight and emphasize the dual, symbiotic, and complex interchanges that occur when external agents (neurotoxins) are introduced into a readily identifiable environment—the womb. The womb is an environment of considerable importance to the healthy development of humans. It is the environment where our cells divide trillions of times, where our genetic code unwinds to create our brains and nervous systems, and where we take on the appearance of human beings. The womb is our first environment, an environment that, for the most part, protects us from outside influences while at the same time connects us to the broader world. The womb, moreover, is a clearly definable environment that lends itself to scientific study.

Understanding the 10 months of human **fetal development** in the womb also helps us understand and explain sources of variation in human traits, dispositions, and characteristics. Although much human variation originates from minor differences in our **DNA** (deoxyribonucleic acid), some portion of variation in human differences can be found by examining differences in in utero development. Many neonates, for example, enjoy a normal period of in utero development, free from exposure to **teratogens**. Others are not so fortunate and develop with cocaine metabolites coursing through their blood, binding to the receptors in their growing brains, and experience classic withdrawal symptoms at birth. Understanding experiences in the womb, it appears, helps to understand the origins of human differences, and thus connect the social world to the biological.

MECHANISMS OF PRENATAL INSULTS TO THE FETUS

The human womb has been subject to thousands of years of natural selection and refinement, with each refinement typically associated with a net reproductive advantage. Because of this lengthy process, the womb can be reliably reproduced across females, regardless of culture or socioeconomic status. In terms of embryonic growth, the womb represents a highly stable, highly efficient, and highly protective environment through which ontogenic development of offspring can occur. Because the womb has been subject to the processes of natural selection, it affords the developing embryo all necessary requirements to sustain embryonic growth and physical maturation. It is not enough, however, to provide sustenance to a growing body of cells. The womb must also offer sufficient protection against an array of potential threats. The threats are varied, but most have the potential to destroy the developing embryo or, at a minimum, to create permanent birth defects. The womb must be able to control temperature, to provide for nutrients to be extracted from maternal stores, and to protect the embryo from the day-to-day bumps and jars. The point is, the female womb not only provides an optimal environment for growth, but it also has evolved to help ensure offspring survival.

You can think of the womb as a series of filters. These filters help block out radiant energy and damaging sound waves, for example, but they also filter bacteria in the mother's blood and possibly prevent dangerous chemicals from reaching the embryo. This active filtration system is composed of the mother's placenta, which produces hormones and allows for the exchange of nutrients and waste byproducts; the umbilical cord, which connects the embryo to the placenta; and the amniotic sac, which contains amniotic fluid.

Going one step further, another filtration system exists within the embryo. The hematoencephalic barrier, better known as the blood–brain barrier, is a semipermeable cellular structure containing more than 400 miles of capillaries filled with endothelial cells. The spacing of the endothelial cells prevents bacteria and other foreign organisms from being transported from the mother to the child. This is why, for instance, infections of the brain are exceedingly rare. The blood–brain barrier prevents large, non–fat-soluble molecules from entering the brain of the embryo. Smaller molecules, such as those composing methyl-alcohol or cocaine, easily pass through. When the molecules pass through the blood–brain barrier, they bind to receptor sites in the brain and disrupt normative growth.

Variation in Effects of Neurotoxins on Development

It is important to keep this filtering system in mind during any discussion of the effects associated with insult due to teratogens. Research makes it clear that the placenta and blood–brain barrier are remarkably effective, and that the physiology of the embryo is, in some instances, highly adaptive. This helps to explain why studies on the behavioral outcomes associated with teratogens show that the influence of these agents is highly variable. Some children exposed in utero to nicotine, lead, alcohol, heroin, or cocaine will be adversely affected, but some will not.

The magnitude of the effects produced by teratogen exposure is likely moderated by minute variances in the placenta and possibly the blood–brain barrier. Some embryos may simply receive a greater dose of the neurotoxin than others. However, the magnitude of effect is also likely moderated by several other biological variables. The genotype of the fetus may be one such variable. Studies into the effects of lead (Pb) exposure in children find that variants in the vitamin D gene receptor gene (VDR-FOK1) modify the amount of lead in blood plasma, as do genes associated with iron metabolism.[6] The gender of the fetus may also be a modifying influence because the central nervous system of males appears more sensitive to insult than that of females.[7] Moreover, the timing of the introduction of the toxin into the developmental sequence and the duration of the exposure may also be important predictors. Illicit drugs taken repeatedly over the first two trimesters of pregnancy, for example, may be associated with more adverse outcomes than an illicit drug taken sporadically later in pregnancy.

Perhaps not surprisingly, given the number of potential modifiers and filters, research into the short- and long-term effects of prenatal insults makes a distinction between "soft" and "hard" effects. "Hard" effects are those where there is precise measurement, such as birth weight, or where observable differences between exposure groups is clear, such as in the case of fetal alcohol syndrome and birth defects. Other effects associated with prenatal insults are less obvious, at least initially. These "soft" effects are typically associated with disturbances in neurological functioning related to specific behavioral traits. Self-control and other brain-based executive functions may be compromised, to varying degrees, by teratogen exposure, but the causal connection is less clear and more difficult to establish.

Tobacco

Tobacco, usually in the form of a cigarette, is a commonly abused drug that generates billions of dollars in associated costs each year. The good news about tobacco is that its use has declined to its lowest point in history, with 20 percent of adults reporting being active smokers. This is down from 42.4 percent in 1966.[8] Major changes in tobacco use have occurred since the early 1990s as the social movement against smoking altered laws and created a social stigma against tobacco use. Today, tobacco smokers are more likely to be found among the less educated, among those living on welfare, and among those with other addictions.[9]

National surveys have found that the prevalence of smoking is lower among pregnant women (16.4 percent) compared with nonpregnant women (28.4 percent), showing that pregnancy causes some women to stop smoking. However, for pregnant females ages 15 to 17, the prevalence of smoking is actually higher than for nonpregnant females (24.3 percent compared with 16 percent).[10] Unfortunately, this group is also more likely to give birth to multiple children, with each successive birth exposed to nicotine. Regardless, estimates indicate that each year upward of 500,000 newborn children were produced by mothers who smoked over the entire term of their pregnancy.[11]

A large literature base exists documenting the varied deleterious effects associated with in utero exposure to nicotine and other tobacco-based compounds. The evidence comes from a variety of sources using a range of assessment techniques. For instance, much research has been conducted on animals, largely because experiments into the effects of neurotoxins on human embryos are inherently unethical. Animal studies converge to show that in utero exposure to nicotine adversely affects the growth, structure, and functioning of pup brains. Rat pups exposed in utero to tobacco smoke, for instance, generally show deficits in cerebral cortex metabolism and demonstrate dysfunction in the dopaminergic and serotonergic systems within the brain.[12] These same studies also find that in utero exposure also appears to condition the brain to make it more susceptible to addiction later in life.

A large body of literature also exists that documents a positive correlation between maternal cigarette smoking and deleterious offspring outcomes. It is important to note that maternal smoking is usually just one of a range of risk factors that tend to co-occur in women who smoke while pregnant. Many pregnant smokers are also relatively young, may be overweight or obese, and may suffer from other medical conditions. These conditions may confer additional risk to the developing fetus. Nonetheless, mothers who smoke during the entire term of their pregnancy are at an increased risk of producing youth born prematurely and substantially underweight.[13] They are also more likely to have children who score significantly lower on tests of IQ and who have a significantly increased number of behavioral problems, including criminal offending in adulthood.

Until a few years ago, conventional scientific wisdom held that maternal smoking was a causal factor in the etiology of these outcomes—that is, nicotine and other tobacco compounds adversely affected child neurological functioning and hence IQ and conduct problems. This view, however, has recently been challenged. Several studies now show that genetic differences between mothers who smoke during pregnancy and those who do not may account for the observable differences between children exposed in utero and those not exposed. In one such study, Wang et al. analyzed 741 mothers from a Boston area hospital, of whom 174 smoked at some point during their pregnancy. Mothers were genotyped on two genes, CYP1A1 and GSTT1, which are involved in the metabolism of drugs. Consistent with prior evidence, maternal smoking was associated with an average birth weight reduction of 377 grams. Genes, however, appeared to moderate the effect of smoking on birth weight. For example, one genetic variant was associated with an average birth weight reduction of only 252 grams, implying a modest protective influence. However, substantial average birth weight reductions were found for mothers who smoked and who had a combination of genetic risk variants. For these mothers, the average birth weight reduction was 1285 grams.[14]

These results were recently replicated by Tsai and colleagues.[15] Their analysis of 1749 mothers found that mothers who smoked the entire length of their pregnancy and who had specific genetic risks were substantially more likely to experience a preterm delivery (birth before 32 weeks' gestation). In the Hokkaido Study of Environments and Children's Health, Kishi and colleagues found that the joint contribution of maternal smoking and genes that regulate drug metabolism was associated with substantial reductions in infant birth weight. Clearly, the results of these studies imply that some infants are at an increased risk of low birth weight and preterm delivery regardless of whether or not the mother smokes. However, maternal smoking appears to interact with specific genes that, in turn, substantially increase the likelihood of preterm delivery and low birth weight.[16]

Assuming these findings are valid, could the same sort of **gene × environment interactions** also explain the association between maternal smoking and offspring behavioral problems? Genetic influences have moderate to large effects on conduct problems early in life—and serious behavioral problems later in life—and have been found to interact with a range of environmental variables.[17] Research to date, however, has produced mixed results. In a study of 538 twin boys, Silberg et al. found that once maternal genetic risks for conduct disorder were accounted for, maternal smoking during pregnancy no longer predicted variation in offspring conduct disturbances.[18] Analysis of the Environmental Risk Longitudinal Twin Study of 1116 twin pairs by Maughan et al. found that genetic influences reduced the effect of prenatal smoking on child conduct problems by 75 to 100 percent. Mothers who smoked during pregnancy were significantly more likely to be antisocial, to have multiple children from multiple antisocial men, to be poor, and to be depressed. These differences make it difficult to link causally prenatal smoking and offspring behavioral problems.[19]

Further evidence demonstrating the negative effects of maternal smoking comes from brain imaging studies of youths whose mothers smoked during pregnancy. **Magnetic resonance imaging** studies of children whose mothers smoked during pregnancy found that the brains of exposed youth have a significantly reduced volume of cortical gray matter and a significantly smaller head circumference.[20] Cortical gray matter is strongly associated with intelligence and overall brain health. In utero exposure to nicotine also appears to alter the brain in ways that make addiction to nicotine and other drugs more likely in the future. However, it is important to note that gray matter differences are under strong genetic control and that these differences have been associated with adult criminal behavior.[21]

Tobacco smoke contains more than 4000 compounds. When these compounds enter the bloodstream of a pregnant woman, they interact with a range of maternal organs that influence the development of the fetus. Maternal smoking constricts the vascular system, reducing blood flow and oxygenation to the fetus. This may be why the most consistent effects associated with maternal smoking are preterm delivery and low birth weight. Infants born preterm and low in birth weight are at an increased risk of mortality and other regulatory defects.

The available evidence cautions against making sweeping generalizations about the behavioral effects associated with prenatal exposure to tobacco smoke. Important differences exist between women who smoke while they are pregnant and those who do not, and these differences may account for the positive correlation between prenatal smoking and child conduct problems. Although it may not be possible to make strict causal statements about the influence of prenatal smoking on conduct problems of children exposed prenatally to nicotine, it appears safe to say that prenatal smoking interacts with maternal genotypes that increase the risk of serious health and behavioral problems. Whether a causal agent or an indicator of maternal antisocial behavior, maternal smoking during pregnancy serves as an important risk indicator. Children born to mothers who smoke may be at an increased genetic, biological, and environmental risk of developing a range of cognitive and behavioral problems.

Alcohol

Unlike trends in smoking by pregnant mothers, trends in alcohol use during pregnancy have not changed in recent years. On average, 10 to 12 percent of all pregnant American women report drinking alcohol at least once during the last 30 days. Similar to the distribution of prenatal smoking, however, 16 percent of pregnant girls aged 15 to 17 reported drinking during the last 30 days. Girls in this age range report drinking, on average, 6 days per month and report consuming, on average, four drinks per sitting when they drink. Extrapolated out, this means that pregnant females aged 15 to 17 consume approximately 24 alcoholic beverages per month over the term of their pregnancy.[22]

In general, studies find a "dose–response" relationship between prenatal drinking and the number and pervasiveness of the developmental problems experienced by offspring. Women who drink frequently and who consume relatively more alcohol at each sitting are more likely to have offspring with serious developmental deficiencies. Stated differently, women who do not drink during pregnancy generally have the healthiest children with the fewest number of developmental problems. Women who drink intermittently generally produce children who fall in the middle of the distribution. Women who suffer from alcoholism, who drink intensely throughout their pregnancy, generally have children with the worst outcomes. It is important to note, however, that no safe level of prenatal alcohol consumption exists.

Heavy and prolonged drinking causes perhaps the best known outcome associated with prenatal alcohol exposure: **fetal alcohol syndrome** (FAS). FAS is one end of a broad spectrum of disorders, referred to as fetal alcohol spectrum disorders (FASD), associated with prenatal alcohol exposure. FAS is characterized by facial disfigurement, asymmetrical growth, and pervasive cognitive deficits. FAS is one of the leading causes of birth defects and mental retardation. The Centers for Disease Control and Prevention estimates that FAS affects 0.5 to 2.0 per 1000 births, with prevalence rates of 3.0 to 5.0 per 1000 births in minority groups. Of the 4 million live births each year in the United States, between 1000 and 6000 will be diagnosed with FAS. FAS is completely preventable.[23]

In utero exposure to alcohol, even at relatively low levels, has also been associated with a wide range of developmental problems that fall under the FASD umbrella. Usually less pervasive and sometimes less obvious than FAS, prenatal exposure to alcohol has been linked to multiple developmental problems, including attention deficit hyperactivity disorder (ADHD), learning deficits, reduced IQ, conduct disorder, delinquency, and criminal behavior in adulthood. One thing is clear about the effects of prenatal alcohol consumption: They are lifelong and affect every aspect of the youth's social, behavioral, and psychological functioning. Individuals exposed in utero to alcohol find it significantly more difficult to make accurate, reliable judgments concerning social interactions and complex social systems. They frequently have problems with expressive and receptive language, which makes it difficult to understand the unique and sometimes subtle differences in communication exchanges between two or more people, and they frequently encounter problems with self-regulation and self-control.[24]

The dangers of consuming alcohol during pregnancy are widely known, which explains why most women quit drinking once they become aware they are pregnant. Those who continue to drink are not randomly distributed—that is, they tend to share certain characteristics. Women who consume alcohol during their pregnancy are significantly more likely to also smoke cigarettes, to be dependent on alcohol, to be antisocial, to have multiple children with a variety of antisocial men, and to use other illicit substances.[25] Studies find, however, that even when these factors are controlled in statistical analyses, prenatal alcohol consumption continues to have a significant, direct effect on the problem behaviors of exposed youth.

That women who drink during pregnancy are different from those who do not brings up the thorny issue of genetic factors. Similar to maternal smoking, genetic factors shared by the mother and her offspring may account for the positive correlation between maternal alcohol consumption and the developmental problems of her children. Empirical tests of this possibility tend to show that genetic influences do account for a large proportion of variance in cognitive and behavioral outcomes of exposed youth, but, unlike the research on cigarette smoking, maternal alcohol consumption still predicts variation in these outcomes. D'Onofrio and colleagues found that, even with genetic and other environmental factors controlled, maternal alcohol consumption continued to predict variation in childhood behavioral problems.[26] Findings from Disney et al.'s analysis of data from the Minnesota Twin Family Study concur with prior studies.[27] Disney and colleagues examined 1252 adolescents and their parents. Structured interviews were used to assess maternal alcohol use during pregnancy and adolescent conduct problems. Even after controlling for the effects of parental substance abuse history, concurrent substance dependency, and a host of other potential confounders, alcohol consumption during pregnancy continued to predict variation in adolescent conduct problems. Mothers reported drinking an average of only three drinks per week over the term of their pregnancy, indicating that prenatal alcohol consumption has the potential for negative effects even at very low doses.

To summarize, prenatal exposure to alcohol can have pervasive, life-altering, negative influences. Although the degree of effect depends in part on the frequency and intensity of maternal drinking, a dose–response effect, negative outcomes have been detected even at very low doses.

Genetic factors likely moderate the effect of alcohol consumption on the developing embryo, with some of these modifiers extending a protective effect and some amplifying the negative effect. Current research is examining these relationships. Nonetheless, consumption of alcohol during pregnancy appears to be a substantive risk factor for a host of developmental problems.

Other Drugs

Images of infants born addicted to drugs undergoing withdrawal have served as a powerful motivator to curb maternal drug use. In the 1980s, it was the image of the "crack baby" that caused much social concern. Exposed in utero to the powerful effects of concentrated levels of cocaine, these youth were predicted to develop into high-rate criminals, to be forever handicapped intellectually and socially, and to cost the state millions of dollars in health care. Overall, illicit drug use by pregnant women is exceedingly rare. National estimates indicate that less than 4 percent of pregnant women use any form of illicit substance. The most commonly used illicit substance is marijuana, which most marijuana-using mothers smoke intermittently during pregnancy. There are, however, substantive racial differences in illicit drug use among pregnant women. Caucasians have the lowest rates (1.3 percent), followed by Hispanics (3.0 percent), and African Americans (7.2 percent). For Hispanics and Caucasians, illicit drug use declines once the mother realizes she is pregnant. The same does not hold true for African Americans, whose prevalence of illicit drug use does not differ between pregnant and nonpregnant status.[28]

Although illicit drug use by pregnant women is rare across the United States, it is, like other behavioral variables, concentrated in certain geographical areas. Addiction to drugs, mental health problems, and violence tend to co-occur in urban, inner-city neighborhoods. These patterns help to explain the dramatic differences in infant mortality. In cities such as Washington, DC or Cincinnati, Ohio, African American infant mortality rates approach 20 percent, whereas Caucasian infant mortality rates have declined to 5 percent. Illicit drug use helps to explain the substantial differences in infant mortality rates between these groups.[29]

Another consistent finding is that pregnant women who use illicit drugs are significantly more likely to also drink alcohol in excess across the term of their pregnancy and to smoke cigarettes throughout their pregnancy. These women are also more likely to be addicted to substances, such as heroin, cocaine, or alcohol; to be relatively young (between 15 and 19 years old); to be antisocial; and to have an above average number of prior born children. Of course, the large number of risk factors in the lives of these women makes it difficult to isolate the potentially harmful effects associated with any single risk factor. This is likely why studies conducted on animals show clearly the harmful effects of illicit drugs on offspring, but that some studies on humans show weak and inconsistent effects.

The most consistent evidence of the effects of illicit substances on embryonic development shows that drug-using mothers are significantly more likely to experience spontaneous abortions and complications during birth. Their children are significantly more likely to die of these complications, to be born premature and with low birth weight, and to die of sudden infant death syndrome. Infants born addicted to substances have the additional burden and health risks associated with drug withdrawal.[30]

Evidence of the long-term effects of illicit substances on the development of exposed children has produced mixed results. Most studies show a correlation between prenatal illicit drug use and reduced intellectual functioning,[31] whereas other studies find that drug-exposed youth can have normal levels of intellectual functioning.[32] Several studies find that prenatal exposure to illicit substances elevates the likelihood of future behavioral problems in exposed youths,[33] but other studies find that the effect is indirect, operating through deficits in language, self-control, and other personality characteristics.[34] No study finds that prenatal exposure to any illicit substance benefits the developing fetus or promotes positive human development.

Although it may be impossible to say with any scientific certainty that illicit drug use by pregnant women "causes" the soft neurological vulnerabilities witnessed in many drug-exposed children, it is only because drug-using women engage in so many risky ventures that isolating any one behavioral effect may be impossible. What is important to understand is not that research cannot always isolate unique causal effects of maternal drug use on the fetus, infant, and child, but that the combined effects of illicit and legal drug use, as well as a host of other pathological behavioral variables, place the exposed youth at high risk of multiple developmental problems—including uncontrolled aggression and violence across the life course.

Lead

Lead (Pb) is a naturally occurring heavy metal that has a wide range of industrial uses. It is also a potent neurotoxin. Since the 1960s, the Centers for Disease Control and Prevention has continually revised downward estimates of "blood lead levels of concern," meaning the level of exposure where deleterious clinical outcomes were thought to occur. In 1960 the level was 60 micrograms per deciliter (μg/dL) of blood. Today the threshold is established at 10 μg/dL of blood; however, the accumulated evidence appears to indicate that "soft" damage can occur at levels below the 10 μg/dL threshold.[35]

Until changes in the laws banned the use of lead in paint and in gasoline, it was a widely distributed pollutant in the air and over the ground. Changes in environmental laws, however, have significantly reduced overall exposure levels, especially in children. In Ohio, for instance, almost 17 percent of children in 1997 had blood lead levels in excess of 10 μg/dL. By 2006 that number had plummeted to only 2.3 percent. This pattern holds true in most states. Today, lead exposure is largely restricted to areas with very old housing stock, where the vestiges of lead-based paint remain. For most places, this translates into inner cities.

Lead mimics a calcium ion so the body cannot tell the difference between a lead particle and a particle of calcium derived from the ingestion of milk. Because it mimics calcium, lead can be stored in the bones. This is particularly problematic for pregnant women. During the first trimester, the fetus requires calcium for the formation of the skeletal system and other neurological functioning. One mechanism through which it obtains calcium is the maternal stores, largely from the soft areas of the mother's bone. Lead stored in the mother's bone becomes bioavailable, passing not only the calcium but also the lead from the mother to the fetus. Calcium is intimately involved in a range of neurological functions. It is implicated in neurotransmission, in the creation and pruning of synaptic pathways, and in the demyelination of axonal branches. When lead binds to sites in the brain, it disrupts calcium-dependent processes and thus may also disrupt a broad range of brain-based functions. High doses of lead, usually above 50 μg/dL, can lead to spontaneous abortion, birth defects, and a wide range of health-related problems.

The effects of lead exposure are pervasive. Studies clearly show that prenatal lead exposure is predictive of lower IQ.[36] Levels of exposure below 10 μg/dL also appear to be associated with lower IQ scores, ranging from 2 to 6 points, although disagreement still exists about the effects of lead at low levels.[37] Prenatal and postnatal exposure has also been linked to self-regulatory problems, to problems with hearing and vision, and to deficits in fine and gross motor abilities.[38]

Of primary concern, a series of studies shows a consistent link between pre- and postnatal lead ingestion and delinquency in adolescence and criminal behavior in adulthood.[39] These studies used a variety of measurement techniques, including k-line x-ray fluorescence spectroscopy and direct assessment of whole blood, across a range of samples with varying levels of exposure. The consistency of results strongly implicates prenatal and postnatal lead exposure in a range of conduct problems that appear stable across the life course. Of course, prenatal ingestion of lead cannot directly influence behavior later in life. Scientists believe that lead compromises areas of the brain associated with behavioral regulation. Direct evidence in support of this hypothesis is in short supply; however, recent brain-imaging studies found that lead exposure is associated with reduced gray matter volume in the prefrontal cortex and the anterior cingulate gyrus, especially for males.[40] These areas of the brain aid in self-regulation, delayed gratification, and other executive functions.

POLICY CONSEQUENCES

A biosocial understanding of the development of persistent offending focuses on the interconnections between internal biological factors, such as genes and heritability; external biological forces, such as lead exposure; and how these factors play out in the social environment. Prenatal exposure to heavy metals, pesticides, drugs, and maternal stress is structured by social forces. Even so, these factors clearly have many detrimental effects on the development of the fetus's central nervous system. Thus, understanding biological influences alone is not sufficient to understand the totality of the problem, just as understanding the geographical distribution of exposure variables tells us nothing of their etiological influences.

There is absolutely no reason why children should continue to be exposed to elevated levels of lead and other environmental neurotoxins. Although the rates of lead poisoning have declined dramatically since the 1980s, much work remains to be done. Lead is pervasive in many inner-city communities where the housing stock is old and blighted. Cities, such as Cincinnati, Ohio; Pittsburgh, Pennsylvania; and Washington, DC, retain some of the oldest housing stock in the country and thus continue to permit pregnant mothers and their children to be exposed to very high levels of lead. In a Cincinnati sample of children measured prenatally through the first 6.5 years of life, for example, elevated blood lead levels (above 10 µg/dL of blood) were a normal part of their development. Every child in the sample had an elevated blood lead level throughout his or her childhood. The clear policy recommendation is to remove the lead from the areas or to remove the people from areas of concentrated lead exposure.

It is past time to take seriously maternal drug use and, especially, drug abuse. The evidence of a general deleterious effect of intrauterine exposure to drugs and other neurotoxins on the developing fetus should be cause for focused interventions. Efforts designed to identify drug-abusing pregnant women and to respond with state-enforced interventions have been met with little success. The available evidence shows that many factors condition or modify the effect neurotoxins have on the central nervous system of the fetus. By last count there were 35 states with some form of smoking ban. Many of these bans came about because of the dangers posed by secondhand smoke. Any observer will note the obvious contradiction: Advocates of smoking bans point to the alleged connection between secondhand smoke and cancer as sufficient reason to justify state intrusion into the operation of legitimate business. The evidence on the health effects of secondhand smoke is less consistent and less compelling than the evidence linking maternal smoking or maternal drug use to health and behavioral problems in children, yet there are no calls to ban such behavior in expectant mothers, and no calls to isolate the pregnant addict so she can get the medical help she needs and no longer jeopardize the health and welfare of her developing child.

CONCLUSION

Addiction to substances is a horrible affliction that destroys lives and families. Many recovering addicts and alcoholics struggle in their day-to-day lives to resist continued use, but the real hardship comes when they have to pick up the pieces of their lives. Because addiction hijacks the brain and short circuits rational thinking, the lives of addicts are littered with damaged and irreparable relationships, with sometimes lengthy arrest and incarceration records, and, for some, with children they have never seen or gotten to know. For pregnant addicts the costs are even greater: HIV infection and AIDS are rampant in this population, and many have their newborn children placed in foster care. Addiction impairs rational decision making and, as such, brings with it life-altering consequences. Yet, the consequences faced by the addict and alcoholic pale in comparison with the consequences experienced by the fetus or newborn child. Addiction is a treatable disease, and intervention may save not only the baby's life, but also the rest of the life of the mother.

GLOSSARY

DNA—a chemical code found in the nucleus of cells that is responsible for guiding the development and functioning of all living organisms

Fetal alcohol syndrome—one end of a broad spectrum of disorders, referred to as fetal alcohol spectrum disorders (FASD), associated with prenatal alcohol exposure

Fetal development—the period of 10 months in which the fetus is developing inside the mother's uterus

Gene × environment interactions—the process whereby an environmental stimulus triggers a genetic effect or vice versa

Magnetic resonance imaging—an imaging technique that uses radio waves and magnetic fields to map organ structure (e.g., brain structure)

Teratogens—any type of toxin that could cause a birth defect, including defects to the developing brain

NOTES

1. Walsh, A., & Ellis, L. (2003). *Biosocial criminology: Challenging environmentalism's supremacy*. Hauppauge, NY: Nova Science Publishers; Walsh, A., & Ellis, L. (2004). Ideology: Criminology's Achilles' heel. *Quarterly Journal of Ideology, 27*, 1–25.
2. Elliott, D. S., Wilson, W. J., Huizinga, D., Sampson, R. J., Elliott, A., & Rankin, B. (1996). The effects of neighborhood disadvantage on adolescent development. *Journal of Research in Crime and Delinquency, 33*, 389–426; Sampson, R. J., Morenoff, J. D., & Earls, F. (1999). Beyond social capital: Spatial dynamics of collective efficacy for children. *American Sociological Review, 64*, 633–660.

3. Garbarino, J. (1992). *Children and families in the social environment* (2nd ed.). New York: Aldine; Guterman, N. B., Cameron, M., & Staller, K. (2000). Definitional and measurement issues in the study of community violence among children and youths. *Journal of Community Psychology, 28*, 571–587.

4. Walsh, A. (2002). Essay review: Companions in crime: A biosocial perspective. *Human Nature Review, 2*, 169–178; Walsh, A. (2002). *Biosocial criminology: Introduction and integration*. Cincinnati, OH: Anderson.

5. Beaver, K. M., Wright, J. P., DeLisi, M., & Vaughn, M. G. (2008). Genetic influences on the stability of low self-control: Results from a longitudinal sample of twins. *Journal of Criminal Justice, 36*, 478–485; Beaver, K. M., Wright, J. P., DeLisi, M., & Vaughn, M. G. (2008). Desistance from delinquency: The marriage effect revisited and extended. *Social Science Research, 37*, 736–752; Haberstick, B. C., Schmitz, S., Young, S. E., & Hewitt, J. K. (2006). Genes and developmental stability of aggressive behavior problems at home and school in a community sample of twins aged 7–12. *Behavior Genetics, 36*, 809–819; Kim-Cohen, J., Caspi, A., Taylor, A., Williams, B., Newcombe, R., Craig, I. W., et al. (2006). MAOA, maltreatment, and gene–environment interaction predicting children's mental health: New evidence and a meta-analysis. *Molecular Psychiatry, 11*, 903–913; Lorber, M. F. (2004). Psychophysiology of aggression, psychopathy, and conduct problems: A meta-analysis. *Psychological Bulletin, 130*, 531–552; Lynam, D. R., Loeber, R., & Stouthamer-Loeber, M. (2008). The stability of psychopathy from adolescence into adulthood: The search for moderators. *Criminal Justice and Behavior, 35*, 228–243.

6. Haynes, E. N., Kalkwarf, H. J., Hornung, R., Wenstrup, R., Dietrich, K., & Lanphear, B. P. (2003). Vitamin D receptor Fok1 polymorphism and blood lead concentration in children. *Environmental Health Perspectives, 111*, 1665–1669; Hopkins, M. R., Ettinger, A. S., Hernández-Avila, M., Schwartz, J., Téllez-Rojo, M. M., Lamadrid-Figueroa, H., et al. (2008). Variants in iron metabolism genes predict higher blood lead levels in young children. *Environmental Health Perspective, 116*, 1261–1266; Rezende, V. B., Barbosa, F., Montenegro, M. F., Sandrim, V. C., Gerlach, R. F., & Tanus-Santos, J. E. (2007). An interethnic comparison of the distribution of vitamin D receptor genotypes and haplotypes. *Clinica Chimica Acta, 384*, 155–159.

7. Craig, I. W., Harper, E., & Loat, C. S. (2004). The genetic basis for sex differences in human behaviour: Role of the sex chromosomes. *Annals of Human Genetics, 68*, 269–284.

8. American Lung Association Research and Program Services. (2008). Trends in tobacco use. *Epidemiology and Statistics Unit*. Washington, DC: American Lung Association.

9. Child Trends DataBank. (2012). *Adolescents: Health & well-being status*. Retrieved from www.childtrendsdatabank .org/?q=node/42.

10. Substance Abuse and Mental Health Services Administration. (2012). *Find substance abuse and mental health treatment*. Retrieved from www.samhsa.gov/treatment.

11. Nishimura, B. K., Adams, E. K., Melvin, C. L., Merritt, R. K., Tucker, P. J., Stuart, G., et al. (2007). *Prenatal smoking databook*. Washington, DC: Centers for Disease Control and Prevention.

12. Abreu-Villaca, Y., Seidler, F. J., & Slotkin, T. A. (2004). Does prenatal nicotine exposure sensitize the brain to nicotine-induced neurotoxicity in adolescence? *Neuropsychopharmacology, 29*, 1440–1450; Muneoka, K., Ogawa, T., Kamei, K., Muraoka, S., Tomiyoshi, R., Mimura, Y., et al. (1997). Prenatal nicotine exposure affects the development of the central serotonergic system as well as the dopaminergic system in rat offspring: Involvement of route of drug administrations. *Developmental Brain Research, 102*, 117–126.

13. Wang, X., Zuckerman, B., Pearson, C., Kaufman, G., Chen, C., Wang, G., et al. (2002). Maternal cigarette smoking, metabolic gene polymorphism, and infant birth weight. *Journal of the American Medical Association, 287*, 195–202.

14. Batty, G. D., Der, G., & Deary, I. J. (2006). Effect of maternal smoking during pregnancy on offspring's cognitive ability: Empirical evidence for complete confounding in the U. S. national longitudinal survey of youth. *Pediatrics, 118*, 943–950; Brennan, P. A., Grekin, E. R., & Mednick, S. A. (1999). Maternal smoking during pregnancy and adult male criminal outcomes. *Archives of General Psychiatry, 56*, 215–219; Gibson, C. L., & Tibbetts, S. G. (1998). Interaction between maternal cigarette smoking and Apgar scores in predicting offending behavior. *Psychological Reports, 83*, 579–586; Piquero, A. R., Gibson, C. L., Tibbetts, S. G., Turner, M. G., & Katz, S. H. (2002). Maternal cigarette smoking during pregnancy and life-course-persistent offending. *International Journal of Offender Therapy & Comparative Criminology, 46*, 231; Rasanen, P., Hakko, H., Isohanni,

M., Hodgins, S., Jarvelin, M. R., & Tiihonen, J. (1999). Maternal smoking during pregnancy and risk of criminal behavior among adult male offspring in the northern Finland 1966 birth cohort. *American Journal of Psychiatry, 156,* 857–862; Weitzman, M., Gortmaker, S., & Sobol, A. (1992). Maternal smoking and behavior problems of children. *Pediatrics, 90,* 342–349; Wang et al. (2002), see Note 13.

15. Tsai, H., Lui, X., Mestan, K., Yu, Y., Zhang, S., Fang, Y., et al. (2008). Maternal cigarette smoking, metabolic gene polymorphisms, and preterm delivery: New insights on GxE interactions and pathogenic pathways. *Human Genetics, 123,* 359–369.
16. Kishi, R., Sata, F., Yoshioka, E., Ban, S., Sasaki, S., Konishi, K., et al. (2008). Exploiting gene-environment interaction to detect adverse health effects of environmental chemicals on the next generation. *Basic & Clinical Pharmacology & Toxicology, 102,* 191–203.
17. Jaffee, S. R., Caspi, A., Moffitt, T. E., Dodge, K. A., Rutter, M., Taylor, A., et al. (2005). Nature X nurture: Genetic vulnerabilities interact with physical maltreatment to promote conduct problems. *Development and Psychopathology, 17,* 67–84; Jaffee, S. R., Belsky, J., Harrington, H., Caspi, A., & Moffitt, T. E. (2006). When parents have a history of conduct disorder: How is the caregiving environment affected? *Journal of Abnormal Psychology, 115,* 309–319.
18. Silberg, J. L., Parr, T., Neale, M. C., Rutter, M., Angold, A., & Eaves, L. J. (2003). Maternal smoking during pregnancy and risk to boys' conduct disturbance: An examination of the causal hypothesis. *Biological Psychiatry, 53,* 130–135.
19. Maughan, B., Taylor, A., Caspi, A., & Moffitt, T. E. (2004). Prenatal smoking and early childhood conduct problems testing genetic and environmental explanations of the association. *Archives of General Psychiatry, 61,* 836–843.
20. Rivkin, M. J., Davis, P. E., Lemaster, J. L., Cabral, H. J., Warfield, S. K., Mulkern, R. V., et al. (2008). Volumetric MRI study of brain in children with intrauterine exposure to cocaine, alcohol, tobacco, and marijuana. *Pediatrics, 121,* 741–750.
21. Durston, S., Hulshoff Pol, H. E., Casey, B. J., Giedd, J. N., Buitelaar, J. K., & van Engeland, H. (2001). Anatomical MRI of the developing human brain: What have we learned? *Journal of the American Academy of Child and Adolescent Psychiatry, 40,* 1012–1020; Sowell, E. R., Thompson, P. M., & Toga, A. W. (2004). Mapping changes in the human cortex throughout the span of life. *The Neuroscientist, 10,* 372–392; Raine, A., Lencz, T., Bihrle, S., LaCasse, L., & Colletti, P. (2000). Reduced prefrontal gray matter volume and reduced autonomic activity in antisocial personality disorder. *Archives of General Psychiatry, 57,* 119–127.
22. National Survey on Drug Use and Health. (2008). *Alcohol use among pregnant women and recent mothers: 2002 to 2007.* Retrieved from www.samhsa.gov/data/2k8/pregnantAlc/pregnantAlc.htm.
23. Centers for Disease Control and Prevention. (2004). *Fetal alcohol syndrome: Guidelines for referral and diagnosis.* Atlanta, GA: Author.
24. Kelly, S. J., Day, N., & Streissguth, A. P. (2000). Effects of prenatal alcohol exposure on social behavior in humans and other species. *Neurotoxicology and Teratology, 22,* 143–149; Kodituwakku, P. W., Kalberg, W., & May, P. A. (2001). The effects of prenatal alcohol exposure on executive functioning. *Alcohol Research & Health, 25,* 192–199; Mattson, S. N., Riley, E. P., Gramling, L., Delis, D. C., & Jones, K. L. (1997). Heavy prenatal alcohol exposure with or without physical features of fetal alcohol syndrome leads to IQ deficits. *Journal of Pediatrics, 131,* 718–721; Mattson, S. N., Schoenfeld, A. M., & Riley, E. P. (2001). Teratogenic effects of alcohol on brain and behavior. *Alcohol Research & Health, 25,* 185–192; Olson, H. C., Streissguth, A. P., Sampson, P. D., Barr, H. M., Bookstein, F. L., & Thiede, K. (1997). Association of prenatal alcohol exposure with behavioral and learning problems in early adolescence. *Journal of the American Academy of Child & Adolescent Psychiatry, 36,* 1187–1194.
25. Disney, E. R., Iacono, W., McGue, M., Tully, E., & Legrand, L. (2008). Strengthening the case: Prenatal alcohol exposure is associated with increased risk for conduct disorder. *Pediatrics, 122,* e1225–e1230.
26. D'Onofrio, B. M., Hulle, C. A., Waldman, I. D., Rodgers, J. L., Harden, K. P., Rathouz, P. J., et al. (2008). Smoking during pregnancy and offspring externalizing problems: An exploration of genetic and environmental confounds. *Development and Psychopathology, 20,* 139–164.
27. Disney et al. (2008), see Note 25.

28. National Household Survey on Drug Abuse. (2001). *Pregnancy and illicit drug use*. Retrieved from www.oas .samhsa.gov/2k2/pregDU/pregDU.htm.

29. El-Mohandes, A., Herman, A. A., Nabil El-Khorazaty, M., Katta, P. S., White, D., & Grylack, L. (2003). Prenatal care reduces the impact of illicit drug use on perinatal outcomes. *Journal of Perinatology, 23*, 354–360.

30. Bauer, C. R., Langer, J. C., Shankaran, S., Bada, H. S., Lester, B., Wright, L. L., et al. (2005). Acute neonatal effects of cocaine exposure during pregnancy. *Archives of Pediatrics and Adolescent Medicine, 159*, 824–834; Fares, I., McCulloch, K. M., & Raju, T. N. (1997). Intrauterine cocaine exposure and the risk for sudden infant death syndrome: A meta-analysis. *Journal of Perinatology, 17*, 179–182; Klonoff-Cohen, H., & Lam-Kruglick, P. (2001). Maternal and paternal recreational drug use and sudden infant death syndrome. *Archives of Pediatrics and Adolescent Medicine, 155*, 765–770.

31. Pulsifer, M. B., Butz, A. M., O'Reilly Foran, M., & Belcher, H. M. E. (2008). Prenatal drug exposure: Effects on cognitive functioning at 5 years of age. *Clinical Pediatrics, 47*, 58–65.

32. Singer, L. T., Minnes, S., Short, E., Arendt, R., Farkas, K., Lewis, B., et al. (2004). Cognitive outcomes of preschool children with prenatal cocaine exposure. *Journal of the American Medical Association, 291*, 2448–2456.

33. Bendersky, M., Bennett, D., & Lewis, M. (2006). Aggression at age 5 as a function of prenatal exposure to cocaine, gender, and environmental risk. *Journal of Pediatric Psychology, 31*, 71–84; Chatterji, P., & Markowitz, S. (2001). The impact of maternal alcohol and illicit drug use on children's behavior problems: Evidence from the children of the national longitudinal survey of youth. *Journal of Health Economics, 20*, 703–731.

34. Paschall, M. J., & Fishbein, D. H. (2002). Executive cognitive functioning and aggression: A public health perspective. *Aggression and Violent Behavior, 7*, 215–235.

35. Lanphear, B. P., Hornung, R., Khoury, J., Yolton, K., Baghurst, P., Bellinger, D. C., et al. (2005). Low-level environmental lead exposure and children's intellectual function: An international pooled analysis. *Environmental Health Perspectives, 113*, 894–899.

36. Barth, A., Schaffer, A. W., Osterode, W., Winker, R., Konnaris, C., Valic, E., et al. (2002). Reduced cognitive abilities in lead-exposed men. *International Archives of Occupational and Environmental Health, 75*, 394–398; Dietrich, K. N., Succop, P. A., Berger, O. G., Hammond, P. B., & Bornschein, R. L. (1991). Lead exposure and the cognitive development of urban preschool children: The Cincinnati lead study cohort at age 4 years. *Neurotoxicology and Teratology, 13*, 203–211.

37. Bellinger, D. C., Needleman, H. L., Eden, A. N., Donohoe, M. T., Canfield, R. L., Henderson, C. R., et al. (2003). Intellectual impairment and blood lead levels. *New England Journal of Medicine, 349*, 500–502; Bellinger, D., Leviton, A., Alfred, E., & Rabinowitz, M. (1994). Pre- and postnatal lead exposure and behavior problems in school-age children. *Environmental Research, 66*, 12–30; Ernhart, C. B. (2006). Effects of lead on IQ in children. *Environmental Health Perspectives, 114*, A85–A86.

38. Needleman, H. L., McFarland, C., Ness, R. B., Fienberg, S. E., & Tobin, M. J. (2002). Bone lead levels in adjudicated delinquents: A case control study. *Neurotoxicology and Teratology, 24*, 711–717; Dietrich, K. N., Succop, P. A., Berger, O. G., & Keith, R. W. (1992). Lead exposure and the central auditory processing abilities and cognitive development of urban children: The Cincinnati lead study cohort at age 5 years. *Neurotoxicology and Teratology, 14*, 51–56.

39. Mendelsohn, A. L., Dreyer, B. P., Fierman, A. H., Rosen, C. M., Legano, L. A., Kruger, H. A., et al. (1998). Low-level lead exposure and behavior in early childhood. *Pediatrics, 101*, e10; Needleman, H. L., Riess, J. A., Tobin, M. J., Biesecker, G. E., & Greenhouse, J. B. (1996). Bone lead levels and delinquent behavior. *Journal of the American Medical Association, 275*, 363–369; Wright, J. P., Dietrich, K. N., Ris, M. D., Hornung, R. W., & Wessel, S. D. (2008). Association of prenatal and childhood blood lead concentrations with criminal arrests in early adulthood. *PLoS Medicine, 5*, 732–740.

40. Cecil, K. M., Brubaker, C. J., Adler, C. M., Dietrich, K. N., & Altaye, M. (2008). Decreased brain volume in adults with childhood lead exposure. *PLoS Medicine, 5*, e112.

CHAPTER FOUR

Developmental Neurobiology from Embryonic Neuron Migration to Adolescent Synaptic Pruning: Relevance for Antisocial Behavior

Anthony Walsh and Ilhong Yun

KEY TERMS

Age–crime curve
Autonomic nervous system
Functional magnetic resonance imaging
Neurons
Prefrontal cortex
Synapses

INTRODUCTION

The brain is the seat of all that we know, feel, and experience; it is the repository of what we were, what we are, and what we may become; and it is the most devilishly complex organization of matter in the universe. This 2-pound, wrinkly, gelatinous blob of wonder takes about 60 percent of our genes to construct, and although it is only 2 percent of total body weight, it consumes 20 percent of the body's energy as it perceives, evaluates, and integrates input from the environment and directs our responses to it.[1] For all the brain's importance, social scientists tend to dismiss it on the assumption that it is a constant, and we all know that constants cannot explain variables. But brains are as variable as environments, perhaps even more so because environments, good or bad, conspire with our genes to build our brains. If negative environmental circumstances and life events constitute an elevated risk for antisocial behavior, we would like to know why at a fundamental neurobiological level. As John Wright and his colleagues point out, social scientists fail to realize that "If deprivation causes or promotes behavioral maladaptation, it has to do so through its influence on the central nervous system."[2]

According to neurobiologist Bruce Perry, there are three key brain systems relevant to the ultimate evolutionary concerns of all sexually reproducing species: neural mechanisms that (1) facilitate responses to threats to wellbeing, (2) facilitate the protection and nurturing of the young, and (3) facilitate mate selection and reproduction. We are only concerned with mechanisms 1 and 2 here.[3]

As developmentalists, we take issue with implications that childhood experiences deserve no special place in the pantheon of behavioral causes. The earliest years, when the brain is most pliable, chisel neural networks in ways that are not easily altered later in life. Later experiences (e.g., "turning points" such as educational, job, and marriage opportunities) are interpreted differently by different neural networks because the influence of contemporary experience is channeled along physical pathways laid down during childhood in response to environmental experiences.

Sampson and Laub accuse developmental theories of preformationism, which implies a latent predetermined form—of waiting only for a developmental process to make it apparent, like dollops of dough waiting for the oven to bake them into cookies.[4] To call someone a preformationist in the old days was akin to calling someone today a genetic determinist (i.e., everything a person is and does is predetermined by genetic inheritance). This is not what we or any other developmental theorist mean by developmental. Biosocial theorists do believe that genes set us on a particular developmental trajectory, but they are aware that the vagaries of the environment, particularly in the earliest days of central nervous system development, can send this trajectory askew.

Terrie Moffitt's developmental theory of criminal behavior serves as our guide for discussing the link between developmental neurobiology and antisocial behavior. The theory posits a dual pathway to delinquent/criminal offending, one in which offending is limited to adolescence and the other which begins before adolescence and continues well into adulthood. The first pathway is called the adolescent-limited pathway and the second is called the life-course-persistent pathway.

Moffitt proposes that neuropsychological and temperamental impairments initiate a cumulative process of negative person–environment interactions for life-course-persistent offenders that results in a life-course trajectory that propels them toward ever-hardening antisocial attitudes and behaviors. The temperamental and neuropsychological deficits most often mentioned—low IQ, hyperactivity, inattentiveness, negative emotionality, and low impulse control—are consistently and robustly found to be correlated with criminality. The problems arise from a combination of genetic and environmental effects on central nervous system development, such as maternal substance abuse during pregnancy, poor nutrition, birthing difficulties, and poor nurturing.[5]

Adolescent-limited offenders have a different developmental history that puts them on a prosocial trajectory that is temporarily derailed at adolescence. These youths are statistically normal, and we may view their offending as adaptive responses to conditions and transitional events that temporarily divert them from their basically prosocial life-course trajectories. In other words, they are responding to the social and biological changes of adolescence in a totally normative way, including disengaging from parents and bonding with peers. These processes are discussed later in this chapter; for now, we concentrate on the specifics of neurological deficits that arise during the early developmental process that give rise to elevated risk of antisocial behavior.

EARLY POSTNATAL BRAIN DEVELOPMENT

After the neurons have migrated and settled in their appointed spots, the brain starts wiring itself together in a process known as synaptogenesis. Synaptogenesis involves the lifelong collaboration of genetic and environmental influences, with wiring patterns differing from person to person according to the kinds of physical, social, and cultural environments they have encountered, and according to the genotype they have inherited. The neonate's cerebral cortex consists of small and underdeveloped neurons and relatively unmyelinated axons with few dendrites. During the first few months, dendrites proliferate and specialized glial cells wrap around axons to begin the process of myelination, working bottom up from the most evolutionarily primitive areas of the brain to the most recent and most human of areas, the **prefrontal cortex** (PFC). The PFC is not completely myelinated until late adolescence/early adulthood.[6]

The most important issue in the process of synaptogenesis is not the birth of **synapses**, but rather which ones survive the competition for synaptic space. Although the brain creates and eliminates synapses throughout life, creation exceeds elimination in the first 2 or 3 years. Production and elimination are roughly balanced thereafter up until adolescence, after which elimination exceeds production.[7] The density of synaptic connections is higher than it will ever be again at around 2 years of age, and about half of them are eventually eliminated in use-dependent fashion. This process of selective production and elimination has been termed "neural Darwinism" by Nobel Prizewinner Gerald Edelman. Edelman posits a selection process among competing brain modules (populations of synapses) with the winners being decided according to how functionally viable they prove to be in the organism's environment. The brain's neuronal populations thus evolve in somatic time very much like species evolve in geological time by the selective elimination and retention of genes according to how well they served their organism's adaptive concerns.[8]

Retention of neural networks is very much a use-dependent process, with those that exchange information frequently and strongly being favored. In the competition for scarce synaptic space, the game "is biased in favor of the [neuron] populations that receive the greatest amount of stimulation during early development."[9] Experiences with strong emotional content are accompanied by strong electrical impulses, and if these impulses are frequently made, the **neurons** involved become more sensitive and responsive to similar stimuli in the future.[10] Frequently activated neurons are thus primed to fire at lower stimulus thresholds once voltage-dependent neurological tracks have been laid down. This process is captured in the neuroscientist's pithy saying: "The neurons that fire together, wire together; those that don't, won't."[11]

Neuroscientists distinguish between two brain developmental processes: experience expected and experience dependent.[12] Experience-expected mechanisms reflect the phylogenic history of the species and are hard-wired in the sense that every organism inherits species-typical brain structures that function identically because they are produced by a common pool of genetic material.[13] Experience-expected processes evolved as neural readiness during certain "critical" or "sensitive" developmental periods to incorporate information that is vital to an organism and ubiquitous in its environment. Over eons of evolutionary time, natural selection has recognized that acquiring survival mechanisms such as sight, speech, depth perception, aversion to insects and waste products, mobility, sexual maturation, and affectionate bonds are vital and has provided for mechanisms designed to take advantage of experiences occurring naturally within the normal range of human environments. Preexperiential brain organization frames or orients experiences so that all members of the species respond consistently and stereotypically to vital stimuli.[14]

Experience-dependent mechanisms, on the other hand, reflect the brain's ontogenic plasticity (the brain's ability to functionally calibrate itself in response to internal and external experiences) in that its further development depends on the experience of the individual organism. Language acquisition helps us to differentiate between experience-expected and experience-dependent processes. The capacity to speak is entirely a genetically hard-wired experience-expected trait of all members of the human species, although it requires aural exposure to language to activate that capacity. What language(s) an individual speaks, on the other hand, is entirely an experience-dependent cultural phenomenon.

EARLY ENVIRONMENTAL INFLUENCES ON BRAIN WIRING

Rowe remarked, "the affection dimension of child rearing appears to pull in more correlates with child behavior than any other dimension."[15] Human infants are adapted to demand the formation of strong affectionate bonds via contact stimuli with caregivers, and there are many negative outcomes associated with the failure to form them, including antisocial behavior.[16] The trajectory toward selective evolutionary pressures for the formation of affectionate bonds was probably precipitated by simultaneous pressures posed by the rapidly (in an evolutionary sense) increasing cranial capacity (because increasing hominid intelligence required bigger brains) of our species from *Australopithecus* (M = 450 cc) to *Homo sapiens* (M = 1350 cc). Because early hominid female pelvises were probably shaped by natural selection to satisfy upright posture and bipedalism (thus narrowing the birth canal) more than for increased fetal cranial size, an evolutionary conflict arose between the obstetrical and postural requirements of early hominid females.[17]

The evolutionary solution to the obstetrics/posture conflict was for infants to be born at even earlier stages of development as cerebral mass increased. Human infants experience 25 percent brain growth inside the womb and 75 percent growth outside the womb, and are thus extremely dependent on caregivers for an extended period of developmental time. The period of brain growth in the womb is called uterogestation and the period outside the womb is called exterogestation. The high degree of incompleteness of the human brain at birth ensures a greater role for the extrauterine environment in its development than is true of any other species.

A species burdened with extremely altricial young must have experienced strong selection pressures for mechanisms designed to ensure the young would be nurtured for an extended period of time. Montagu stresses the importance of a "continuous symbiotic relationship between mother and child" (across the two gestation periods), and calls this relationship love. He goes on to add: "It is, in a very real and not in the least paradoxical sense, even more necessary to love than it is to live, for without love there can be no healthy growth or development, no real life." Modern science tells us that love confers enormous benefits in terms of healthy functioning in human beings through its impact on central and peripheral nervous system functioning, and that there are negative consequences from not receiving it during the earliest period of life.[18]

The proximate mechanisms of mother–infant bonds of affection are probably organized by the birthing process itself and consolidated by mother–infant contact and interaction during the immediate postpartum period. Oxytocin, a neuropeptide that has been called the "cuddle chemical," underlies these mechanisms. It is synthesized in the hypothalamus and is stimulated by environmental events such as the birthing process, infant distress, and breastfeeding. Breastfeeding, which combines the panoply of sight, sound, smell, touch, and the tangible evidence in the mother's arms that affirms her womanhood, stimulates the release of oxytocin, which intensifies the warm feeling that released it in a felicitous feedback loop. The oxytocin released by breastfeeding is related to mothers' reduced sensitivity to environmental stressors, which allows for greater sensitivity to the infant. Lactating mothers show significantly fewer stress responses to infant stimuli, as determined by skin conductance and cardiac response measures, than nonlactating mothers, and the oxytocin-generated sense of emotional warmth motivates significantly greater desire to pick up their infants in response to infant-presented stimuli.[19]

Breastfeeding has many positive developmental effects to recommend itself to us, including superior cognitive development. In a large randomized trial of 13,889 breastfeeding women, half were provided with incentives to continue prolonged and exclusive breastfeeding and the other half simply followed their usual hospital and outpatient follow-up care. Following these groups for 6 years, it was found that the children breastfed for an extended period had a mean IQ almost six points higher than the control group children and also received higher academic ratings from teachers.[20] The randomized experimental design allowed researchers to measure breastfeeding effects on children's cognitive development without possible biasing confounds such as the known positive relationship between mothers' IQ and the probability of prolonged breastfeeding.

Although the researchers were confident that the breastfeeding–IQ relationship was causal, it could not be determined if it was due to the nutritional constituents of breastmilk or to mother–child interactions, including the tactile stimulation experienced during the process of breastfeeding. Tactile stimulation has many positive effects on brain development via its ability to release the quiescent features of oxytocin and the reward features of dopamine. The medical literature is unanimous in its praise of tactile stimulation for infants; it confers enormous benefits on them and is recommended by physicians for the optimal brain development of low-birth-weight infants and infants who have suffered some kind of head trauma. Breastfeeding and the tactile stimulation accompanying it is an experience-expected feature of human brain development, and the failure of the infant to experience it may be viewed as the deprivation of important experience-expected developmental input.[21]

There are many studies of orphanage-reared children that bring home the importance of early tactile stimulation for developing healthy attachment, but perhaps the most interesting is one that looked at Russian and Romanian orphans who had been in orphanages an average of 16.6 months before being fostered to American families. After being with their American families for an average of 34.6 months they showed significantly lower base levels of the neuropeptides vasopressin and oxytocin compared with a control group of American children reared by their biological parents.[22] They also showed significantly lower levels of these peptides after being experimentally exposed to the kinds of interaction with their mothers that normally increases these levels. In effect, these children had been reared during a period in life without the frequent tactile comfort that is very likely an experience-expected feature of human rearing. Being so deprived, they become vulnerable to difficulties in forming secure relationships with caregivers and in forming social bonds with the wider society. Thus, the patterns of neuropeptide release offer palpable evidence that the neurohormonal machinery of such children has not been calibrated to expect those bonds and relationships.

The neuroscience literature is replete with imaging and electroencephalographic studies documenting abnormal brain connectivity (and therefore functioning) of children subjected to severe early socioemotional (contact comfort) deprivation. These connection abnormalities tend to be primarily those between the higher cognitive areas such as the PFC and emotional areas such as the amygdala. The lack of frequent bodily contact with the mother has to be interpreted by the infant's experience-expected mechanisms as abandonment, because the infant can only "think" with its skin. The infant's contact comfort derived from the sensitive responses of the mother (and other caregivers) during times of duress tells it that everything is okay: "She's there for me," "I'm safe," "All's right in my world." If this contact input is not forthcoming, the infant's underspecified and miswired circuits interpret otherwise.[23]

THE BRAIN AND EXPOSURE TO VIOLENCE

One of the major areas of concern for criminologists is violence. Violence is rightly condemned as morally repugnant, but because it is painfully and obviously in the behavioral repertoire of our species, it must have served some useful evolutionary purpose. In the Pleistocene epoch in which our most human traits evolved, one could not just dial 911 to get others to take care of threats to one's self or one's property. It would thus be beneficial to have the wherewithal to "take care of business" one's self and to cultivate a reputation of being a person who could. In some environments today where social control has broken down, a "bad ass" reputation is a most valuable asset, and taking violent action may be quite rational (adaptive) when one's person or property is threatened.

From an evolutionary point of view, the major long-term factor in violence instigation is how much violence a person has been exposed to in the past. When a person has observed and experienced many acts of violence there is a feedback effect; each violent act observed makes observers feel more at risk and therefore more likely to resort to preemptive violence themselves. Impulsive aggression is often the proximate behavioral expression of a brain wired by frequent exposure to violence. Individuals whose brains are wired in response to violent environments naturally come to expect hostility from others and behave accordingly, but by doing so they invite the very hostility for which they are on guard. When further hostility is forthcoming, it confirms the belief that the world is a dangerous and hostile place and sets in motion a vicious circle of negative expectations and confirmations.[24]

Chronic exposure to violence, even vicarious exposure via violent video games, also desensitizes viewers to it and makes them callous and indifferent to the suffering of others. Desensitization has been assessed by measures such as heart rate and galvanic skin responses when exposed to depictions of violence. **Functional magnetic resonance imaging** (fMRI) studies show decreased activity in brain structures that regulate aggression (e.g., anterior cingulate gyrus, and PFC areas) and increased activity in structures associated with increased aggression, such as the amygdala. Experimental studies such as these are conducted with subjects exposed to chronic vicarious violence, not real violence aimed at themselves or at loved ones. Witnessing and experiencing violence "upfront and personal" is presumably captured on the neural circuitry of children more strongly, and plants a warning sign there that the world is a dangerous place in which they must be prepared to protect their interests by violent means if necessary.[25]

ABUSE, NEGLECT, STRESS, AND BRAIN DEVELOPMENT

If the loving touch exemplified by breastfeeding positively affects brain development, the evil touch of maltreatment negatively affects brain development in an experience-dependent fashion. Maltreatment, whether it is experienced as physical, sexual, or psychological abuse, or as neglect, is stressful, particularly if experienced at an age in which children's brains are most plastic (plasticity unfortunately implies vulnerability as well as opportunity). Stress is a state of psychophysiological arousal accompanying perceived challenges to an organism's wellbeing. It is a normal adaptive part of life without which we would be seriously handicapped in our ability to meet these challenges, but protracted and toxic stress damages vital memory storage and behavioral regulatory regions such as the amygdala and hippocampus. Individuals differ in their ability to manage stress, partly for genetic reasons and partly for experiential reasons that have been captured in the brain's circuitry. Thus, frequent activation of stress response mechanisms may lead to their dysregulation, leading in turn to a number of psychological, emotional, and behavioral problems.[26]

The stress response is mediated by two separate but interrelated systems controlled by the hypothalamus: the **autonomic nervous system** (ANS) and the hypothalamic-pituitary-adrenal (HPA) axis. The ANS has two complementary systems: sympathetic and parasympathetic. When an organism perceives a threat to its wellbeing the hypothalamus directs the sympathetic system to mobilize the body for vigorous action by dilating the pupils, accelerating the heart and lungs, stopping digestion, and a number of other things that are aided by pumping out the hormone epinephrine (adrenaline) from the adrenal glands. When the organism perceives the threat is over, the parasympathetic system restores the body to homeostasis (the return of physiological functions to their "set points" of acceptable range values).

Although the ANS response is instantaneous, the HPA axis response is much slower because it occurs through changes in gene expression. It is activated in situations that call for a prolonged series of thoughtful responses rather than the visceral immediacy of the ANS's fight-or-flight response. The HPA axis response begins with the hypothalamus feeding various chemical messages to the pituitary gland (master of the endocrine system), which leads to further chemicals that stimulate the adrenal glands to release the hormone cortisol. Unlike epinephrine, which does not cross the blood–brain barrier, the brain is a major target for cortisol.[27] All of this is adaptively normal and enables us to respond appropriately to environmental contingencies.

However, frequent and chronic HPA axis arousal may lead to hypercortisolism (overproduction of cortisol) or hypocortisolism (underproduction). Hypercortisolism leads to anxiety and depressive disorders, is most likely to be found in maltreated females, and can lead to posttraumatic stress disorder. Hypercortisolism suggests a failure of the system to adjust to chronic environmental stressors, whereas hypocortisolism suggests an adaptive adjustment to chronic aversive circumstances. Frequent stressful encounters habituate the organism such that it does not perceive further encounters as quite so stressful, thus both the HPA axis and ANS responses become blunted. Hypocortisolism has been linked to early onset of aggressive antisocial behavior, to criminal behavior in general, and is more likely to be found in maltreated males. Consistent with previous studies, O'Leary, Loney, and Eckel's study of cortisol and psychopathic personality traits found that males high in psychopathic traits lacked stress-induced increases in cortisol displayed by males low in psychopathic traits.[28] Blunted arousal means a low level of anxiety and fear, something that is very useful for those committing or contemplating antisocial behavior.

The development of hyper- or hypocortisolism is an example of the process of allostasis. Whereas homeostasis (homeo = "same"; stasis = "balance") maintains the body's equilibrium by returning physiological set points to their previous state when stress is terminated, allostasis (allo = "other" or "different") describes the body's attainment of equilibrium by altering the acceptable range of physiological set points to adapt to extreme acute stress or to chronic stress. Conceptually, allostasis means to achieve physiological stability through change. According to Goldstein and Kopin, "Adaptations involving allostasis to cope with real, simulated, or imagined challenges are determined by genetic, developmental, and previous experiential factors. While they may be effective for a short interval, over time the alterations may have cumulative adverse effects."[29]

Frequent or prolonged stress-inducing events leading to allostatic responses is termed *allostatic load*. This is like having heating and cooling systems running simultaneously in the same apartment, a situation guaranteed to waste energy and to contribute to the wear and tear of both systems. In addition to affecting neurohormonal functioning by down-regulating the stress response systems and thus affecting behavioral response, allostatic load compromises the cardiovascular and immune systems, and therefore goes a long way to explain the poor health outcomes of individuals who reside in high-stress environments.[30]

The take-home message is that child maltreatment can lead to the down-regulation of the child's stress responses (although there are also genetic reasons for differential ANS and HPA axis responsiveness). The hyporeactivity of these stress mechanisms can lead to poor development of the conscience due to low levels of fear and anxiety, and thus to an elevated risk of norm violation.[31]

THE BRAIN AT ADOLESCENCE

The other pathway in Terrie Moffitt's dual pathway model is the adolescent-limited pathway. Although the youths who take this pathway are not burdened with the neurological deficits that life-course-persistent offenders are, it is still vital to understand what is going on in the brain during adolescence to understand why antisocial behavior is so common "twixt twelve and twenty."

The **age–crime curve** (the sudden sharp increase in offending around the onset of puberty and the steady decline in offending in young adulthood) has been noted around the world for as long as age-based statistics have been kept. Sociologists have long admitted that they are not able to explain the age–crime curve by any known set of sociological variables.[32] Other sociologically oriented criminologists have made feeble attempts to explain the curve, but end up merely describing situations that correlate with the onset and desistence of juvenile offending.[33] Those who attempt such explanations forget one of the first lessons of statistics: Just as we cannot explain a variable by appealing to a constant, we cannot explain a constant by appealing to a variable. The invariance of the age–crime curve must be explained by something that is itself invariant. This is not to say that the relative peaks of the curves (revealing differential prevalence of offending) cannot be explained in terms of social, political, and economic variables; they must be so explained. The invariance is in the presence of the age–crime curve, not in variation in its peak.

The constant across time and place is puberty, the developmental stage marking the onset of the transition from childhood to adulthood that prepares us for procreation. Puberty initiates a cascade of physical, hormonal, and neurological events that can dramatically affect the behavior of adolescents. Adolescence is a normal and necessary period in the human lifespan. There is much to learn about being an adult, and adolescence is a time to experiment with a variety of social skills before putting them into practice in earnest. It is a time of sorting out what aspects of the previous generation to retain and what to discard. If teens are to become capable of adapting to new situations, it is necessary for them to temporarily strain the emotional bonds with parents that served their purpose well in childhood. If they do not assert themselves, they will be hindered in their quest for independence and in their efforts to bond and mate with their own generation. Leaving the nest is an evolutionary design feature of all social primates as males seek out sexual partners from outside the rearing group, and fighting with parents and seeking age peers "all help the adolescent away from the home territory."[34]

Natural selection has provided adolescent males with a huge increase in testosterone to help them in all their experimentation. Testosterone (aromatized to estradiol) is a hormone that organizes the male brain during the second trimester of pregnancy so that it will respond to the pubertal surge of testosterone in male-typical ways. After testosterone organizes the male brain there is little difference between the sexes in levels of testosterone until puberty, when the second surge of testosterone—dramatically higher for males than for females—activates the male brain to engage in male-typical behavior. In addition to preparing us for procreation, the pubertal testosterone surge facilitates behaviors such as risk taking, sensation seeking, sexual experimentation, dominance contests, and self-assertiveness. Although none of these behaviors is necessarily antisocial in and of itself, it can easily be pushed in that direction in antisocial environments. No one claims a causal role for testosterone in these behaviors; testosterone is more of a facilitating accomplice than a perpetrator. Testosterone levels are highly heritable, but at least 40 percent of the variance in testosterone is attributable to environmental factors because levels rise and fall in response to environmental challenges that the organism confronts. The "need" to conform to risky behavioral patterns, to seek dangerous sensations, and to engage in dominance competitions with other males certainly qualify as challenges that would require raising testosterone levels to meet them.[35]

The changes that occur in the brain during adolescence are more salient than the hormonal surge in explaining the teenage jump in antisocial behavior, although the surge itself initiates some of these changes. Risky adolescent behavior appears to be a function of an imbalance between areas in the brain associated with approach–avoidance behavior. Much of the behavior characterizing adolescence is rooted in biology intermingling with environmental influences to cause teens to conflict with their parents, take more risks, and experience wide swings in emotion. The lack of synchrony between a physically mature body and a still maturing nervous system may explain these behaviors. Adolescents' sensitivities to reward appear to be different from adults', prompting them to seek higher levels of novelty and stimulation to achieve the same feeling of pleasure. With the right dose of guidance and understanding, adolescence can be a relatively smooth transition.

Functional MRI studies have shown that relative to children and adults, adolescents have exaggerated nucleus accumbens activity feeding into the PFC.[36] The nucleus accumbens is implicated in reward-seeking behaviors, and the PFC is an inhibitor and guide of impulses. Findings such as these suggest what parents have long suspected: Adolescents are hypersensitive to reward cues (approach), and are deficient in their ability to reign in their pursuit of such rewards (avoidance). Here is the biological answer, mingling with our disinhibited cultural milieu, to today's high prevalence of teen pregnancy, drug and alcohol abuse, and antisocial behavior in general.

This approach–avoidance imbalance is facilitated by a shift in the ratios of excitatory and inhibitory neurotransmitters during adolescence. The excitatory neurotransmitters dopamine and glutamate peak during adolescence, and the inhibitory neurotransmitters gamma-aminobutyric acid and serotonin are reduced.[37] Adolescents are thus provided with the biological tools in the form of increased testosterone and excitatory chemicals and decreased inhibitory chemicals needed to increase novelty seeking, sensation seeking, status seeking, and competitiveness, all of which strongly suggest they are evolutionary adaptations. Mid-adolescence and early adulthood comprise a period of intense competition among males for dominance and status among many primate species, aimed ultimately at securing more mating opportunities than other males. As Daly put it: "There are many reasons to think that we've been designed [by natural selection] to be maximally competitive and conflictual in young adulthood."[38]

The pubertal hormonal surge also prompts the increase of gene expression in the brain to slowly refine the adolescent neural circuitry to its adult form. From a criminological perspective, the most important part of the cerebral cortex is the PFC, which is the most uniquely human of all brain structures. This vital area has extensive connections with other cortical regions and with deeper structures in the limbic system. Because of these interstructural connections, it is considered to play the major integrative and the major supervisory role in the brain. The PFC is vital to the forming of moral judgments, mediating affect, and social cognition.

Although the adolescent brain as a whole undergoes a wave of synaptic overproduction just before puberty, followed by a period of pruning during adolescence and early adulthood, the PFC is undergoing some very dramatic changes. As Steinberg points out: "Significant changes in multiple regions of the prefrontal cortex [occur] throughout the course of adolescence, especially with respect to the processes of myelination and synaptic pruning."[39] The incompleteness of PFC myelination has been implicated in the weaker decision-making skills of adolescents and accounts for the "time lapse" between the adolescent perception of an event in the emotional limbic system and the PFC's rational judgment of it.[40] A number of functional MRI studies have shown that adolescents exhibit greater limbic activation in response to angry and fearful faces than to happy and neutral faces, whereas adults show the opposite.[41] Adolescents thus pay greater attention to negative emotional states than do adults.

Putting all this together, we may conclude from the neurobiology of adolescents that there are physical reasons for their immature and often antisocial behavior. The immaturity of the teen brain combined with a teen physiology on "fast forward" facilitates the tendency to assign faulty attributions to the intentions of others. In other words, a brain on "go slow superimposed on a physiology on fast forward" explains why many teenagers "find it difficult to accurately gauge the meanings and intentions of others and to experience more stimuli as aversive during adolescence than they did as children and will do so when they are adults."[42] As Restak so aptly put it: "The immaturity of the adolescent's behavior is perfectly mirrored by the immaturity of the adolescent's brain."[43]

In early adulthood, the ratio of excitatory transmitters to inhibitory transmitters becomes more balanced as the former start to decrease and the latter start to increase. A more bio-balanced brain signals the emergence of more adult-like personality traits, and thus more adult-like behavior. Findings from five different countries show age-related decreases in personality traits positively related to antisocial behavior (e.g., neuroticism, extraversion) and increases in personality traits positively related to prosocial behavior (e.g., agreeableness, conscientiousness).[44] The fine-tuning of neurohormonal systems, correlated worldwide with steep declines in antisocial behavior, lays the foundations for the acquisition of responsible social roles that help us stay on the straight and narrow.

CONCLUSION

From this brief review of the literature it is plain that although the brain is the stupendously complex magnum opus of natural selection, it can be compromised by exposure to environmental teratogens that did not exist in 99.99 percent of our evolutionary history. The brain can also be compromised by depriving its owner of experience-expected developmental input such as that supplied by tactile stimulation received while suckling the milk of human kindness from its mother's breast. Individuals who begin offending at an early age are often found to possess many of the deficits linked to unexpected perturbations in brain development. Unfortunately, these individuals cluster in similar socioeconomic groups in which exposure to the risk factors for compromised brain development are elevated—that is, among those living in poverty. It is in these groups that pregnancy out of wedlock is most likely to occur, and where maternal smoking and drinking during pregnancy is likely to be more prevalent. It is also among these groups that long-term breastfeeding is less likely to occur and where abuse, neglect, and frequent exposure to violence is common.

Sociological criminologists will also tell us that social pathologies cluster together in lower socioeconomic areas, but are incapable of telling us why without taking off into flights of fancy by appealing to the ineffable, the implausible, and the unlikely. Without a solid foundation in genetics, neuroscience, and evolutionary biology, areas in which the most exciting work in criminology is currently being done (although most professional criminologists are unaware of it), they will forever be incapable. It is for this reason that no less a criminological luminary of the sociological paradigm as Francis T. Cullen has written that he has become "persuaded that sociological criminology has exhausted itself as a guide for the future study on the origins of crime. It is a paradigm for the previous century, not the current one." Cullen sees that the paradigm that will take the place of sociology as the guiding model for 21st century criminology is one that embraces the fundamental biological sciences, a biosocial paradigm that he describes as "a broader and more powerful paradigm."[45] Traditional criminologists have scared themselves silly over such fictions as "biological determinism." All they have to do to assuage this fear is learn a little biology, and when they do they will find that our biology (genes, brains, endocrine system, and everything else) has been exquisitely designed by natural selection to adapt us to the variety of environments in which we find ourselves. When or if they learn this, they will come to the conclusion that to consider any class of behavior without considering the evolutionary history of the species, the genetics of individuals, and the developmental context in which genes and environments combine to construct our brains, is pure folly.

GLOSSARY

Age–crime curve—the inverted-U, age-graded distribution of crime, where crime begins to increase at around 12 years of age, peaks around the ages of 19 and 20, and then begins to decrease steadily thereafter

Autonomic nervous system—part of the peripheral nervous system that is largely responsible for controlling involuntary processes, such as heart rate

Functional magnetic resonance imaging—a brain-imaging technique that assesses brain function via brain blood flow

Neurons—nerve cells found in the brain

Prefrontal cortex—the area of the brain located behind the forehead that is responsible for judgment, thinking, planning, and other higher order cognitive processes

Synapses—the small gaps that exist between neurons

NOTES

1. Mitchell, K. (2007). The genetics of brain wiring: From molecule to mind. *PLoS Biology, 4*, 690–692.
2. Wright, J., Boisvert, D., Dietrich, K., & Ris, M. (2009). The ghost in the machine and criminal behavior: Criminology for the 21st century. In A. Walsh & K. Beaver (Eds.), *Biosocial criminology: New directions in theory and research* (pp. 73–89). New York: Routledge, p. 85.
3. Perry, B. (2002). Childhood experience and the expression of genetic potential: What childhood neglect tells us about nature and nurture. *Brain and Mind, 3*, 79–100.
4. Sampson, R. J., & Laub, J. H. (2005). A life-course view of the development of crime. *American Academy of Political & Social Sciences, 602*, 12–45.
5. Moffitt, T. (1993). Adolescent-limited and life-course-persistent antisocial behavior: A developmental taxonomy. *Psychological Review, 100*, 674–701; Moffitt, T., & Walsh, A. (2003). The adolescence-limited/life-course persistent theory of antisocial behavior: What have we learned? In A. Walsh & L. Ellis (Eds.), *Biosocial criminology: Challenging environmentalism's supremacy* (pp. 125–144). Hauppauge, NY: Nova Science.
6. Sowell, E., Thompson, P., & Toga, A. (2004). Mapping changes in the human cortex throughout the span of life. *Neuroscientist, 10*, 372–392.
7. Giedd, J. (2004). Structural magnetic resonance imaging of the adolescent brain. *Annals of the New York Academy of Science, 1021*, 77–85.
8. Edelman, G. (1992). *Bright air, brilliant fire*. New York: Basic Books.
9. Levine, D. (1993). Survival of the synapses. *The Sciences, 33*, 46–52, p. 52.

10. Shi, S., Cheng, T., Jan, L., & Jan, Y. (2004). The immunoglobin family member dendrite arborization and synapse maturation 1 (Dasm1) controls excitatory synapse maturation. *Proceedings of the National Academy of Sciences, 101*, 13246–13351.

11. Penn, A. (2001). Early brain wiring: Activity-dependent processes. *Schizophrenia Bulletin, 27*, 337–348, p. 339.

12. Black, J., & Greenough, W. (1997). How to build a brain: Multiple memory systems have evolved and only some of them are constructivist. *Behavioral and Brain Sciences, 20*, 558–559; Schon, R., & Silven, M. (2007). Natural parenting—back to basics in infant care. *Evolutionary Psychology, 5*, 102–183.

13. Gunnar, M., & Quevedo, K. (2007). The neurobiology of stress and development. *Annual Review of Psychology, 58*, 145–173.

14. Geary, D. (2005). *The origin of mind: Evolution of brain, cognition, and general intelligence.* Washington, DC: American Psychological Association; Nelson, C. (2007). A neurobiological perspective on early human deprivation. *Child Development Perspectives, 1*, 13–18.

15. Rowe, D. (1992). Three shocks to socialization research. *Behavioral and Brain Sciences, 14*, 401–402, p. 402.

16. Lee, V., & Hoaken, P. (2007). Cognition, emotion, and neurobiological development: Mediating the relation between maltreatment and aggression. *Child Maltreatment, 12*, 281–298.

17. Bromage, T. (1987). The biological and chronological maturation of early hominids. *Journal of Human Evolution, 16*, 257–272; Buck, R. (1999). The biological affects: A typology. *Psychological Review, 106*, 301–336; van As, A., Fieggen, G., & Tobias, P. (2007). Severe abuse of infants: An evolutionary price for human development? *South African Journal of Children's Health, 1*, 54–57.

18. Montagu, A. (1981). *Growing young.* New York: McGraw-Hill, p. 93; Esch, T., & Stefano, G. (2005). Love promotes health. *Neuroendocrinology Letters, 3*, 264–267.

19. Curtis, J., & Wang, Z. (2003). The neurochemistry of pair bonding. *Current Directions in Psychological Sciences, 12*, 49–53; Nair, H., & Young, L. (2006). Vasopressin and pair-bond formation: Genes to brain to behavior. *Physiology, 21*, 146–152; Hiller, J. (2004). Speculations on the links between feelings, emotions and sexual behaviour: Are vasopressin and oxytocin involved? *Sexual and Relationship Therapy, 19*, 1468–1479.

20. Li, R., Zhao, Z., Mokdad, A., Barker, L., & Grummer-Strawn, L. (2003). Prevalence of breastfeeding in the United States: The 2001 National Immunization Survey. *Pediatrics, 111*, 1198–1201; Kramer, M., Aboud, F., Mironova, E., Vanilovich, I., Platt, R. W., Matush, L., et al. Promotion of Breastfeeding Intervention Trial (PROBIT) Study Group. (2008). Breastfeeding and child cognitive development: New evidence from a large randomized trial. *Archives of General Psychiatry, 65*, 578–584.

21. Adams, B., & Moghaddam, B. (2000). Tactile stimulation activates dopamine release in the lateral septum. *Brain Research, 858*, 177–180; Elbert, T., & Rockstroh, B. (2004). Reorganization of human cerebral cortex: The range of changes follow use and injury. *The Neuroscientist, 10*, 129–141; Weiss, S., Wilson, P., & Morrison, D. (2004). Maternal tactile stimulation and neurodevelopment of low birth weight infants. *Infancy, 5*, 85–107.

22. Wismer Fries, A., Ziegler, T., Kurian, J., Jacoris, S., & Pollak, S. (2005). Early experience in humans is associated with changes in neuropeptides critical for regulating social behavior. *Proceedings of the National Academy of Sciences, 102*, 17237–17240.

23. Chugani, H., Behan, M., Muzic, O., Juhasz, C., Nagy, F., & Chugani, C. (2001). Local brain functional activity following early deprivation: A study of postinstitutionalized Romanian orphans. *Neuroimage, 14*, 1290–1301; Eluvathingal, T., Chugani, H., Behan, M., Juhasz, C., Muzic, O., Maqbool, M., Chugani, D., & Makki, M. (2006). Abnormal brain connectivity in children after early severe socioemotional deprivation: A diffusion tensor imaging study. *Pediatrics, 117*, 2093–2100; Nelson, E., McClure, E., Monk, C., Zahran, E. Leibenluft, E., Pine, D., & Ernst, M. (2003). Developmental differences in neuronal engagement during implicit encoding of emotional faces: An event-related fMRI study. *Journal of Child Psychology and Psychiatry and Allied Disciplines, 44*, 1015–1024.

24. Niehoff, D. (2003). A vicious circle: The neurobiological foundations of violent behavior. *Modern Psychoanalysis, 28*, 235–245.

25. Cooley-Quille, M., Boyd, R., Frantz, E., & Walsh, J. (2001). Emotional and behavioral impact of exposure to community violence in inner-city adolescents. *Journal of Clinical Child Psychology, 30*, 199–206; Carnagey, N.,

Anderson, C. A., & Bushman, B. J. (2007). The effect of video game violence on physiological desensitization to real violence. *Journal of Experimental Social Psychology, 43,* 489–496; Murray, J., Liotti, M., Mayberg, H., Pu, Y., Zamarripa, F., & Liu, Y. (2006). Children's brain activations while viewing televised violence revealed by fMRI. *Media Psychology, 8,* 25–37; Sterzer, P., Stadler, C., Krebs, A. Kleinschmidt, A., & Poustka, F. (2003). Reduced anterior cingulate activity in adolescents with antisocial conduct disorder confronted with affective pictures. *NeuroImage, 19*(Suppl 1), 123.

26. Perry, B., & Pollard, R. (1998). Homeostasis, stress, trauma, and adaptation: A neurodevelopmental view of childhood trauma. *Child and Adolescent Psychiatric Clinics of America, 7,* 33–51.

27. van Voorhees, E., & Scarpa, A. (2004). The effects of child maltreatment on the hypothalamic-pituitary-adrenal axis. *Trauma, Violence, & Abuse, 5,* 333–352.

28. Ellis, L. (2005). A theory explaining biological correlates of criminality. *European Journal of Criminology, 2,* 287–315; McBurnett, K., Lahey, B., Rathouz, P., & Loeber, R. (2000). Low salivary cortisol and persistent aggression in boys referred for disruptive behavior. *Archives of General Psychiatry, 57,* 38–43; van Goozen, S., Fairchild, G., Snoek, H., & Harold, G. (2007). The evidence for a neurobiological model of childhood antisocial behavior. *Psychological Bulletin, 133,* 149–182; O'Leary, M., Loney, B., & Eckel, L. (2007). Gender differences in the association between psychopathic personality traits and cortisol response to induced stress. *Psychoneuroendocrinology, 32,* 183–191.

29. Goldstein, D., & Kopin, I. (2007). Evolution of the concept of stress. *Stress, 10,* 109–120, p. 111.

30. Bremner, D. (2002). *Does stress damage the brain? Understanding trauma-related disorders from a neurological perspective.* New York: Norton.

31. Alink, L., van Ijzendorn, M., Bakermans-Kranenburg, M., Mesman, J., Juffer, F., & Koot, H. (2008). Cortisol and externalizing behavior in children and adolescents: Mixed meta-analysis evidence for the inverse relation of basal cortisol and cortisol reactivity with externalizing behavior. *Developmental Psychobiology, 50,* 427–450.

32. Gottfredson, M., & Hirschi, T. (1990). *A general theory of crime.* Stanford, CA: Stanford University Press; Hirschi, T., & Gottfredson, M. (1983). Age and the explanation of crime. *American Journal of Sociology, 89,* 552–584; Shavit, Y., & Rattner, A. (1988). Age, crime, and the early lifecourse. *American Journal of Sociology, 93,* 1457–1470.

33. Akers, R. (1998). *Social learning and social structure: A general theory of crime and deviance.* Boston: Northeastern University Press; Warr, M. (2002). *Companions in crime: The social aspects of criminal conduct.* New York: Cambridge University Press.

34. Powell, K. (2006). How does the teenage brain work? *Nature, 442,* 865–867, p. 867.

35. Booth, A., Granger, D., Mazur, A., & Kivligan, K. (2006). Testosterone and social behavior. *Social Forces, 85,* 167–191; Mazur, A. (2005). *Biosociology of dominance and deference.* Lanham, MD: Rowan & Littlefield.

36. Eshel, N., Nelson, E., Blair, R., Pine, D., & Ernst, M. (2007). Neural substrates of choice selection in adults and adolescents: Development of the ventrolateral prefrontal and anterior cingulated cortices. *Neuropsychologia, 45,* 1270–1279; Galvan, A., Hare, T., Parra, C., Penn, J., Voss, H., Glover, G., & Casey, B. (2006). Earlier development of the accumbens relative to orbitofrontal cortex might underlie risk-taking behavior in adolescents. *Journal of Neuroscience, 26,* 6885–6892.

37. Collins, R. (2004). Onset and desistence in criminal careers: Neurobiology and the age-crime relationship. *Journal of Offender Rehabilitation, 39,* 1–10; Walker, E. (2002). Adolescent neurodevelopment and psychopathology. *Current Directions in Psychological Science, 11,* 24–28.

38. Spear, L. (2000). Neurobehavioral changes in adolescence. *Current Directions in Psychological Science, 9,* 111–114; Daly, M. (1996). Evolutionary adaptationism: Another biological approach to criminal and antisocial behavior. In G. Bock & J. Goode (Eds.), *Genetics of criminal and antisocial behaviour* (pp. 183–195). Chichester, England: Wiley, p. 193.

39. Steinberg, L. (2005). Cognitive and affective development in adolescence. *Trends in Cognitive Sciences. 9,* 69–74, p. 70.

40. Beckman, M. (2004). Crime, culpability, and the adolescent brain. *Science, 305,* 596–599.

41. Monk, C., Klein, R., Telzer, E., Schroth, E., Mannuzza, S., Moulton, J., et al. (2008). Amygdala and nucleus accumbens activation to emotional facial expressions in children and adolescents at risk for major depression. *American Journal of Psychiatry, 165,* 90–98.
42. Walsh, A. (2002). *Biosocial criminology: Introduction and integration.* Cincinnati, OH: Anderson, p. 143.
43. Restak, R. (2001). *The secret life of the brain.* New York: copublished by Dana Press and Joseph Henry Press, p. 76.
44. McCrae, R., Costa, P., Ostendorf, F., Angleitner, A., Hrebickova, M., Avia, M., et al. (2000). Nature over nurture: Temperament, personality, and life span development. *Journal of Personality and Social Psychology, 78,* 173–186.
45. Cullen, F. (2009). Foreword. In A. Walsh & K. Beaver (Eds.), *Biosocial criminology: New directions in theory and research* (pp. xvi–xvii). New York: Routledge.

CHAPTER FIVE

The Intergenerational Transmission of Antisocial Behavior: Parents, Genes, and Unsettling Realities

Brian B. Boutwell

KEY TERMS

Gene–environment correlation (rGE)
Gene–environment interaction (G×E)
Intergenerational transmission of criminality
Plasticity genes

INTRODUCTION

At its core, the discipline of criminology exists to better understand the origins of antisocial, delinquent, and criminal behavior. Centuries of research along these lines have produced a bewildering body of findings, all purporting to offer some insight into the etiological underpinnings of aberrant conduct. Wading through this vast body of knowledge in order to search out a vein of consistency can represent a daunting task. To be sure, some lines of inquiry have represented dead ends and others have produced only limited returns. The good news, however, is that some very clear patterns have emerged over decades of study and these trends may help to illuminate the origins of criminal behavior. Perhaps one of the most striking features of deviant, antisocial, and criminal conduct is that it runs in families.[1]

To illustrate the point, consider the findings from a study conducted by Farrington and colleagues.[2] This group of researchers examined data collected from a sample of more than 1000 families living in the city of Pittsburgh. Once the analyses were complete, the results were indeed stark: Almost half of the individuals who had been arrested belonged to less than 10 percent of the participating families. Although it may be tempting to dismiss such a finding as a statistical artifact, the results dovetail closely with other studies. Research analyzing data drawn from the Cambridge Study in Delinquent Development, for instance, has arrived at similar conclusions concerning the aggregation of crime in families.[3]

In a more recent study, Frisell and colleagues examined the concentration of violent crime in an impressively large sample of more than 12.5 million individuals living in Sweden.[4] Similar to the findings mentioned above, the results clearly pointed toward an aggregation of criminal violence in families. The full siblings of individuals convicted of a violent crime were four times more likely to have also reported a conviction. Interestingly, the unrelated adopted siblings of violent offenders were not noticeably more likely to report violent convictions. The relevance of this last finding should become more apparent as we progress. The important point is that there is a voluminous body of research pertaining to the **intergenerational transmission of criminality**—the transfer of antisocial traits and behaviors across generations.

BAD FAMILIES, BAD PARENTS, AND BAD CHILDREN

The findings discussed in the introduction raise an obvious question: What is it about some families that make them so criminogenic? Farrington has broadly classified and reviewed six family-level risk factors that consistently correlate with antisocial behavior: (1) criminal and antisocial parents, (2) large family size, (3) inadequate child-rearing methods (e.g., poor supervision; harsh, inconsistent, or coercive disciplinary techniques), (4) exposure to abuse and neglect, (5) parental conflict and disrupted families, and (6) other parental features (e.g., young maternal age, substance use, working mothers).[5]

Upon examining the list of possible culprits, at least two common themes emerge rather quickly. First, not all of the categories are mutually exclusive; the use of harsh discipline, for example, might escalate into abuse, thus blurring together two categories. Second, and most relevant to this discussion, practically every category implicates the actions of parents when describing child outcomes. Parents, either directly or indirectly, are thought to influence aberrant behavior in their offspring. This is perhaps logical given the vast amount of research linking exposure to physical discipline, abuse, neglect, and lack of parental supervision to adverse child development.

A critical question, however, is not if parenting factors are *related* to negative outcomes in childhood. Rather, the more pertinent question is whether the effect is *causal*. To better illustrate the point, we will borrow an example articulated by the biologist Stephen Jay Gould and drawn from the world of architecture.[6] When two arches are placed beside one another, the area in between forms a distinctive triangular shape. The vacant space between the neighboring arches is referred to as a spandrel. The spandrel itself is a complete accident—a byproduct of placing two arches next to one another. Importantly, spandrels do not *do* anything; it is the arch itself that is the relevant design feature. It would be a fallacy to say that two arches were *caused* by erecting a spandrel (the exact opposite would be true). Applied to the current topic, then, could parental influence be a kind of spandrel? Put differently, it is possible that parent and child behavior coincide, not because the parent directly affects the child, but because some other factor influences both parent and child behavior.

With this in mind, the purpose of this chapter is to entertain two possibilities. The first is that the correlation between parental criminality and child criminality is entirely spurious (i.e., a spandrel). As should become very clear, the standard methodologies employed by most social scientists will be virtually useless for resolving this issue.[7] Alternative lines of research, which employ various analytical strategies, will have to be consulted. The second possibility to be considered is that parental criminality plays a role in influencing the criminal behavior in children. However, we postulate very different pathways by which this might take place. Regardless of what conclusion is ultimately reached, it should become apparent that the current understanding regarding the intergenerational transmission of antisocial and delinquent behavior is woefully inadequate. Ultimately, research on the concentration of crime in families is going to need an extensive overhaul.

PARENTING EFFECTS: CRIMINOLOGY'S SPANDREL?

Perhaps one of the least surprising, and indeed rather banal, insights from research concerning inter-generational criminality is that exposure to adverse parenting corresponds to criminal behavior in children. This finding shocks almost no one, either in the academy or the general public. Indeed it has become almost "common sense" to most social science researchers. Parenting effects have been elevated to the status of near "social fact" not only in criminology, but also in the fields of developmental and child psychology. Indeed, if one makes inquiries as to why children behave in certain ways, both layperson and academic will almost invariable invoke some aspect of parental influence.[8]

Almost without fail, studies examining the relationship between parental criminality and child criminality will find a positive correlation. What is important to realize, however, is that much of the existing research on intergenerational transmission tends to overlook an important fact. Specifically, parents generally pass along two things to their children, environment *and* genes. While genetic factors may receive some mention, they are often glossed over in favor of more environmentally oriented explanations. This would not be problematic if genes exerted little or no impact on criminogenic involvement. As we will see, however, such a conclusion is simply not supported.

Do Genes Really Matter?

Evidence concerning the genetic underpinnings of behavior has reached a critical mass. A voluminous body of scholarship on the topic has revealed with remarkable consistency that behaviors run in families partly because they are influenced by genes. Hundreds of studies analyzing tens of thousands of subjects continue to arrive at the same conclusion: Human variation, for virtually every measurable outcome, is the combined product of genes and the environment. Moreover, similar findings have been garnered regardless of the analytical approach (i.e., twin studies, adoption studies, family studies, monozygotic twins reared apart studies, and molecular genetic studies) or the demographic composition of the sample.[9] In short, genes impact all behaviors and generally account for around half of the observed variance.

At first blush, such a conclusion seems unremarkable. As Cohen and others have pointed out, the immediate rebuttal of most social scientists goes something like the following: "Of course genes matter, we never really said that they didn't. But the environment matters too and your studies clearly demonstrate that." This reply on the part of researchers, however, is disingenuous, because when one looks closer, the interpretation becomes far less intuitive. While behavior genetic studies unequivocally demonstrated that environments matter, it was not the type of environment that most social researchers had in mind.[10]

Most behavior genetic studies are designed to partition variance in a given outcome (i.e., antisocial behavior) into a genetic and environmental component. However, behavior geneticists go one step further and subdivide the environmental variance component into two distinct types: shared environments and nonshared environments. Nonshared environments represent the unique experiences of children raised in the same household. Put differently, nonshared experiences function to make siblings different from one another. Shared environments, on the other hand, refer to environmental experiences (perhaps including parenting factors, socioeconomic status, etc.) that function to make two siblings similar. The bombshell finding that emerged from behavior genetic research was that shared environments matter very little as a source of variation for most behaviors. Put more directly, the family environment consistently accounts for practically zero variance in behavior, including antisocial behavior.[11]

To be sure, studies that fail to find parenting effects on antisocial and criminal behaviors continue to emerge within the discipline of criminology. Wright and Beaver, for example, were among the first to show that parenting effects on levels of self-control fell from statistical significance once genetic influences were held constant.[12] Barnes and Boutwell recently demonstrated that genetic and nonshared environmental influences accounted for most of the variance in measures of adolescent delinquency and adult criminal behavior.[13] The findings of Barnes and Boutwell dovetail closely with Wright and colleagues, who reported negligible parenting influences on measures of delinquent involvement during adolescence.[14]

Ultimately, the question of whether genes influence behavior seems almost an absurdity in the post genomic era. Even so, criminologists persist in their rejection of biosocial explanations of criminal behavior.[15] This ideologically charged boundary maintenance is unfortunate, because as Moffitt points out, "The family concentration of antisocial behavior could be explained by a genetic influence on antisocial behavior, but it could just as easily be explained by nongenetic social transmission of antisocial behavior within families. Again, causation is not well understood. Studies that cannot disentangle genetic and environmental influences cannot help."[16]

Families, Genes, and Criminal Behavior

Thus far, behavior genetic research has converged on two conclusions: (1) that antisocial and criminal behavior are highly heritable, and (2) shared environments, including those shared by parents and families, have little to do with creating variation in adverse behavior. Upon perusing the findings from behavior genetics further, one might retort that children may experience very different environments, despite being raised in the same home, and these differences in parental treatment may have important ramifications.[17] Consider, for example, a situation where one child is abused and his or her sibling is not. This could open the door for family influences to impact intergenerational criminality. The argument is certainly logical and, more importantly, testable.

Beaver examined this issue by testing whether differences in the environmental exposure of identical twins explained differences in their involvement in delinquent outcomes during adolescence. Using data drawn from the National Longitudinal Study of Adolescent Health (Add Health), Beaver looked at several parental socialization variables, all of which have been implicated in the intergenerational transmission of antisocial behavior, including parental permissiveness, parental disengagement, parental involvement, and parental attachment. The findings were indeed stunning in that differences in parenting experiences were virtually unrelated to delinquency, levels of self-control, and contact with delinquent peers.[18] The outcome of the study becomes less shocking, however, when one considers that it accords closely with prior research. In a classic study, Plomin and Daniels concluded that the only reason children from the same family were similar is because of shared genetic factors.[19]

At this juncture, it may appear as if parenting, and the influence of the family, is being shoved aside in favor of a more "deterministic" explanation of intergenerational effects (i.e., a purely genetic explanation). This is simply not the case for two reasons. First, genetic influences on behavior are probabilistic, not deterministic. There is no such thing as a "crime gene." Second, and equally relevant, a lack of shared environmental influence does not necessarily indicate that family influences are unimportant and can be dismissed out of hand.[20] What it does mean is that criminologists need to refocus their view of how family factors either promote or discourage intergenerational continuities.

RETHINKING THE ROLE OF FAMILIES

In light of behavior genetic evidence, we are left to reconsider why the family factors described earlier influence criminality. In this regard, there may be two possibilities: (1) family- (and parent-) level factors might moderate the effect of genes on behavior in children, and (2) family-level factors may, themselves, be influenced by genes. In the parlance of biosocial research, the first scenario is referred to as a gene–environment interaction (G×E) and the second is known as a gene–environment correlation (rGE). Currently, there is a growing line of evidence concerning both possibilities and we explore this research in more detail as follows.

Understanding Why Environments Matter: Gene–Environment Interaction

Perhaps one of the most perplexing questions social scientists try to tackle is why some individuals who are exposed to very similar environments often experience divergent outcomes. For intergenerational research, one might wonder why one child raised by criminal parents fails to ever lead a life of crime, while his or her sibling—raised by the same parents—becomes criminal. In the case of **gene–environment interaction (G×E)** the environment acts to amplify the effects of genes (and vice versa) so that genetic predispositions become most pronounced when certain environmental conditions are present. In other words, environmental effects may condition the effects of a child's unique genome.

To best understand the logic of G×E, consider the often utilized agricultural example of two seeds, each of the same variety, planted in two different types of soil. Perhaps one plot of soil is arid and lacks the appropriate nutrient profile for proper growth. The other patch of soil is moist and contains the appropriate nutrients. Although the seeds possess the same set of genetic instructions, the differences between their environments may moderate the expression of those genetic tendencies. One seed may flourish in the enriched environment while the other may fail to reach its full growth potential. This same logic may be applied to humans in that the environment might either facilitate or dampen the effects of certain genetic proclivities.

Some of the first evidence of G×Es in humans was detected using adoption studies designed to compare adopted children to both their biological and adoptive parents.[21] Cadoret and colleagues were among the first to demonstrate that adopted children, reared in an adverse environment, were most likely to evince behavioral problems when they also reported having a biological mother who exhibited adverse behavior.[22] As with all statistical techniques, however, adoption studies are not without methodological limitations, and thus alternative approaches are helpful in validating the findings of adoption research. Importantly, a similar pattern of findings has emerged utilizing alternative behavior genetic designs.[23]

Analyzing a sample of twins, Boutwell and colleagues recently uncovered evidence that physical discipline (i.e., spanking) conditioned genetic risk factors for externalizing problems in children.[24] In a landmark study, Caspi and his team were the first to detect evidence of a G×E using molecular genetic data. Specifically, Caspi and colleagues reported evidence that a functional polymorphism in the monoamine oxidase A (MAO-A) gene interacted with exposure to abuse and neglect to increase conduct problems in children.[25]

In a study directly pertaining to intergenerational criminality, DeLisi and his team applied the logic of G×E in a sample of African American females participating in the Add Health study. Specifically, these researchers tested whether being raised by a criminal father (a family condition mentioned earlier) moderated the effects of a polymorphism in the dopamine receptor gene (DRD2).

The results were striking in that being raised by a criminal father interacted with a variant of DRD2 to significantly increase scores on five different measures of antisocial behavior: serious delinquency (measured at two points in time), violent delinquency (measured at two points in time), and contact with the police.[26]

An emergent area of G×E research that is also important to mention concerns the proposition that putative genetic risk factors may actually represent markers for plasticity. Put differently, when coupled with prosocial environments, certain alleles may work to increase prosocial functioning in the individual. These same **plasticity genes**, when coupled with a negative environment, may further exacerbate negative outcomes. In a recent study, Belsky and Beaver uncovered evidence that an index of genetic plasticity (measured by assessing the presence of specific dopaminergic and seroto-nergic alleles) interacted with adverse forms of parenting to predict increased levels of impulsivity and low self-control. Importantly, as parenting effects shifted toward the prosocial end of the spec-trum (i.e., parenting became warm and supportive), levels of self-control increased based on levels of genetic plasticity.[27]

Other studies on the topic of plasticity have also uncovered similar findings.[28] What this means is that parenting effects may work to either blunt or exacerbate the influence of genes on behavior. A note of caution is necessary, however, concerning research on genetic plasticity. The concept of plasticity alleles is both compelling and potentially important, but there are specific theoretical limitations that need to be addressed in future studies. For example, there is yet to be a clearly articulated reason to expect plasticity from an evolutionary perspective. In other words, why would natural selection either react favorably, or at least neutrally, to genes that increase plasticity? Moreover, genetic factors generally function to delineate set points, creating an envelope of sorts for environmental effects to function within.[29] It remains to be seen whether the effects of plasticity are as widespread and encompassing as has been previously suggested.[30]

Ultimately, the logic of G×E suggests that some criminogenic family environments may exac-erbate genetic vulnerabilities for behavior problems in some children. To borrow a term coined by Jaffee and colleagues, and referenced by DeLisi and his team, certain children may receive a "double whammy" of risk—stemming from both environmental and genetic factors.[31] Importantly, the argument for genetic plasticity, despite potential limitations, may also point toward a potential avenue for breaking the cycle of intergenerational continuity. Moving forward, research in the area of G×E will no doubt continue to shed light on the question of why certain families remain mired in cycles of crime and interpersonal violence.

Pulling Back the Curtain on Environments: Gene–Environment Correlation

Despite the promise of G×E research to inform the study of intergenerational transmission, it does nothing to address the question of why some children encounter adverse environments in the first place. Why are some children born to very young mothers and raised in broken homes? Why are some children exposed to harsh and coercive forms of discipline or subjected to abuse and neglect? Rarely do researchers examining intergenerational questions pause to consider that adverse parenting and disrupted homes might be something other than a predictor of child outcomes.[32] This oversight is unfortunate because the child's environment can just as easily be considered a dependent variable.

The line of inquiry known as **gene–environment correlation (rGE)** examines the genetic and envi-ronmental origins of exposure to deleterious circumstances. A review of the literature conducted by Kendler and Baker revealed that genetic factors accounted for a significant proportion of the variance in a range of environmental experiences.[33] On average, heritability estimates for parenting behaviors ranged from low (12–17 percent) to moderate (34–37 percent). Family environmental variables, such as levels of cohesion, parental control, and parental conflict were also found to be under genetic influence. In this case, genes accounted for between 18 and 30 percent of the variance in different types of family environments. A similar pattern of findings emerged concerning marital quality. On average, genetic factors consumed between 13 and 28 percent of the variance in measures assessing levels of satisfaction, conflict, and warmth within a marriage.

Children raised in broken homes have consistently been found to exhibit higher levels of antisocial and criminal behavior.[34] Even so, much of the research in this area has failed to fully account for genetic factors and selection biases that may predict both family structure and behavioral problems in children. Cleveland and colleagues found evidence that while genes accounted for a significant proportion of the variance in childhood behavioral problems, they also corresponded to mean level differences in behavior problems for children raised in broken homes versus children raised by two parents. In short, the findings presented by Cleveland et al. serve as evidence that parents select into certain family structures based, in part, on genetic factors. In other words, even the structure of families is not beyond genetic influence.[35]

Recently, Boutwell and colleagues examined the association between maternal disengagement and externalizing problem behaviors in children. There is no shortage of evidence exploring the role that parent–child attachment plays in the development of antisocial tendencies early in life. What Boutwell et al. reported, however, was that genetic risk factors in children increased the likelihood of experiencing disengaged relationships with primary caregivers (i.e., an rGE). Interestingly, the findings also suggested that maternal disengagement moderated genetic effects on behavioral problems (i.e., a G×E). This is important because it further illustrates the close interplay between genes, environments, and child development. Put differently, it is likely the case that rGEs and G×Es both play an integral role in the cycle of familial criminality.[36]

CONCLUSION

Perhaps one of the most important topics that criminologists grapple with concerns the reasons why crime and antisocial behavior tend to concentrate in certain families. In this regard, parenting effects continue to occupy a lofty position in the pantheon of explanatory variables. Indeed, there is no shortage of evidence that researchers can point to in order to implicate bad parenting in the origins of antisocial behavior. The problem with the vast majority of this research, however, is that it is not interpretable. Unless a study is capable of disentangling genetic from environmental influences, it can shed no light on causal associations between families, parenting, and the transmission of antisocial behavior.

To illustrate the point, consider a study by Boutwell and Beaver.[37] These researchers found that levels of self-control appear to run in families. Analysis of the Fragile Families and Child Wellbeing Study revealed that maternal and paternal levels of self-control correlated with childhood levels of self-control (despite controlling for a host of important covariates). Although Boutwell and Beaver posited that a genetic mechanism might be involved, it is entirely possible (albeit unlikely) that environmental factors can fully account for their findings. Until the analyses are replicated using samples capable of controlling for genetic influences, no definitive conclusion can be reached.

The findings outlined in this chapter are intended to provide a way forward. In this regard, we reviewed several approaches to studying family-level dynamics that may yield more in the way of causal insight than previous research was capable of providing (i.e., G×E and rGE). Conversely, there is also a growing body of research, not covered in this chapter, regarding factors that children experience outside the home that may also inform intergenerational research. To illustrate, Barnes and Morris provided evidence that the deleterious effects of young maternal age were mediated by the child's exposure to delinquent peers.[38] Exposure to delinquent peers represents a potential foothold to further test questions of intergenerational transmission. The most important point to recognize in this regard is that the putative risk factor of young motherhood operated through an environmental experience *outside* of the home.

In closing, criminologists have a history marked by contentious debate over the introduction of new ideas into the study of criminal and delinquent behavior. The finding that behaviors are heritable, however, is not a new idea and it is time that researchers in the discipline of criminology accept this reality despite how unsettling it may be. What this means is that research on the topic of intergenerational transmission cannot progress without incorporating the insights of behavior genetics and biosocial criminology.[39] Data sets that permit genetically sensitive analysis are becoming more widely available, yet there remains a need to collect more rigorous data capable of addressing questions regarding intergenerational effects. It is perhaps the most exciting time ever to be a criminologist. However, unless researchers in the field can remove their ideological blinders, we run the very real risk of becoming irrelevant, at least where questions regarding intergenerational transmission are concerned.

GLOSSARY

Gene–environment correlation (rGE)—behaviors caused by the correlation between genetic factors and environmental factors

Gene–environment interaction (G×E)—behaviors caused by an interaction between genetic factors and environmental factors

Intergenerational transmission of criminality—the transfer of antisocial traits and behaviors across generations

Plasticity genes—genetic variants that increase risk for antisocial behavior when paired with negative environments and that protect from antisocial behavior when paired with prosocial environments

NOTES

1. Avakame, E. F. (1998). Intergenerational transmission of violence, self-control, and conjugal violence: A comparative analysis of physical violence and psychological aggression. *Violence and Victims, 13*, 301–316; Boutwell, B. B., & Beaver, K. M. (2010). The intergenerational transmission of low self-control. *Journal of Research in Crime and Delinquency, 47*, 174–209; DeLisi, M., Beaver, K. M., Vaughn, M. G., & Wright, J. P. (2009). All in the family: Gene × environment interaction between DRD2 and criminal father is associated with five antisocial phenotypes. *Criminal Justice and Behavior, 36*, 1187–1197; Frisell, T., Lichtenstein, P., & Langstrom, N. (2011). Violent crime runs in families: A total population study of 12.5 million individuals. *Psychological Medicine, 41*, 97–105.

2. Farrington, D. P., Jolliffe, D., Loeber, R., Stouthamer-Loeber, M., & Kalb, L. M. (2001). The concentration of offenders in families, and family criminality in the prediction of boy's delinquency. *Journal of Adolescence, 24*, 579–596.

3. Rowe, D. C., & Farrington, D. P. (1997). The familial transmission of criminal convictions. *Criminology, 35,* 177–201; Farrington, D. P., & Welsh, B. C. (2007). *Saving children from a life of crime: Early risk factors and effective interventions.* Oxford, UK: Oxford University Press.

4. Frisell et al. (2011), see Note 1.

5. Farrington, D. P. (2010). Family influences on delinquency. In D. W. Springer & A. R. Roberts (Eds.), *Juvenile justice and delinquency* (pp. 203–222). Sudbury, MA: Jones & Bartlett.

6. Gould, S. J. (1991). Exaptation: A crucial tool for an evolutionary psychology. *Journal of Social Issues, 47,* 43–65.

7. Beaver, K. M. (2008). Nonshared environmental influences on adolescent delinquent involvement and adult criminal behavior. *Criminology, 46,* 341–369.

8. Harris, J. R. (1998). *The nurture assumption: Why children turn out the way they do.* New York: Free Press

9. Ferguson, C. J. (2010). Genetic contributions to antisocial personality and behavior: A meta-analytic review from an evolutionary perspective. *The Journal of Social Psychology, 150,* 160–180; Rhee, S. H., & Waldman, I. D. (2002). Genetic and environmental influences on antisocial behavior: A meta-analysis of twin and adoption studies. *Psychological Bulletin, 128,* 490–529.

10. Harris (1998), see Note 8; Cohen, D. B. (1999). *Stranger in the nest: Do parents really shape their child's personality, intelligence, or character?* New York: Wiley.

11. Ferguson (2010), Rhee, & Waldman (2002), see Note 9; Barnes, J. C., & Boutwell, B. B. (2012). On the relationship of past to future involvement in crime and delinquency: A behavior genetic analysis. *Journal of Criminal Justice, 40,* 94–102.

12. Wright, J. P., & Beaver, K. M. (2005). Do parents matter in creating self-control in their children? A genetically informed test of Gottfredson and Hirschi's theory of low self-control. *Criminology, 43,* 1169–1202.

13. Barnes, & Boutwell (2012), see Note 11.

14. Wright, J. P., Beaver, K. M., DeLisi, M., & Vaughn, M. G. (2008). Evidence of negligible parenting influences on self-control, delinquent peers, and delinquency in a sample of twins. *Justice Quarterly, 25,* 544–569.

15. Walby, K., & Carrier, N. (2010). The rise of biosocial criminology: Capturing observable bodily economies of 'criminal man.' *Criminology and Criminal Justice, 10,* 261–285.

16. Moffitt, T. E. (2005). The new look of behavioral genetics in developmental psychopathology: Gene-environment interplay in antisocial behaviors. *Psychological Bulletin, 131,* 533–554, p. 533.

17. Plomin, R., & Daniels, D. (1987). Why are children in the same family so different from each other? *Behavioral and Brain Sciences, 10,* 1–16.

18. Beaver (2008), see Note 7.

19. Plomin, & Daniels (1987), see Note 17.

20. Cohen (1999), see Note 10.

21. Bohman, M., Cloninger, R., Sigvardsson, S., & von Knorring, A. (1982). Predisposition to petty criminality in Swedish adoptees: I. Genetic and environmental heterogeneity. *Archives of General Psychiatry, 39,* 1233–1241; Mednick, S. A., & Christiansen, K. O. (1977). *Biosocial bases of criminal behavior.* New York: Gardner Press.

22. Cadoret, R. J., Cain, C. A., & Crowe, R. R. (1983). Evidence for gene–environment interaction in the development of adolescent antisocial behavior. *Behavior Genetics, 13,* 301–310.

23. Moffitt (2005), see Note 16.

24. Boutwell, B. B., Franklin, C., Barnes, J. C., & Beaver, K. M. (2011). Physical punishment and childhood aggression: The role of gender and gene-environment interplay. *Aggressive Behavior, 37,* 559–568.

25. Caspi, A., McClay, J., Moffitt, T. E., Mill, J., Martin, J., Craig, I. W., Taylor, A., & Poulton, R. (2002). Role of genotype in the cycle of violence in maltreated children. *Science, 297,* 851–854; Kim-Cohen, J., Caspi, A., Taylor, A., Williams, B., Newcombe, R., Craig, I. W., & Moffitt, T. E. (2006). MAOA, maltreatment, and gene–environment interactions predicting children's mental health: New evidence and a meta-analysis. *Molecular Psychiatry, 11,* 903–913.

26. DeLisi et al. (2009), see Note 1.

27. Belsky, J., & Beaver, K. M. (2010). Cumulative genetic plasticity, parenting, and adolescent self-regulation. *Journal of Child Psychology and Psychiatry, 52*(5), 619–626.

28. Belsky, J., Jonassaint, C., Pluess, M., Stanton, M., Brummett, B., & Williams, R. (2009). Vulnerability genes or plasticity genes? *Molecular Psychiatry, 14,* 746–754.

29. Rowe, D. C. (2001). Do people make environments or do environments make people? *Annals of the New York Academy of Sciences, 935,* 62–74.

30. Belsky et al. (2009), see Note 28.

31. Jaffee, S. R., Moffitt, T. E., Caspi, A., & Taylor, A. (2003). Life with (or without) father: The benefits of living with two biological parents depend on the father's antisocial behavior. *Child Development, 74,* 109–126.

32. Cleveland, H. H., Wiebe, R. P., van den Oord, E., & Rowe, D. C. (2000). Behavior problems among children with different family structures: The influence of genetic self-selection. *Child Development, 71,* 733–751.

33. Kendler, K. S., & Baker, J. H. (2007). Genetic influences on measures of the environment: A systematic review. *Psychological Medicine, 37,* 615–626.

34. Boutwell, B. B., & Beaver, K. M. (2010). The role of broken homes in the development of self-control: A propensity score matching approach. *Journal of Criminal Justice, 38,* 489–495.

35. Cleveland et al. (2000), see Note 32.

36. Boutwell, B. B., Beaver, K. M., Barnes, J. C., & Vaske, J. (2012). The developmental origins of externalizing behavioral problems: Parental disengagement and the role of gene-environment interplay. *Psychiatry Research, 197*(3), 337–344.

37. Boutwell, & Beaver (2010), see Note 1.

38. Barnes, J. C., & Morris, R. G. (2012). Young mothers, delinquent children: Assessing mediating factors among American youth. *Youth Violence and Juvenile Justice, 10*(2), 172–189.

39. Moffitt (2005), see Note 16.

CHAPTER SIX

The Impact of Biosocial Criminology on Public Policy: Where Should We Go From Here?

J. C. Barnes

KEY TERMS

Adaptability
Biosocial criminology
Phenotype
Reproductive fitness
Standard social science method (SSSM)

INTRODUCTION

Sociological explanations of human behavior have traditionally dominated criminology. These theories primarily focus on environmental/social causes of crime, criminality, and criminal propensity, leaving little room for alternative explanations that include biological, genetic, and/or evolutionary origins.[1] Thus, it should come as no surprise that **biosocial criminology** (research that explores the biological and genetic underpinnings to criminality) has, until very recently, remained on the fringes of the criminological landscape.

Since about 2000, biosocial criminology has gained momentum as an empirical perspective. Biosocial criminology encompasses several "subperspectives," each raising different questions and utilizing different methodological techniques. The existence of different perspectives often leads to confusion among criminologists as to what biosocial criminology *is*, how it can be used to understand human behavior, and whether it can inform humane and pragmatic public policies. This chapter addresses each of these issues. In doing so, a brief primer on biosocial criminology is offered first, followed by a discussion of the policy implications that stem from this line of research.

BIOSOCIAL CRIMINOLOGY: A PRIMER

The term *biosocial criminology* can best be viewed as a blanket concept that covers five different subperspectives or subcategories (note that these categories are not an established rubric and other scholars may situate things differently): evolutionary psychology, biological criminology, behavior genetics, molecular genetics, and neurocriminology. Though these five perspectives are not mutually exclusive, separating them in this manner will facilitate an understanding of their interrelatedness and, eventually, will assist in determining whether biosocial research can be applied to public policy. This section provides a brief overview of the five subperspectives, research that has developed out of them, and a summary of their interconnectedness.

Evolutionary Psychology

Evolutionary psychology applies the principles of evolution to human behavioral phenotypes. A **phenotype** can be defined as any measurable trait or behavior such as levels of self-control or involvement in delinquency.[2] To oversimplify a bit, evolutionary theory states that a species will adapt (i.e., change) to changes in the environment by means of natural selection. This process occurs at the phenotypic level, but the selection actually occurs at the genetic level; meaning that selection is *for* genes that lead to the observable phenotype.[3] Thus, adaptations occur at the genetic level such that various selection pressures operating in the environment function to alter the distribution of certain genes in the population.

Note that I have discussed adaptation in terms of a genetic "response" to the environment. This is a useful framework from which to understand the logic of evolutionary theory, but it must be noted evolution is not directed by goal-oriented outcomes; there is no omnipotent oversight of species' adaptations directing them in one way versus another. Rather, to borrow from Richard Dawkins, a blind watchmaker capitalizes upon random genetic variations, selectively retaining those few mutations that increase the fitness of certain organisms.[4] In other words:

> *Evolution is a* description of a process *that lacks consciousness, intentions, morals, and goals. Tigers do not have stripes because evolution wanted them to blend in against a forested background where the sun highlights one area but leaves an adjacent area in deep shade. Instead, tigers have stripes because at some point in their past striped tigers outreproduced other tigers.*[5]

Evolutionary psychology offers an explanation of the etiology of a wide range of human phenotypes. These explanations are often built upon two key pieces of information: *adaption* and *reproductive fitness*. Natural selection and, therefore, human evolution hinges on reproductive fitness and humans' adaptability.[6] **Adaptability**, as described earlier, refers to a species' ability to change its genetic distribution in response to shifts in environmental conditions. **Reproductive fitness**—simply defined—indexes reproduction success (i.e., the number of offspring contributed to the next generation). In short, genes (or the phenotypic outcomes of those genes) that increase the chances an individual will successfully reproduce (or reproduce more often than other members of the same species) will have a greater probability of being passed along to the next generation. Over time, these genes will proliferate in the population. Thus, to the extent that human behaviors are influenced by genetic factors (see the discussion of behavioral and molecular genetics that follows), evolutionary psychology offers one lens through which we can view the ultimate causes of antisocial outcomes.

Evolutionary psychology offers a unique perspective on human development and human behaviors. Where criminological research often looks to the recent past (relatively speaking) for illuminating causal processes, evolutionary psychology recognizes that the etiology of human behavior—including antisocial behavior—may have roots in our remote ancestral history. Toward this point, anthropological evidence suggests that modern humans (*Homo sapiens*) evolved somewhere between

250,000 and 50,000 years ago.[7] We evolved as a social species and, therefore, major portions of our lives are governed by our ability to interact with and navigate the social environment. These selection pressures have shaped human behaviors and must be considered if we are to fully understand antisocial behaviors.

In terms of the relevance for criminological research, evolutionary psychology brings three main points into focus. First, evolutionary psychology offers an explanation of why certain characteristics appear to be universal (shared by nearly all human cultures). Research has revealed that, while cultures differ in their tolerance of violence and aggression, all human cultures recognize certain acts, like intra-group homicide, to be antisocial and beyond the scope of permissible group behavior.[8] Thus, humans have evolved a mechanism to recognize wrongdoing in others and to respond to such misbehavior via the creation of laws and taboos. These findings also indicate that all cultures contain members who display antisocial behavior. Perhaps evolutionary psychology can shed light on the issue of why antisocial behavior persists across all human cultures.

The second way that evolutionary psychology applies to criminology is that it may offer an explanation for some of the most aggravating, elusive, and consistent findings known to criminologists. For example, the gender gap in offending is one of the most consistent findings—perhaps even to the level of being considered a "law" of criminology. The gender gap has been shown to be invariant across cultures and across time. Yet, criminology has failed to reach a plausible explanation of why males are overwhelmingly more likely to be involved in antisocial behavior as compared to females.[9] Evolutionary psychology, on the other hand, has the potential to address these shortcomings and offer a cogent explanation of the gender gap in offending. Specifically, evolutionary biologists/psychologists have consistently shown that, in sexual species, the sex bearing the largest parental investment (parental investment refers to both biological investment [i.e., gestation, sexual resources] and temporal investment during child rearing, though biological and temporal investment almost always go hand-in-hand) will tend to display fewer "risky" behaviors and will, instead, be more passive and more discriminating in their selection of potential mates. Since human females bear the larger parental investment, evolutionary psychology would predict that they will be less involved in antisocial behavior as compared to males. Thus, evolutionary psychology offers one avenue by which we can build an understanding of the gender gap in offending.

Third, evolutionary psychology asks new questions. Daly and Wilson offer a visible example of the "new" questions that evolutionary psychology poses to seemingly "old" issues.[10] Specifically, Daly and Wilson approached the issue of homicide (why do humans kill each other?) from an evolutionary framework. By doing so, they unlocked numerous realities that had before gone unnoticed; for example, stepchildren are more likely to be victims of child abuse than biological children. Other examples of new questions offered by evolutionary psychology relate to the sexual behavior of offenders and whether we can ever eliminate crime.[11]

Behavior Genetics

A large body of criminological research analyzes individual-level data to test hypotheses about the causes of crime. To gather the data necessary to analyze these types of hypotheses, researchers commonly rely on the **standard social science method (SSSM)**.[12] The SSSM can be defined as any method of data gathering and/or data analysis that does not allow the researcher to account for genetic influences. Just because SSSMs are the most common technique should not give the impression that they are the best method for every purpose. In fact, SSSMs may generate results that are systematically biased.[13] The reason for this systematic bias is that SSSMs do not allow the researcher to control for genetic influences on the outcome variable (i.e., omitted variables bias). Because genetic factors are likely to influence nearly every human behavioral phenotype, omitting genetic factors as an explanatory variable runs a high risk of introducing bias into researchers' models. In other words, the problem is one of spuriousness with genetic factors acting as the omitted third variable.

Unlike SSSMs, behavior genetic research allows for the study of genetic as well as environmental influences on human phenotypes such as behaviors and personality traits. Researchers are often interested in predicting differences—or variation—in phenotypes within their samples. For example, a researcher may wish to explore the variation in levels of self-control across a sample of adolescents. The researcher would calculate the variance in self-control that exists in the sample and then use statistical analyses to determine the factors—both genetic and environmental—that account for, or explain, the observed variance. To do so, variance in the phenotype (self-control) is decomposed into three latent components: heritability (h^2), shared environment (c^2), and nonshared environment (e^2). The heritability component (h^2) measures the amount of variance in the phenotype that can be attributed to genetic differences in the sample. For example, a heritability estimate of 0.25 would mean that one-fourth of the variance in self-control is attributable to differences in genetic material between the respondents in the sample. For ease of interpretation, researchers often transform heritability estimates into percentages by multiplying the estimate by 100. Thus, an estimate of 0.25 can be interpreted as 25 percent of the variance being explained by differences in genetic factors. The environmental components (i.e., c^2 and e^2) estimate the amount of variance in the phenotype that can be attributed to environmental factors.

It is important to point out that behavioral geneticists distinguish between two types of environmental influences: shared (c^2) and nonshared (e^2). The shared environment captures any environmental influence that makes two siblings more alike. Growing up in poverty and experiencing a parental divorce might be two examples of shared environmental influences. Nonshared environmental influences capture environmental effects that make siblings *different* from one another. Typically, nonshared environments are conceptualized as events that are not experienced by both siblings. Nonshared environments may also arise, however, when siblings experience the same objective event but have different subjective interpretations or perceptions of the event.[14] Examples of nonshared environments might include exposure to different peer groups, exposure to different teachers, and experiencing different parenting strategies (i.e., parents treating their children differently). Measurement error is also captured by the nonshared environmental component.

In quantitative behavioral genetics, the three components (h^2, c^2, and e^2) are estimated simultaneously as latent traits. In this way, behavioral genetic models do not suffer from the specification problems that afflict SSSMs (i.e., omitted variable biases) because all factors that can influence variation in the phenotype are included in the model. Specifically, 100 percent of the phenotypic variance is always explained by behavioral genetic methods.

In order to obtain estimates of the three components, researchers must analyze pairs of individuals with known levels of genetic relatedness. Twin studies have emerged as one of the most common methodologies used by behavioral genetic researchers, because the genetic differences between monozygotic (MZ) twins and dizygotic (DZ) twins can be seen as a "natural experiment." Specifically, the logic of twin studies rests on one key piece of information: MZ twins share 100 percent of their genetic material while DZ twins, on average, share 50 percent of their distinguishing genetic material. Both types of twins, however, share an equal part of their environments (referred to as the equal environments assumption).[15] Thus, if MZ siblings resemble one another more closely than DZ twins, genetic factors are likely to play a role in the etiology of the phenotype under examination.

In terms of the sources of antisocial behavior, five meta-analyses have summarized the role that genetic and environmental factors play. To be brief, all five meta-analyses have concluded that genetic factors play a role in the etiology of aggression, delinquency, criminality, and other related phenotypes.[16] Further, these studies were surprisingly consistent in their estimate of the amount of variance that was explained by genetic factors. Indeed, all five meta-analyses suggest that antisocial behavior is around 50 percent heritable.

Molecular Genetics

As already discussed, behavioral genetic research typically decomposes the variance in a phenotype into three separate components (i.e., heritability, shared environment, and nonshared environment). One limitation of this research design is that the individual factors comprising each of the three components remain unidentified. For example, a research study may estimate that genetic factors account for 50 percent of the variance in antisocial behavior. Although this estimate is informative, it does not tell us anything about *which* genes are driving the effect. In other words, behavioral genetic research designs estimate latent traits that cannot inform scholars about specific risk factors.

With the mapping of the human genome in the early 2000s, researchers are beginning to pull back the "heritability curtain" and identify links between *measured* genes and phenotypic outcomes. Although an overview of human molecular genetics is well beyond the scope of this chapter, a few of the more important points are worth noting. First, it is important to keep in mind that there is no "crime gene." To be sure, complex human phenotypes like behavioral tendencies are not governed by a single gene. Instead, a complex mixture of genetic and environmental influences is necessary to explain the emergence of antisocial behavior. Second, genes do not directly "cause" antisocial behavior. Genes code for amino acid production, amino acids are strung together to form proteins, proteins are then put to work in various capacities throughout the body. In short, any genetic influence on antisocial behavior is mediated by a complex biological system. Third, any genetic influence on antisocial behavior is almost certainly mediated by the brain. The brain controls our thoughts and emotions, and is built largely based on the blueprint laid out in our genetic code. Thus, molecular geneticists often look for genes known to influence brain structure or functioning when hypothesizing about the genetic pathways to antisocial behavior.[17]

Molecular genetics research has already produced a wealth of knowledge. Certain genetic polymorphisms have been linked to well known predictors of antisocial behaviors such as attention deficit hyperactivity disorder (ADHD), childhood conduct disorder, and gang membership.[18] Perhaps more importantly, molecular genetics research has demonstrated the importance of the environment in "triggering" genetic effects: a process known as gene–environment interaction (G×E). Findings from G×E research show that certain genetic effects are more likely to manifest when combined with environmental risk factors.[19] This line of work has also shown that genetic influences can change throughout the life course as a result of differential environmental exposure. In other words, changes in the environment can alter the effects of genetic factors.

Biological Criminology

Biological criminology recognizes that a host of physiological factors, not just genetic factors, may be related to antisocial behavior. Moreover, the genetic effects identified by behavioral and molecular geneticists must work *through* physiological factors. Although many areas of research could be pigeonholed into the biological criminology concept, three foci have garnered much interest from behavioral scientists: (1) the influence of hormone levels, (2) the influence of resting heart rate, and (3) the influence of environmental factors on biological functioning. Each of these three areas are briefly discussed in this section.

Hormones

An interesting stream of research has reported a correlation between products of the endocrine system (i.e., hormones) and human behavior. Testosterone is one hormone that has received much attention. Testosterone is an androgen, meaning that it is a steroid hormone that is produced to masculinize the body. Both females and males produce testosterone. The primary difference between the two sexes, however, lies in the *amount* of testosterone that is produced—males produce testosterone at much greater levels.[20] Because testosterone is often conceptualized as the "male hormone," it is easy to understand why scholars have searched for a link between testosterone levels and antisocial behavior. Research coming out of this area, however, has struggled to explain the idiosyncratic association between testosterone and male behavior. To be brief, research has supported the notion that testosterone levels are linked with aggressive and dominant behavior, but it is unclear exactly which way the causal relationship works. Because testosterone levels vary throughout the day, it is nearly impossible to determine whether increased testosterone causes aggression or vice versa. The most likely scenario is that testosterone and aggression/dominance enjoy a reciprocal relationship with one another. Supporting this argument is research showing that testosterone levels vary according to marital status and fatherhood—testosterone has been shown to decrease during times of marriage and fatherhood and increase after a divorce.[21] In short, testosterone has been linked with aggressive and dominant behavior, but there remains much room for future research.

Resting Heart Rate

Resting heart rate has been shown to be a robust (if not the most robust) biological correlate of antisocial behavior. It is thought that resting heart rate influences autonomic arousal levels which, in turn, influence sensation-seeking behaviors. To be specific, scientists have shown that individuals with low resting heart rates engage in risk-seeking behaviors to a greater degree as compared to people with normal resting heart rates. Indeed, Ortiz and Raine's meta-analysis uncovered a strong correlation between resting heart rate and antisocial behavior.[22] Furthermore, recent work has revealed that college students who had low resting heart rates (operationalized as being one or more standard deviations below the mean) were less likely to be deterred from antisocial behavior. Specifically, Armstrong and Boutwell reported that students who had low resting heart rates perceived fewer disadvantages (e.g., perceived sanctions, perceived feelings of guilt/shame) from committing an assault during a hypothesized confrontation.[23] In short, individuals with low resting heart rates may be *less* deterrable from offending.

Environmental Factors

Many "environmental" influences are thought to impact human behavior via their influence on biological/physiological functions. An environmental influence, in this sense, can be thought of as anything that is *not* genetic or inherently biological in origin. For instance, a long history of psychological research has revealed that persons who are subject to head trauma are at an increased risk of showing personality changes.[24] Others have shown that persistent offending is predicted by the individual's experiences with birth complications.[25] Building on these findings, it is not surprising that some research has reported that life-course-persistent offenders are more likely to display reduced cognitive functioning and that they are at greater risk of having experienced head trauma.[26] Though the causal impact of head trauma works via physiological factors in the brain (either reduced functioning or altered structure), the event itself (i.e., the experience of head trauma) is an environmental influence. Biological criminology reveals, therefore, the interconnectedness between biological/physiological functioning (or structure) and environmental influences on human behavior.

A telling line of research has reported that individuals exposed to high amounts of lead early in development are more likely to display antisocial behavior later in life. Lead is not naturally produced by the body, thus, lead exposure would be considered an "environmental" influence. In one of the most rigorous tests of the influence of lead on antisocial behavior, Wright et al. measured blood lead levels at three different time points: during prenatal development (i.e., when the participant was still in the womb), during early childhood, and at 6 years of age. In short, the authors found that blood lead levels—whether measured prenatally or at age 6—predicted involvement in crime. The relationship was positive such that a higher concentration of lead in the blood was predictive of a greater involvement in crime.[27]

Neurocriminology

Neurocriminology is the final subperspective of biosocial criminology. Just because we have saved it for last should not give the impression that it is less important, that the research is less conclusive, or that this approach is inferior to the others in any way. To be sure, neurocriminology has received a large amount of attention, it relies on extremely advanced techniques and technologies, and it has produced an overwhelmingly consistent body of evidence.[28] Instead, we saved neurocriminology for last for one reason: Researchers are beginning to understand the importance of the brain for controlling human behavior, and it is believed that many of the influences discussed earlier (i.e., genes, hormones, environment) impact behavior *via* their impact on the brain.[29] As with the preceding sections, the literature surrounding neurocriminology is far too vast to review in a few paragraphs. Thus, this space is reserved for offering a quick glimpse into the interconnectedness between genes, hormones, the environment, and the brain.

The brain is the epicenter for all human behavior and emotions. Neuroscientists have linked emotional responses to environmental stimuli with various regions of the brain (see Wright, Tibbetts, and Daigle for an overview of the brain written for social scientists).[30] At the same time, neuroscience research has shown that human behaviors are controlled by the brain and, in particular, that certain regions of the brain appear to be critically important for understanding the etiology of antisocial behavior.[31] Juxtaposing these findings with those from evolutionary psychology, behavior genetics, molecular genetics, and biological criminology, a causal model begins to emerge. Specifically, as Raine explained, it is likely that most of the evolutionary, genetic, and biological risk factors that have been linked to antisocial behavior are mediated by the brain.[32] In other words, when a molecular geneticist finds a specific gene to be correlated with antisocial behavior, it is likely that this gene impacts antisocial behavior through its impact on either the structure or the function of the brain.

Summary

Biosocial criminology encompasses a vast literature. Although this perspective is gaining traction among criminologists, it has yet to enjoy majority support within the discipline. As can be seen from this review, the field of criminology can no longer excuse biosocial research on grounds of too limited evidence. There is now an abundance of research showing that genetic factors influence antisocial behavior, that they interact with the environment to predict criminal offending (i.e., G×E), and that the synergistic relationship between genes and the environment is played out in the brain, which translates physiological and environmental inputs into thoughts and behavior.[33] In short, the age of biosocial criminology is upon us. Those who continue to ignore its existence or its contributions to criminological research risk being left by the wayside.

BIOSOCIAL CRIMINOLOGY AND PUBLIC POLICY

One of the most important issues stemming from any social science research is whether and how that research can be used to create policies that will better society. In addressing this topic, the remainder of this chapter is organized around three questions that are frequently asked when considering the policy implications of biosocial criminology. Perhaps the most common question raised when navigating these issues is a philosophical one: *Should biosocial criminology inform public policy?* Note that this is a philosophical question, meaning that there is no way to reach a "right" or "wrong" answer with empirical data. Instead, the question challenges scientists to consider whether we *should* or *ought* to do something. When pondering this question, we must consider the implications of action and the implications of inaction. In other words, ethics and morals must play a major role in answering this question.

The second question considered in this section is closely related to the first: *Are inhumane policies an inevitable byproduct of biosocial research?* At the risk of spoiling the discussion that follows, the answer is unequivocally "no." Finally, the third question naturally flows from the conclusions of the previous two. Specifically, if biosocial criminology can inform policies that are humane and progressive, what do these policy implications look like? In other words, *what are the policy implications of contemporary biosocial criminology?*

Should Biosocial Criminology Inform Public Policy?

One of the first questions a student of biosocial criminology will face is whether biosocial research should inform public policy. As noted earlier, this is a philosophical question that requires a philosophical answer. In other words, it is easy to imagine two scholars: one who argues the merits of biosocial criminology and why it should be used to develop policy, and another who argues against biosocial criminology's influence on public policy. The scholar who supports the use of biosocial criminology in policy formation might begin by noting that all social science research should aim to better humanity, and new policy is but one avenue that can be used to do so.

The biosocial proponent might also point out that biosocial research is already being used to shape policy. A good example comes from the Supreme Court's decision in *Roper v. Simmons* (2005).[34] Based on extensive testimony of developmental psychologists and neuroscientists, the Supreme Court ruled that imposing the death penalty on persons under the age of 18 is unconstitutional because it violates the principle that the punishment must match the culpability of the offender. This is not to say that juvenile defendants are *less* guilty than their adult counterparts; on the contrary, the Supreme Court ruled this way because research has shown that the teenage brain is not fully developed in areas known to control logical reasoning and decision making (the prefrontal cortex). In other words, teenagers may be more prone to risky behavior due to the different "weights" they place on the risks and rewards of such behavior.[35]

Though proponents of biosocial research may point to the positive policy decisions such as the one noted above, critics will certainly have plenty of "bad" policies they can point to in order to argue against biosocial influences on policy. One of the most common arguments is that biosocial research has the potential to lead to dangerous, "Nazi-like" policies. Unfortunately, this is a valid argument that must be taken into serious consideration. The next section considers this side of the discussion.

Are Inhumane Policies an Inevitable Byproduct of Biosocial Research?

The argument that biosocial research can lead to dangerous policy implications is true. As such, it is important that we frequently pause and remind ourselves of the tainted history of "biosocial" policy (the term *biosocial* is in quotations here because it has often been argued that policies of the past, such as the eugenics movement, were based on such a warped version of empirical reality that they can barely be described as biosocial).[36] Scholars often point to the eugenics movement when condemning biosocial research.[37] Briefly, the term *eugenics* means "wellborn" and, thus, the eugenics movement sought to change society at the genetic level by selectively allowing (or disallowing) certain individuals to reproduce.[38] For example, in the Supreme Court case, *Buck v. Bell* (1927), the court upheld a law that allowed for the sterilization of mentally challenged persons.[39] This is just one example of a "eugenics-like" policy. There are certainly other examples, but the point is clear: Eugenics sought to change society by controlling and manipulating the distribution of genetic factors within the population.

The eugenics movement is a failed experiment and, in the words of Rowe, "eugenics has been discredited, and it is unlikely to be embraced again."[40] Indeed, eugenics has been discredited on several grounds. First, and most importantly, eugenics is immoral and should be eliminated for that reason alone. Second, eugenics is impractical from a scientific standpoint; it would take far too long to change the genetic distribution of an entire population enough for it to have any impact. On the contrary, behavioral geneticists and population geneticists have argued extensively that changing the environment is a much quicker and more manageable way to impart large-scale changes on society. Third, eugenics works against the nature of evolution. Evolution seeks to maintain population variation. Without variation, a species would be more susceptible to threats and have little chance of long-term survival.[41] Thus, eugenics is incompatible with the principle laws of nature and, therefore, there is no reason to expect it would work.

Though biosocial research has the potential to lead to harmful policies, this does not mean that *non*biosocial research always leads to altruistic policy. Put differently, "social" research (e.g., sociology and traditional criminology) can also lead to dangerous policies. Perhaps the best example comes from one of the mainstays of social research: Marx's theory of socialism.

I do not mean to digress into a conversation about the merits of socialist versus capitalist policies. Instead, I only intend to show that *all* social science research (including biosocial research), despite good intentions, can lead to dangerous policies. Marxist socialism is based, fundamentally, on the idea that human nature can be changed if society is subjected to a large-scale shift in ideology and structure. Marx saw staunch inequalities between members of different societal classes (for simplicity, think of the ruling class and the [much larger] lower class). Marx noted that a capitalist system will inevitably lead to this dichotomy and, as a result, the only solution is to reshape society by distributing property equally (or according to need) across all members of society. This theory was based—in the words of Steven Pinker—on a Utopian Vision of society and human nature.[42] In short, the key argument is that the ills of capitalism can be remedied with a radical shift in economic policy.

Socialism has been implemented in numerous places across the globe over the past century. Despite the good intentions of scholars and policymakers, however, socialism appears to have been a failed experiment—at least in regard to its actual implementation.

It is likely a useless endeavor to extend this argument to one of body counts, so suffice it to say that many human lives have been lost in the name of socialism (as have many been lost in the name of other political movements). The point to take away is that socialism is, at its roots, a sociologically based theory. Given the ills that have been suffered in the name of socialism, it should be clear that any social scientist who claims biosocial research must be ignored due to the possibility of dangerous policy suffers from historical myopia and academic hypocrisy.[43]

Aside from eugenics, another concern for biosocial research is that it will justify differential treatment (i.e., discrimination) based on genetic differences. This criticism is rooted in the moralistic fallacy—the idea that anything explained by biology is legitimated as desirable. This is a fallacy precisely because biology and evolution have no conception of morality. As described earlier, evolution capitalizes on traits that give the individual a survival and/or reproductive advantage. There is no room for morality in this equation. Instead, it is up to humans to determine what fits within the bounds of legitimate, moral, and desirable behavior. In other words, these issues bring one point into focus: It is critically important that researchers and policymakers distinguish between what *can be done* and what *ought to be done*. Simply because science allows something to be carried out does *not* justify the behavior. For instance, many have examined the genetic underpinnings to sex differences in behavior. It is interesting to note that sex is, at its root, a genetic difference: Males have a Y chromosome and females do not (females have two Xs). The differences between males and females are stark on many issues, such as physical characteristics, but just because we understand why males and females are different does *not* justify differential treatment. Fairness does *not* hinge on sameness.[44]

This is a long-winded way of saying that biosocial research can be used to develop humane policies, but we must be cautious that they do not backslide into the horrible policies littering human history. The challenge, therefore, is to overcome certain fallacies about human nature, about how scientific evidence should be used, and about how the world "ought" to be. Any research that has bearing on human behavior can be used in harmful or destructive ways. It is up to society (meaning researchers, policymakers, and the public) to ensure that this does not occur by setting clear boundaries between acceptable and proscribed policies/ideas. This requisition applies to *all* types of research, not just biosocial scholarship.

In short, biosocial research can be dangerous. This does not mean, however, that it will *always* be dangerous. Biosocial research—just like social research—must be cautiously used when creating and implementing policy. Biosocial research should not be denied nor shunned simply because it *may* lead to harm. All research *may* lead to harm. So, any argument against the infusion of biosocial research into the policy debate simply because it may be harmful falls flat as it is an attack on all science, not just biosocial science.

What Are the Policy Implications of Contemporary Biosocial Criminology?

Social scientists are often surprised to find that biosocial policy implications tend to be similar to, if not exactly the same as, traditional criminology/sociology policies. No serious biosocial researcher would ever argue that policymakers should consider policies like those that are often referenced in popular culture. For example, many popular books and movies have been based around the idea that children should be tested for genetic markers for crime (and other things) at birth. Though there are a number of reasons to attack this policy, one stands most prominent: It is unethical, immoral, and illegal.[45] The fear of genetic testing, however, is often expressed by nonbiosocial scientists. This section, therefore, attempts to assuage these concerns by offering clear examples of modern biosocial policies that are already in place. You may be surprised to find that you have already heard of many of these but have never thought of them as "biosocial policies."

Biosocial criminology offers a more thorough and in-depth look at how the mind and the body work in concert with the environment to produce certain behaviors. Recall from our earlier discussion that biosocial criminology acknowledges the interplay between genes/biology and the environment. This is important because, as sociologists, criminologists, and geneticists have long noted, it is *far* easier to change the environment than it is to change the genetic code/distribution. This means that biosocial policies are *environmental* policies. The only difference between biosocial policies and typical criminological policies is that biosocial criminologists are able to make clearer predictions about what will work, for whom it will work, when it will work, and when it will not work. The focus for the remainder of this section centers on three biosocial policy implications (though there are many more): (1) Eradicate toxins such as lead from the environment (especially poor/disadvantaged neighborhoods); (2) improve pre-, peri-, and postnatal care for all children (especially those born to poor/disadvantaged parents); and (3) utilize targeted intervention for high-risk individuals.

The first policy implication is to eradicate toxins such as lead from the environment. Lead is a naturally occurring metal (though most lead exposure is due to human uses of the material such as in leaded gasoline and lead-based paint) that is highly toxic to the human body and has been linked to decreased brain volume, injury to the internal organs, and increased involvement in criminal activity.[46] Importantly, children, as compared to adults, appear to be more sensitive to lead exposure, showing toxicity at lower levels of exposure, a greater absorption rate, and a lower cleansing rate.[47] Importantly, dietary deficiencies (especially in calcium and iron) appear to regulate lead absorption: Children with dietary insufficiencies are at greater risk of absorbing lead. Because children from disadvantaged neighborhoods are more likely to have dietary deficiencies *and* are more likely to be exposed to lead, the obvious policy implication is to eradicate lead from the environment, with immediate attention being given to disadvantaged communities.[48] A recent analysis estimated that every $1 invested in lead paint hazard control would result in a return of $17–$221 in net savings (savings in terms of health care, crime, and tax revenue costs, among others).[49]

Although a recent report from the Centers for Disease Control and Prevention (CDC) revealed that blood lead levels have been decreasing nationwide, stark contrasts remain between communities of different socioeconomic status. Additionally, CDC data revealed that between 2003 and 2004, non-Hispanic blacks recorded a mean blood lead level of 1.69 μg/dL compared to the mean for non-Hispanic whites of 1.37 μg/dL. Perhaps differential rates of exposure to the environmental toxin, lead, can provide an understanding of the racial gap in offending.[50]

A second policy implication centers on improving pre-, peri-, and postnatal care (PPPC) for all children. Hundreds of studies have shown support for the notion that life-course-persistent offending is often preceded by deficiencies in PPPC that result in structural and/or functional brain abnormalities (i.e., neuropsychological deficits).[51] Many of these neuropsychological deficits are readily avoidable through better health care during the developmental stages both within and outside the womb (i.e., PPPC). Olds is putting this hypothesis—that better health care can improve infants' development—to the test with three ongoing randomized trials. Referred to as the Nurse Home Visitation Program (NHVP), three experimental trials (one in New York, one in Tennessee, and one in Colorado) have focused efforts on educating soon-to-be, first-time mothers about childbirth, child rearing, and healthy child development.[52] All three trials share a common structure, and therefore the remainder of this discussion will simply refer to the NHVP rather than the individual trials.

The NHVP was built on a few simple intervention techniques that proved to have large-scale and long-term impacts on child development, parent–child relationships, and parental outcomes. A full overview of the NHVP is beyond the scope of this chapter; instead, let us focus on three elements from the intervention that best fit the current discussion: (1) prenatal health, (2) child maltreatment, and (3) self-efficacy.[53] A major focus of the NHVP was to improve mothers' health during pregnancy (i.e., prenatal health). In order to achieve this mission, the NHVP sent nurses to expecting mothers' homes. During these nurse home visitations, mothers were educated about the potential harms of substance abuse, about the benefits of healthy eating, and about other factors known to impact the child's health at birth. Further, mothers were educated about how to seek medical attention, who to contact in case of a medical emergency, and how to identify potential pregnancy complications. The second element of the intervention is that expecting mothers were instructed on the importance of proper childrearing, forming attachment bonds with the child, and about the detrimental impacts of child maltreatment, abuse, and neglect. The third element of the intervention that warrants attention here is that the program's curriculum was designed with the mother's self-efficacy in mind. In short, the nurses helped mothers understand that the choices they make are important and will influence not only their lives, but the lives of their unborn children. As a result, the nurses often helped mothers to "establish realistic goals and small achievable objectives that once accomplished, increase parents' reservoir of successful experiences. These successes in turn increase women's confidence in taking on larger challenges."[54]

To briefly summarize, the NHVP provided nurse home visitations free of charge to at-risk, first-time mothers. These visitations were remarkably progressive in that they did not provide any "magic pill," nor did they force any unwanted changes on the mothers. Instead, the NHVP focused on pragmatic outcomes such as improving physical and mental health and on teaching proper parenting skills. In other words, the NHVP provided a form of oversight and a wealth of information that at-risk mothers often lack. So, what were the effects of the NHVP? To be direct, the effects have been astounding. Expecting mothers showed marked improvement in their prenatal health (e.g., smoking during pregnancy was reduced by 25 percent), children were healthier at birth (e.g., babies were less likely to be low birth weight), parents were *much* less likely to abuse/neglect their child, children were seen in the emergency room with less frequency during their second year of life, children showed improved intellectual functioning at ages 3 and 4, and children were less likely to have been arrested by age 15, among many other positive outcomes.[55]

A third policy implication that stems directly from biosocial research is targeted rehabilitation/intervention/reintegration for high-risk individuals. (For brevity, rehabilitation, intervention, and reintegration efforts can simply be termed *rehabilitation*. Note, however, that different meanings are often attached to each term. So, the "collapsing" of the terms is more for ease of demonstration than anything else.) This suggestion should sound familiar, as criminologists have recently begun a resurgence of interest in offender rehabilitation. Though definitions/conceptualizations of the term *rehabilitation* differ, Cullen and Jonson define it as "a planned correctional intervention that targets for change internal and/or social criminogenic factors with the goal of reducing recidivism and, where possible, of improving other aspects of an offender's life."[56]

Clearly, rehabilitation programs can differ on many aspects. One aspect that is of interest for the current purposes, however, is that rehabilitation programs have consistently reported that individuals who are *most* at risk often show the greatest improvement.[57] This finding is worth restating:

Rehabilitation programs tend to work best for those who need them most. This is an extremely important piece of information and, when combined with a biosocial focus, may lead to even greater rehabilitation success. Consider that medical research has long shown drug treatments to have variable effectiveness based on a person's genotype (termed *pharmacogenomics*).[58] Biosocial research has shown that genetic factors play a role in the etiology of antisocial behavior, hinting that gene X rehabilitation interactions may be important.

Oslin et al. examined whether a variant of the OPRM1 gene, which has been implicated in substance dependence, had an impact on individuals' response to alcohol treatment. These authors reported that participants who had a certain variant of the OPRM1 gene were more likely to respond to an alcohol treatment program (which utilized naltrexone—a common treatment aid for alcohol dependence). Specifically, participants who went through treatment *and* had a certain version of the OPRM1 gene were *less* likely to relapse and evinced *longer* times to relapse as compared to participants who went through treatment but had a different version of the gene. In the words of the authors, "If the findings reported here are replicated, OPRM1 genotyping may prove to be an efficient mechanism for identifying patients who are most likely to respond to naltrexone and those for whom other available treatments may be more efficacious."[59]

Brody and colleagues, building on an impressive body of research that had shown a link between a serotonin gene (5-HTTLPR; serotonin is a type of neurotransmitter expressed in the brain to send signals of information between neurons) and risky/antisocial behavior, hypothesized that a gene X rehabilitation effect would exist between genetic risk and exposure to an environmental rehabilitation program (the Strong African American Families Program [SAAF]). Specifically, the hypothesis was that youth who had the genetic risk factor would benefit most from the SAAF.[60] In this way, the study tested the hypothesis that rehabilitation has the greatest benefits for those who need it most. The study's findings supported the hypothesis by revealing that youth who had the genetic risk factor showed the most improvement (by displaying *less* risky behavior) after the intervention program had ended. Beach et al., also analyzing the SAAF data, reported a similar pattern of findings for a dopamine receptor gene (DRD4).[61]

In summary, biosocial research offers policy implications that are very similar (if not identical) to those offered by other lines of scholarship. The main difference between biosocial policy and criminology policy is that biosocial research offers a more in-depth understanding of what works and for whom. It has long been noted that "one-size-fits-all" policies do *not* work. Continuing that analogy, biosocial research offers a way to determine who needs which size (treatment).[62]

CONCLUSION

The purpose of this chapter is to dissolve the mystique that tends to surround biosocial criminology and the policy implications that stem from this perspective. As discussed, biosocial policies need not be different from the policies criminologists have already set in place. Instead, biosocial research offers an exciting opportunity to better evaluate and understand why certain extant policies work and why others do not. Thus, there is nothing inherently "scary," nor is there anything "evil" about biosocial research. Indeed, biosocial research may be the only way to truly understand antisocial behavior and to pull criminology away from the fringes and into the realm of serious science.[63] As any scientist will attest, only after we have reached an understanding of a phenomenon can we begin to influence it in earnest. Thus, our ability to improve society through policy is only as strong as our knowledge about *all* contributors to behavior—both genetic and environmental.

GLOSSARY

Adaptability—a species' ability to change its genetic distribution in response to shifts in environmental conditions

Biosocial criminology—research that explores the biological and genetic underpinnings to criminality

Phenotype—any measurable trait or behavior, such as levels of self-control or involvement in delinquency

Reproductive fitness—reproduction success (i.e., the number of offspring contributed to the next generation)

Standard social science method (SSSM)—any method of data gathering and/or data analysis that does not allow the researcher to account for genetic influences

NOTES

1. Beaver, K. M. (2009). *Biosocial criminology: A primer*. Dubuque, IA: Kendall/Hunt; Udry, J. R. (1995). Sociology and biology: What biology do sociologists need to know? *Social Forces, 73*, 1267–1278; van den Berghe, P. L. (1990). Why most sociologists don't (and won't) think evolutionarily. *Sociological Forum, 5*, 173–185.
2. Tooby, J., & Cosmides, L. (2005). Conceptual foundations of evolutionary psychology. In D. M. Buss (Ed.), *The handbook of evolutionary psychology*. Hoboken, NJ: Wiley.
3. Dawkins, R. (1989). *The selfish gene*. New York: Oxford University Press.
4. Dawkins, R. (1996). *The blind watchmaker: Why the evidence of evolution reveals a universe without design*. New York: W. W. Norton & Company.
5. Carey, G. (2003). *Human genetics for the social sciences*. Thousand Oaks, CA: Sage, pp. 218–219.
6. Dawkins (1989), see Note 3.
7. Wright, J. P. (2009). Inconvenient truths: Science, race, and crime. In A. Walsh & K. M. Beaver (Eds.), *Biosocial criminology: New directions in theory and research*. New York: Routledge.
8. Brown, D. E. (1991). *Human universals*. New York: McGraw-Hill; Pinker, S. (2011). *The better angels of our nature: Why violence has declined*. New York: Viking.
9. Campbell, A. (2009). Gender and crime: An evolutionary perspective. In A. Walsh & K. M. Beaver (Eds.), *Biosocial criminology: New directions in theory and research*, New York: Routledge; Campbell, A. (1995). A few good men: Evolutionary psychology and female adolescent aggression. *Ethology and Sociobiology, 16*, 99–123.
10. Daly, M., & Wilson, M. (1988). *Homicide*. New York: Aldine.
11. Boutwell, B. B., Barnes, J. C., & Beaver, K. M. (2012). Life-course persistent offenders and the propensity to commit sexual assault. *Sexual Abuse: A Journal of Research and Treatment*, Epub.
12. Harris, J. R. (1998). *The nurture assumption: Why children turn out the way they do*. New York: The Free Press.
13. Cleveland, H. H., Beekman, C., & Zheng, Y. (2011). The independence of criminological "predictor" variables: A good deal of concerns and some answers from behavioral genetic research. In K. M. Beaver & A. Walsh (Eds.), *The Ashgate research companion to biosocial theories of crime*. Franham, UK: Ashgate Publishing.
14. Turkheimer, E., & Waldron, M. (2000). Nonshared environment: A theoretical, methodological, and quantitative review. *Psychological Bulletin, 126*, 78–108.
15. Carey (2003), see Note 5.
16. Burt, S. A. (2009). Are there meaningful etiological differences within antisocial behavior: Results of a meta-analysis. *Clinical Psychology Review, 29*, 163–178; Ferguson, C. J. (2010). Genetic contributions to antisocial personality and behavior: A meta-analytic review from an evolutionary perspective. *Journal of Social Psychology, 150*, 160–180; Mason, D. A., & Frick, P. J. (1994). The heritability of antisocial behavior: A meta-analysis of twin and adoption studies. *Journal of Psychopathology and Behavioral Assessment, 16*, 301–323; Miles, D. R., &

Carey, G. (1997). Genetic and environmental architecture of human aggression. *Journal of Personality and Social Psychology, 72*, 207–217; Rhee, S. H., & Waldman, I. D. (2002). Genetic and environmental influences on antisocial behavior: A meta-analysis of twin and adoption studies. *Psychological Bulletin, 128*, 490–529.

17. Moffitt, T. E. (2005). The new look of behavioral genetics in developmental psychopathology: Gene-environment interplay in antisocial behaviors. *Psychological Bulletin, 131*, 533–554.

18. Faraone, S. V., Doyle, A. E., Mick, E., & Biederman, J. (2001). Meta-analysis of the association between the 7-repeat allele of the dopamine D4 receptor gene and attention deficit hyperactivity disorder. *American Journal of Psychiatry, 158*, 1052–1057; Beaver, K. M., Wright, J. P., DeLisi, M., Walsh, A., Vaughn, M. G., Boisvert, D., & Vaske, J. (2007). A gene × gene interaction between DRD2 and DRD4 is associated with conduct disorder and antisocial behavior in males. *Behavioral and Brain Functions, 3*, 30; Beaver, K. M., DeLisi, M., Vaughn, M. G., & Barnes, J. C. (2010). Monoamine oxidase A genotype is associated with gang membership and weapon use. *Comprehensive Psychiatry, 51*, 130–134.

19. Caspi, A., McClay, J., Moffitt, T. E., Mill, J., Martin, J., Craig, I. W., Taylor, A., & Poulton, R. (2002). Role of genotype in the cycle of violence in maltreated children. *Science, 297*, 851–854.

20. Mazur, A. (2009). Testosterone and violence among young men. In A. Walsh & K. M. Beaver (Eds.), *Biosocial criminology: New directions in theory and research*, New York: Routledge.

21. Gettler, L. T., McDade, T. W., Feranil, A. B., & Kuzawa, C. W. (2011). Longitudinal evidence that fatherhood decreases testosterone in human males. *Proceedings of the National Academy of Sciences, 108*, 16194–16199.

22. Ortiz, J., & Raine, A. (2004). Heart rate level and antisocial behavior in children and adolescents: A meta-analysis. *Journal of the American Academy of Child and Adolescent Psychiatry, 43*, 154–162.

23. Armstrong, T. A., & Boutwell, B. B. (2012). Low resting heart rate and rational choice: Integrating biological correlates of crime in criminological theories. *Journal of Criminal Justice, 40*, 31–39.

24. Groom, K. N., Shaw, T. G., O'Connor, M. E., Howard, N. I., & Pickens, A. (1998). Neurobehavioral symptoms and family functioning in traumatically brain-injured adults. *Archives of Clinical Neuropsychology, 13*, 695–711; Kurtz, J. E., Shealy, S. E., & Putnam, S. H. (2007). Another look at paradoxical severity effects in head injury with the personality assessment inventory. *Journal of Personality Assessment, 88*, 66–73.

25. Olds, D. L. (2006). The Nurse-Family Partnership: An evidence-based preventive intervention. *Infant Mental Health Journal, 27*, 5–25.

26. Raine, A., Moffitt, T. E., Caspi, A., Loeber, R., Stouthamer-Loeber, M., & Lynam, D. (2005). Neurocognitive impairments in boys on the life-course persistent antisocial path. *Journal of Abnormal Psychology, 114*, 38–49.

27. Wright, J. P., Dietrich, K., Ris, M. D., Homung, R. W., Wessel, S. D., Lanphear, B. P., Ho, M., & Rae, M. (2008). Association of prenatal and childhood blood lead concentrations with criminal arrests in early adulthood. *PLoS Medicine, 5*, 732–740.

28. Boots, D. P. (2011). Neurobiological perspectives of brain vulnerability in pathways to violence over the life course. In K. M. Beaver & A. Walsh (Eds.), *The Ashgate research companion to biosocial theories of crime*. Burlington, VT: Ashgate Publishing Company.

29. Raine, A. (2008). From genes to brain to antisocial behavior. *Current Directions in Psychological Science, 17*, 323–328.

30. Wright, J. P., Tibbetts, S. G., & Daigle, L. E. (2008). *Criminals in the making: Criminality across the life course*. Thousand Oaks, CA: Sage.

31. Beaver, K. M., Wright, J. P., & DeLisi, M. (2007). Self-control as an executive function: Reformulating Gottfredson and Hirschi's parental socialization thesis. *Criminal Justice and Behavior, 34*, 1345–1361; Raine, A. (1993). *The psychopathology of crime: Criminal behavior as a clinical disorder*. New York: Academic Press.

32. Raine (2008), see Note 29.

33. Pinker, S. (2002). *The blank slate: The modern denial of human nature*. New York: Penguin.

34. *Roper v. Simmons*, 543 U.S. 03-633 (2005).

35. DeLisi, M., Wright, J. P., Vaughn, M. G., & Beaver, K. M. (2010). Nature and nurture by definition means both: A response to Males. *Journal of Adolescent Research, 25*, 24–30; Steinberg, L. (2008). A social neuroscience perspective on adolescent risk-taking. *Developmental Review, 28*, 78–106; Steinberg, L. (2009). Adolescent development and juvenile justice. *Annual Review of Clinical Psychology, 5*, 47–73.

36. Pinker (2002), see Note 33.
37. Rukus, J., & Gibson, C. L. (2011). From petri dish to public policy: A discussion of the implications of biosocial research in the criminal justice arena. In K. M. Beaver & A. Walsh (Eds.), *The Ashgate research companion to biosocial theories of crime*. Burlington, VA: Ashgate Publishing Company.
38. Rowe, D. C. (2001). *Biology and crime*. Los Angeles, CA: Roxbury.
39. *Buck v. Bell*, 274 U.S. 200 (1927).
40. Rowe (2001), see Note 38.
41. Ridley, M. (1993). *The red queen: Sex and the evolution of human nature*. New York: Penguin Putnam.
42. Pinker (2002), see Note 33.
43. Walsh, A., & Yun, I. (2011). Race and criminology in the age of genomic science. *Social Science Quarterly*, *92*, 1279–1296.
44. Pinker (2002), see Note 33.
45. Genetic Information Nondiscrimination Act of 2008, Public Law 110-233; Sankar, P. (2003). Genetic privacy. *Annual Review of Medicine*, *54*, 393–407.
46. Wright et al. (2008), see Note 27.
47. Agency for Toxic Substances and Disease Registry. (2007). *Toxicological profile for lead*. Washington, DC: Department of Health and Human Services.
48. Sankar (2003), see Note 45.
49. Agency for Toxic Substances and Disease Registry (2007), see Note 47.
50. Wright et al. (2008), see Note 27; Walsh, &Yun (2011), see Note 43.
51. Raine et al. (2005), see Note 26; Gould, E. (2009). Childhood lead poisoning: Conservative estimates of the social and economic benefits of lead hazard control. *Environmental Health Perspectives*, *117*, 1162–1167.
52. Cullen, F. T., & Jonson, C. L. (2011). Rehabilitation and treatment programs. In J. Q. Wilson & J. Petersilia (Eds.), *Crime and public policy*. New York: Oxford University Press, p. 295; Olds (2006), see Note 25; Olds, D. L., Henderson, C. R., Chamberlin, R., & Tatelbaum, R. (1986). Preventing child abuse and neglect: A randomized trial of nurse home visitation. *Pediatrics*, *78*, 65–78; Olds, D. L., Henderson, C. R., Cole, R., Eckenrode, J., Kitzman, H., Luckey, D., et al. (1998). Long-term effects of nurse home visitation on children's criminal and antisocial behavior: 15-year follow-up of a randomized controlled trial. *Journal of the American Medical Association*, *280*, 1238–1244; Olds, D. L., Henderson, C. R., & Tatelbaum, R. (1994). Prevention of intellectual impairment in children of women who smoke cigarettes during pregnancy. *Pediatrics*, *93*, 228–233; Rutter, M. (1985). Resilience in the face of adversity: Protective factors and resistance to psychiatric disturbance. *British Journal of Psychiatry*, *147*, 598–611.
53. Olds (2006), see Note 25.
54. Olds (2006), p. 14, see Note 25.
55. Olds et al. (1986), Olds et al. (1994), Olds et al. (1998), see Note 52.
56. Cullen, & Jonson (2011), see Note 52.
57. Rutter (1985), see Note 52.
58. Evans, W. E., & Johnson, J. A. (2001). Pharmacogenomics: The inherited basis for interindividual differences in drug response. *Annual Review of Genomics and Human Genetics*, *2*, 9–39.
59. Oslin, D. W., Berrettini, W., Kranzler, H. R., Pettinati, H., Gelernter, J., Volpicelli, J. R., & O'Brien, C. P. (2003). A functional polymorphism in the μ-opioid receptor gene is associated with naltrexone response in alcohol-dependent patients. *Neuropsychopharmacology*, *28*, 1546–1552, p. 1551.
60. Brody, G. H., Beach, S., Philibert, R. A., Chen, Y., & Murry, V. M. (2009). Prevention effects moderate the association of 5-HTTLPR and youth risk behavior initiation: Gene X environment hypotheses tested via a randomized prevention design. *Child Development*, *80*, 645–661.
61. Beach, S. R., Brody, G. H., Philibert, R. A., & Lei, M. (2010). Differential susceptibility to parenting among African American youths: Testing the *DRD4* hypothesis. *Journal of Family Psychology*, *24*, 513–521.
62. Cullen, F. T. (2011). Beyond adolescence-limited criminology: Choosing our future—The American Society of Criminology 2010 Sutherland Address. *Criminology*, *49*, 287–330.
63. Cullen (2011), see Note 62.

CHAPTER SEVEN

The Heritability of Common Risk and Protective Factors to Crime and Delinquency

Kevin M. Beaver, Michael G. Vaughn, Matt DeLisi, John Paul Wright, Richard Wiebe, H. Harrington Cleveland, and Anthony Walsh

KEY TERMS

Heritability
Nonshared environment
Protective factors
Risk factors
Shared environment

INTRODUCTION

Criminological research is largely concerned with identifying the various factors that may contribute to the development of criminal and delinquent involvement. Much of this research has been guided by a **risk factor** approach, where putative criminogenic risk factors are examined quantitatively to determine whether they are associated with antisocial outcomes. This line of research has revealed that exposure to a wide array of criminogenic risk factors, ranging from poor parental supervision to contact with delinquent peers, appears to confer an increased risk of antisocial outcomes. In general, as the number of risk factors increases, so do the odds of criminal and delinquent behavior. Somewhat paradoxically, however, this line of research has also revealed that some adolescents who are exposed to multiple criminogenic risk factors do not become involved in crime and delinquency.[1]

The reasons why these risk-exposed adolescents do not become delinquent has spurred another line of research that examines the various factors, known broadly as **protective factors**, that appear to shield youth from delinquency and other problem behaviors.[2] Similar to the research on risk factors, research examining protective factors has revealed that a broad spectrum of protective factors, ranging from self-efficacy to academic competence, may act as buffers against delinquency among those adolescents who are *a priori* identified as being at high risk for delinquency.[3] The findings flowing from research on risk and protective factors have been used to guide and inform delinquency prevention and intervention programs.[4] These types of programs, in general, aim to reduce exposure to criminogenic risk factors and, at the same time, to increase the potency of available protective factors.

Unfortunately, this approach to intervention rests on a foundation of questionable internal validity. The correlational and longitudinal studies that underlie most of traditional criminology (including the dataset used in the present study) cannot, by their nature, reveal whether a particular correlate of or precursor to delinquency actually causes delinquency, to say nothing of exactly how it might have its effect. For example, because delinquency increases as parent–child relationships deteriorate, good relations with one's parents are generally thought to be a "protective factor" against delinquency and drug use, while its obverse, family conflict, is a "risk factor." The usual reaction among policymakers and, indeed, researchers is to assume that family conflict directly causes delinquency, and to further assume that improving family functioning will reduce the risk of delinquency. In fact, at least one public service ad campaign has embraced this approach (i.e., see www.theantidrug.com).

Other causal relationships are possible, however. The direct model might be reversed. Family conflict may as easily result from parental reaction to child misbehavior in the first place. Alternatively, a third variable may be involved. Poor parent–child relations might merely indicate the presence of some yet-unidentified latent risk factor, possibly of genetic origin, that functions as a spurious cause of both family conflict and delinquency.

The existence of alternate causal models implies that, until a causal relationship has been established, it may be better to consider variables like poor parent–child relations as risk indicators rather than risk factors.[5] A risk indicator misidentified as an actual cause of delinquency can lead to ineffective interventions. For the purposes of this chapter, we will continue to use the traditional language, mindful that, until internal validity has been established, the terms *risk factor* and *protective factor* should be considered to be hypotheses.

One way to improve the internal validity of causal models of delinquency is to examine the extent to which genetic and environmental factors explain variation in purported risk and protective factors. As shall be seen, this approach can help direct interventions. For example, variance in a family-based risk factor like parent–child conflict might turn out to be split between genetic and unshared environmental effects, with shared environmental effects largely absent. If this were the case, a family-based intervention designed to reduce conflict would probably fail. Rather, energies would better be directed into interventions with peers and other potential sources of unshared environmental effects.

THE ETIOLOGY OF RISK AND PROTECTIVE FACTORS

Risk and protective factors are not randomly distributed across all adolescents. The odds of being exposed to a single risk or protective factor tends to vary across youths, and differential exposure to these single risk or protective factors corresponds to an increase or decrease in the odds of antisocial behaviors. Perhaps even more important than differential exposure to a single risk factor is differential exposure to multiple risk or protective factors. Research has revealed, for example, that some youths are saturated with risk while others possess a full complement of protective factors, and still others are exposed to relatively few risk or protective factors.[6] Of particular importance is that the more risk (protective) factors that are accumulated, the stronger the effect. In general, serious violent juvenile offenders are characterized by possessing multiple risk factors. The opposite is also true. Adolescents who abstain from or are only mildly involved in delinquency (e.g., adolescence-limited offenders) tend to have very few risk factors, while adolescents who are characterized as being resilient to risk exposure tend to be those risk-exposed adolescents who are also exposed to multiple protective factors.

Given that the risk and protective factors framework has produced an impressive amount of knowledge, the next step is to begin to identify the underlying contributors that create variation in exposure to risk and protective factors. Understanding what accounts for differential exposure to risk and protective factors can aid in the development of effective delinquency prevention and intervention programs. At the same time, knowing the potential causes of risk and protective factors can add greater theoretical specificity to existing criminological theories and can aid in the creation of new, or more refined, explanations of crime. Because such a broad array of factors can be considered risk or protective factors, it is difficult to determine the specific variables that might have effects that cut across all risk and/or protective factors. For example, the causes of one specific risk factor might be very different from the causes of a specific protective factor. The end result is that it is very difficult to examine the causes of more than just a couple of risk and protective factors in the same study. One way to overcome this limitation, however, is to apply a behavioral genetic research methodology to the study of risk and protective factors. To understand why this is the case, it is first necessary to provide a brief overview of behavioral genetics.

A Brief Introduction to Behavioral Genetics

Generally speaking, behavioral genetics is a field of study that examines the extent to which variance in a measure (e.g., a risk factor or a protective factor) is the result of genetic effects, shared environmental effects, and nonshared environmental effects. Genetic effects, quantified as a **heritability** estimate, capture the extent to which genetic variance explains variance in the measure of interest. **Shared environmental** effects refer to the proportion of variance in the measure of interest that is the result of environmental factors that are the same between children raised in the same household, and thus can also be characterized as family environmental effects. **Nonshared environmental** effects capture the effects of all environmental factors that are different between siblings, as well as differential reactions to similar environments (plus error). Together, the heritability, shared environmental, and nonshared environmental components account for 100 percent of the variance in the measure of interest.

To estimate these three variance components, behavioral geneticists typically analyze samples of kinship pairs, including full siblings, half siblings, and twins living in the same household, or compare adoptees with their biological siblings, or adoptive siblings or parents. The similarity in scores on the measure of interest is then compared across these kinship pairs of differing levels of genetic relatedness. If the assumptions of behavioral genetic research are met, then genetic influences are inferred when scores on the measure of interest are more similar for kinship pairs that share more genetic material. For example, genetic influences on a risk factor would be detected if the score for a risk factor is more similar between monozygotic (MZ) twin pairs (who share 100 percent of their DNA) than between dizygotic (DZ) twin pairs (who share 50 percent of their dissenting DNA—DNA that varies among individuals of the same species—on average). If, on the other hand, scores were similar regardless of genetic relatedness, this would indicate the presence of shared family environmental effects and the absence of genetic effects. The same logic can be applied to all different types of kinship pairs, not just twin pairs.

The Intersection of Behavioral Genetics with Risk and Protective Factors

Much research has employed behavioral genetic methodologies to examine the genetic and environmental underpinnings to antisocial phenotypes.[7] The results of these studies have provided at least two reasons to suspect that risk and protective factors will be at least partially influenced by genetic factors. First, research has revealed that antisocial behavior and resiliency to antisocial outcomes is partially the result of genetic factors. The results of four meta-analyses have indicated that genetic factors account for about 50 percent of the variance in measures of antisocial behaviors, including violence, aggression, and delinquency.[8] While research on protective factors is limited, preliminary evidence suggests that genetic factors are involved in resiliency, accounting for as much as 50 percent of the variance.[9]

Heritability estimates of antisocial behavior are frequently interpreted to mean that genes directly lead to crime and delinquency. This interpretation, however, is erroneous; genetic factors do not have direct effects on complex behaviors, such as delinquency or resiliency. Instead, genetic factors tend to operate indirectly via intermediary processes and variables.[10] For example, genetic factors have been found to contribute to variation in levels of self-control, which in turn, have been found to predict delinquent and criminal involvement.[11] Part of the genetic influence on crime thus can likely be traced to genetic influences on levels of self-control. It is important to point out, however, that there are likely hundreds of risk factors, each of which may be influenced by genetic factors. Collectively, these genetically influenced risk factors account for genetic influences on antisocial behavior. The same logic also can be extended to resiliency, where genetically influenced protective factors, such as self-efficacy, contribute to the heritability of resiliency. Such an explanation, of course, is contingent on whether salient risk and protective factors are influenced by genetic factors.

The second key finding to emerge from the behavior genetic research adds support for the explanation advanced earlier by revealing that virtually every measure ever examined has been found to be under some level of genetic influence. The precise heritability estimates vary across studies, partially as a function of study-specific characteristics (e.g., the sample being examined), but also as a function of the precise measure being examined. For the purposes of the current study, measures that have been analyzed previously and that are closely aligned to putative risk and protective factors can be broadly divided into two groups: those that are *internal* and those that are *external* to the child or adolescent.

Internal risk and protective factors refer to those factors that are measured at the individual level and are psychologically oriented, such as cognitive abilities, personality traits, and mental illnesses. Levels of self-control, neuropsychological deficits, and self-efficacy are just a few of the many risk and protective factors that would fall under the rubric of internal factors. Behavioral genetic research has a long tradition of estimating the genetic, shared, and nonshared effects on internal factors. Although the precise estimates vary somewhat across the different factors, in general this line of research has revealed that these internal factors are genetically influenced, with most being approximately 50 percent heritable.[12] Against this backdrop, it would seem reasonable to conclude that internal risk and protective factors that are frequently studied by criminologists would also likely be influenced by genetic factors.

External risk and protective factors capture those factors that are frequently labeled as environmental, social, or contextual in criminological research. Examples of external factors include social support, parental negativity, and neighborhood conditions. Behavioral genetic research has often examined whether external factors are influenced by genetic factors.[13] These studies employ the same methodology described earlier by comparing the similarity of kinship pairs. However, instead of employing a behavioral or trait-based measure as the phenotype of interest, the variance of an environmental measure is decomposed. Heritability estimates of external measures vary across studies, but a fairly recent review article sheds some light on the extent to which genetic factors affect environments. Kendler and Baker reviewed the results of 55 studies that had estimated genetic influences on environmental measures, including many of those that are central to criminological theory and research (e.g., exposure to antisocial peers, parent–child conflict). They reported that the weighted heritability of environmental measures was around 0.25, meaning that about one-quarter of the variance in external measures is accounted for by genetic factors. As such, they concluded, "Genetic influences on measures of the environment are pervasive in extent and modest to moderate in impact."[14]

From a criminological vantage point, it might seem somewhat odd to view external factors as being partially influenced by genetic factors. Behavioral geneticists have advanced the concept of gene–environment correlation to elucidate the mechanisms that produce genetic influences on environments. There are three main types of gene–environment correlations: passive gene–environment correlation, active gene–environment correlation, and evocative gene–environment correlation. First, passive gene–environment correlation highlights the fact that biological parents pass along both an environment and a genotype to their children. Because they are produced by the same source (i.e., parents), the two are likely to be correlated. Second, active gene–environment correlation draws attention to the possibility that people can actively be involved in seeking out environments that are compatible with their genetic tendencies. What this necessarily means is that the environments that people find themselves in are created, in part, by their own genetic predispositions. Third, evocative gene–environment correlations are created by people eliciting or evoking certain environmental responses—responses that are correlated with their genetic predispositions. For example, a child with conduct disorder (a genetically influenced disorder) is at risk for eliciting negative reactions from his or her parents, peers, and teachers. And, these negative reactions stem, in part, from the genetic predispositions of the child. Together, these three gene–environment correlations provide the conceptual scaffolding needed to understand how genetic factors are able to influence individuals' exposure to external risk and protective factors.

The purpose of this chapter is to examine empirically the extent to which genetic, shared environmental, and nonshared environmental factors are involved in explaining variance in risk and protective factors frequently studied by criminologists. Due to data limitations, analyses include only variables that were available in the data and that had been consistently linked to antisocial behaviors and resiliency. To do so, we analyzed a sample of kinship pairs drawn from a longitudinal and nationally representative sample of American youths.

METHODOLOGY

Data for this study come from the National Longitudinal Study of Adolescent Health (Add Health). The Add Health is a nationally representative sample of American youths who were enrolled in middle and high schools during the 1994–1995 academic school year. Three waves of data are currently available for analysis. The first wave of data was collected when the adolescents were attending school; more than 90,000 students completed self-report surveys. These surveys, which are referred to as the wave 1 in-school surveys, contained an array of questions that asked youths about their demographics, their social relationships, and their family life. In order to gain more detailed information about some of the youths, a subsample of respondents and their primary caregivers (usually their mother) was then asked to complete follow-up interviews at the adolescent's home (i.e., the wave 1 in-home surveys). During these in-home interviews, respondents were asked questions that were more sensitive in nature, including their involvement in delinquency, their sexual activities, and their use of drugs and alcohol. In total, 20,475 adolescents and approximately 17,700 of their primary caregivers participated in the wave 1 in-home component of the Add Health study.[15]

Approximately a year and a half after the wave 1 surveys were completed, the second wave of data was collected. Because relatively little time had lapsed between the two waves, most of the respondents were still adolescents and thus the questions that were asked at wave 1 were still applicable to respondents at wave 2. As a result, the survey instruments between waves were very similar. For example, respondents were still asked about their delinquent behaviors, their family and peer relationships, and their achievements at school. To reduce costs, there was no attempt to find respondents who were seniors in wave 1 for the wave 2 data collection. Overall, 14,738 respondents participated in the wave 2 component of the Add Health study. Finally, during 2001 and 2002 the third wave of data was collected. Since most of the participants were no longer adolescents, the questions asked of the respondents were amended to be appropriate for young adults. For example, during wave 3 interviews, participants were asked about their lifetime contact with the criminal justice system, their childrearing techniques, and their employment status. Unlike the wave 2 data collection, the wave 3 data collection plan included attempts to follow-up all wave 1 in-home respondents. Altogether, 15,197 young adults were reinterviewed at wave 3.

The Add Health dataset also includes a subsample of kinship pairs. During wave 1 in-school interviews, youths were asked to indicate whether they resided with a co-twin, a half sibling, a stepsibling, or a cousin. If they responded affirmatively, and if their sibling was between the ages of 11 and 20 years old, then they were also added to the sample. A probability sample of full biological siblings was also included in the sample; this sampling process netted more than 3000 "sibling" pairs.[16] The data analyzed in this study included MZ twins, DZ twins, full biological siblings, half siblings, and cousins.

Risk Factors

Family Risk

In general, research has revealed that risk experiences inside families, such as abuse, neglect, and parental apathy, are significantly related to various forms of antisocial behavior. To take this finding into account, we created a family risk-weighted factor score. Previous researchers analyzing the Add Health data have developed three parenting scales drawn from the wave 1 data: a 10-item maternal involvement index, a 5-item maternal disengagement scale, and a 2-item maternal attachment scale. We recreated these scales and then subjected them to a principal components factor analysis with varimax rotation. The results of these analyses indicated that the covariance structure of these three scales could be accounted for by a single construct. So, these three scales were then used to create the family-risk weighted factor score together to create the family risk scale.[17] Descriptive statistics for the family risk scale and all of the other scales used in the analyses are presented in **Table 7-1**.

Delinquent Peers

Exposure and contact with delinquent peers represents one of the strongest and most consistent predictors of adolescent delinquent involvement. As a result, we included a three-item delinquent

TABLE 7-1 Descriptive Statistics for Selected Add Health Sample Variables

Variable	Mean	Median	Mode	SD	Min–Max
Risk Factors					
Family risk	11.61	11.00	11.00	4.82	1–33
Delinquent peers	2.44	2.00	0.00	2.62	0–9
Neuro. deficits	49.89	52.00	69.00	26.77	0–100
Depression	15.69	14.00	11.00	7.68	4–54
Protective Factors					
Social support	31.98	32.00	32.00	4.66	10–40
Academic achievement	11.22	11.00	11.00	3.00	4–16
School attachment	18.52	19.00	20.00	3.74	5–25
Religiosity	0.06	0.52	2.61	2.47	−7.21–2.61

peers scale that has been used previously by Add Health researchers.[18] During wave 1 interviews, respondents were asked to indicate how many of their three best friends smoke cigarettes every day, drink alcohol at least once a month, and smoke marijuana at least once a month. Responses to these items were then summed together to create the wave 1 delinquent peers scale (α = 0.75). Higher values on this scale represent greater exposure to delinquent peers.

Neuropsychological Deficits

A line of research has revealed that measures tapping neuropsychological deficits are associated with a variety of forms of antisocial behaviors, especially chronic violence. As such, we included a single-item neuropsychological deficits item in the analyses. To measure neuropsychological deficits, we used scores on the wave 1 Peabody Picture Vocabulary Test (PPVT). Scores on the PPVT were reverse-coded such that higher values reflect more neuropsychological deficits. This measure is in line with previous research that has used tests of verbal abilities as a measure of neuropsychological deficits.[19]

Depression

Empirical evidence suggests a link between depression and antisocial outcomes, including criminal and delinquent involvement. To examine the extent to which this risk factor is influenced by genetic factors, an 18-item depression scale was created. During wave 1 interviews, respondents were asked a series of questions drawn from the Center for Epidemiological Studies Depression Scale that were designed to tap symptoms of depression.[20] For example, respondents were asked whether they could not shake off the blues, whether they had a poor appetite, and whether they felt depressed. The item responses were then summed together with higher values indicating more symptoms of depression.

Risk Factor Index

Much of the extant risk factor research creates a cumulative risk factor index. To do so, the individual risk factors are dichotomized and summed together. We followed this protocol and created a composite risk factor index. This index was created by standardizing all of the individual risk factors. These standardized items were then dichotomized at the mean: Scores at or below the mean were assigned a value of "0," indicating the absence of that risk factor; scores above the mean were assigned a value of "1," indicating the presence of that risk factor. Then, the dichotomous risk factor variables were summed together to create the composite risk factor index, with scores ranging between 0 and 4.

Protective Factors

Social Support

Social support systems can work to significantly decreased adolescents' involvement in delinquent activities. We examined the possibility that social support was influenced by genetic factors by employing an 8-item social support scale. During wave 1 interviews, adolescents were asked a series of questions tapping the support that they perceive they receive from their family, friends, and teachers. For instance, they were asked how much they felt that: Adults care about them, teachers care about them, people in their family care about them, and they have fun together with their family. Responses to these items were summed together to create the social support scale ($\alpha = 0.71$). Higher values on this scale reflect more social support.

Academic Achievement

All else equal, adolescents who excel at school are less likely to become delinquents compared with their same-age peers who struggle at school. We employed a 4-item academic achievement scale to examine the extent to which genetic factors affect academic achievement. During wave 1 interviews, youths were asked to self-report their grades on their most recent report card for English, mathematics, history, and science. Responses were scored as follows: $1 = D$, $2 = C$, $3 = B$, and $4 = A$. These four grades were summed together to create the academic achievement scale ($\alpha = 0.74$). Higher values on this scale indicate higher grades in school.

School Attachment

Students who are highly bonded and attached to school are at reduced risk for delinquent activities than are students who are not as attached to school. As a result, we included a 5-item school attachment scale in the analyses. During wave 1 interviews, adolescents were asked whether they feel close to people at their school, whether they feel like they are part of the school, and whether they feel safe in their school. Responses were summed together to create the school attachment scale with higher values indicating greater attachment to the school ($\alpha = 0.75$). This scale is identical to one that has been used previously.[21]

Religiosity

Religiosity appears to have some protective effects against delinquency where adolescents who have strong religious beliefs are less likely to engage in antisocial behaviors compared with adolescents lacking religiosity. The Add Health data contain a 3-item religiosity scale. During wave 1 interviews, adolescents were asked how many times in the past 12 months they had attended religious services, how important religion is to them, and how frequently they prayed. The response sets for these three items varied, and so to give them equal weighting, each item was standardized. These standardized items were then summed together to create the religiosity scale ($\alpha = 0.74$). Higher values on this scale reflect more religiosity. This scale is identical to the one that has been used previously by Add Health researchers.[22]

Protective Factor Index

Similar to research examining risk factors, research examining protective factors has tended to create a protective factors index that is the function of the individual protective factors. Following the lead of prior research, we created a composite protective factor index. This index was developed using the same procedure that was used to create the risk factor index. First, all of the individual protective factors were standardized. Next, these variables were dichotomized at the mean, where values at or below the mean were assigned a value of "0" and values above the mean were assigned a value of "1." Finally, all four of the dichotomous variables were summed together to create the protective factor index.

The main objective of this study is to determine the extent to which genetic factors account for the variance in the risk and protective factors. To do so, we estimated a series of DeFries-Fulker (DF) equations. DF analysis is a regression-based statistic that analyzes samples containing kinship pairs to estimate the genetic and environmental effects on outcome measures, such as environmental risk and protective factors.[23] The original DF equation was developed to analyze clinical samples, where kinship pairs were concordant or discordant for a particular disorder. The original DF equation, however, has since been altered to be used with kinship pairs who were drawn from the general population. The DF equation that can be used for nonclinical samples takes the following form:

Equation 1: $K_1 = b_0 + b_1 K_2 + b_2 R + b_3 (R * K_2) + e$

where K_1 represents the score on the outcome measure (e.g., a risk or protective factor) for one sibling, K_2 represents the score on the same outcome measure for their sibling, R is a measure of genetic relatedness ($R = 1.0$ for MZ twins, $R = 0.5$ for DZ twins and full siblings, $R = 0.25$ for half-siblings, and $R = 0.125$ for cousins), and ($R * K_2$) is a multiplicative interaction term between R and K_2. For Equation 1, b_0 = the constant, b_1 = the proportion of variance in K_1 that is the result of shared environmental factors, b_2 is not typically interpreted in the DF equation, b_3 = the proportion of variance in K_1 that is accounted for by genetic factors (i.e., heritability), and e = the proportion of variance in K_1 that is the result of nonshared environmental factors and error.

A newly modified DF equation takes the following form:

Equation 2: $K_1 = b_0 + b_1(K_2 - K_m) + b_2[R * (K_2 - K_m)] + e$

where K_1 remains the score on the outcome measure for one sibling, K_2 remains the value on the same outcome measure for their sibling, and R remains the measure of genetic similarity. Note, however, that this new DF equation introduces a new term, K_m. K_m = the mean value for K_2. Equation 2 also indicates that K_2 is being mean centered, whereas in Equation 1, K_2 was left untransformed. Moreover, the main effect of R has been removed from Equation 2, but remains in the interaction term, $[R * (K_2 - K_m)]$. The meaning and the interpretation of the coefficients in Equation 2 are very similar to those in Equation 1: b_1 = the proportion of variance in K_1 that is explainable by shared environmental factors, b_2 = the proportion of variance in K_1 that is the result of genetic factors (i.e., heritability), and e = the proportion of variance in K_1 that is due to nonshared environmental factors plus error.

One of the issues with using DF analysis is to figure out which sibling from each sibling pair should be used as the dependent variable and which should be used as the independent variable. This problem is often overcome by "double-entering" sibling pairs, where each sibling is included in the data twice: once as the dependent variable and once as the independent variable. While double-entering is advantageous, it also results in nonindependence in observations. Frequently, this problem is addressed by correcting the standard errors by using Huber/White standard errors. We opted to sidestep this issue by randomly selecting one sibling from each sibling pair to be the dependent variable and the other sibling to be the independent variable. As a result, the unit of analysis was the twin pair and since all of the observations were independent of each other, there was no need to correct the standard errors. However, we recalculated all of the models using "double-entry" modeling procedures and the results were virtually identical. All of the models were estimated using Equation 2. To provide a better fit to the data, all nonsignificant coefficients were removed and the models were reestimated retaining only the statistically significant coefficients.

RESULTS

The analysis begins by employing the four criminogenic risk factors as dependent variables in separate DF equations. The results of these models are presented as bar graphs in **Figure 7-1** with the proportion of variance explained by genetic factors, shared environmental factors, and nonshared environmental factors depicted on the y-axis and each of the risk factors located on the x-axis. As can be seen, all four criminogenic risk factors are influenced by genetic factors. For

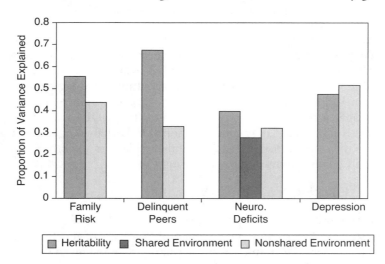

FIGURE 7-1. Genetic and Environmental Influences on Criminogenic Risk Factors

Note: The shared environmental effect was zero (0.00) for all of the criminogenic risk factors except neurological deficits.

example, genetic factors explain 56 percent of the variance in family risk, 67 percent of the variance in delinquent peers, 40 percent of the variance in neuropsychological deficits, and 48 percent of the variance in depression. The remaining variance for family risk, delinquent peers, and depression is accounted for by nonshared environmental factors; the shared environment accounts for none of the variance in these three risk factors. The only risk factor that was influenced by shared environmental factors was neuropsychological deficits, with 28 percent of the variance attributable to the shared environment. The nonshared environment explained 32 percent of the variance in neuropsychological deficits.

The next set of DF models used the four protective factors as dependent variables. As **Figure 7-2** shows, genetic factors accounted for a significant amount of variance in all four of the measures. Specifically, genetic factors explained 48 percent of the variance in social support, 51 percent of the variance in academic achievement, 48 percent of the variance in school attachment, and 42 percent of the variance in religiosity. The nonshared environment also explained a significant amount of variance in all four protective factors. Nonshared environmental factors, for example, explained 52 percent of the variance in social support, 35 percent of the variance in academic achievement, 52 percent of the variance in school attachment, and 21 percent of the variance in religiosity. The shared environment explained a significant amount of variance for only academic achievement and religiosity, accounting for 14 percent and 37 percent of the variance, respectively.

The final two DF equations were estimated for the cumulative risk factor index and the cumulative protective factor index, the results of which are presented in **Figure 7-3**. The pattern of results is in line with those found for the individual risk and protective factors. For instance, 59 percent of the variance in the cumulative risk factor index is explainable by genetic factors and the remaining 41 percent of variance is attributable to the nonshared environment. The shared environment accounts for none of the variance in the cumulative risk factor index. Strikingly similar findings were garnered for the cumulative protective factor index. Fifty-eight percent of the variance in the

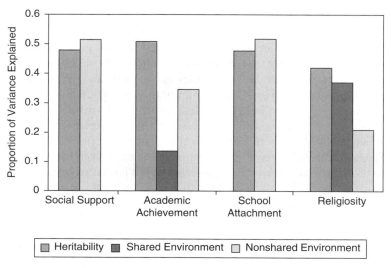

FIGURE 7-2. Genetic and Environmental Influences on Protective Factors

Note: The shared environmental effect for social support and school attachment was zero (0.00).

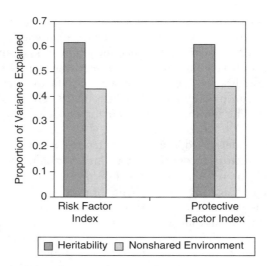

FIGURE 7-3.   Genetic and Environmental Effects on the Cumulative Risk Factor Index and the Cumulative Protective Factor Index

Note: The shared environmental effect was zero (0.00) for both indexes.

protective factor index was the result of genetic factors and the remaining 42 percent of the variance was accounted for by nonshared environmental factors. Once again, the shared environment failed to explain any of the variance in the protective factor index.

CONCLUSION

The study of risk and protective factors has gained a considerable amount of traction among criminologists during the past couple of decades. This line of research has produced a vast array of knowledge identifying specific risk factors that are linked to delinquent and criminal involvement and it has also uncovered certain protective factors that appear to mitigate risk and facilitate resiliency to delinquency among high-risk youths. Of critical importance is that elucidating the underlying etiologic structure of risk and protection holds considerable promise to make clearer the developmental origins to antisocial behaviors. At the same time, research on protective factors has the capability to increase the effectiveness of delinquency prevention and treatment programs. A logical next step in risk/protective factor research, however, is to examine what causes variation in established risk and protective factors. For the most part, criminological research has failed to address this critical issue. The current study addressed this gap in the literature by examining the genetic and environmental basis of a range of risk and protective factors.

Analysis of kinship pairs drawn from the Add Health data revealed three key findings. First, the shared family environment explained a significant amount of variance in only three of the eight risk and protective factors and none of the variance in either the cumulative risk factor index or the cumulative protective factor index. These findings stand in stark contradiction to what criminological theory would predict. Most criminological theories and research highlight the importance of family environmental factors, such as parental socialization, in the etiology of antisocial outcomes as well as the importance of including risk and protective factors. This line of research, however, is likely misspecified because the methodologies employed by criminologists are not able to isolate the effects of the shared environment from the effects of nonshared environmental factors and genetic factors.[24] As a result, analyses that purportedly show an association between a shared environmental variable and a risk or protective factor is likely artificially inflating the true effect of the shared environmental variable. One should not conclude, however, that parenting practices or related family dynamics are unimportant. Greater attention by criminological researchers employing genetically sensitive research methodologies where possible when assessing the effects that shared environmental variables have on risk and protective factors is needed to fully examine the conditions under which these variable exert their influence.

The second main finding to emerge from the analysis is that nonshared environmental effects were large and statistically significant across all of the risk and protective factors and for the cumulative risk factor index and the cumulative protective factor index. These findings are in line with a large pool of behavioral genetic research, indicating that nonshared environmental effects tend to explain roughly 50 percent of the variance in behaviors, traits, and environments.

These findings can also be viewed as consistent with criminological perspectives that emphasize children and adolescents' experiences that can vary across siblings within the same family, such as differential school and peer exposures. Because mainstream criminological research does not make use of genetically informative research designs, it does not use the term *nonshared environmental influences*, but viewed through a behavioral genetic lens, it is clear that many criminological theories invoke causal mechanisms that would fall into the nonshared environment. Future criminological research that examines the causes of variance in risk and protective factors needs to be cognizant of employing a research design that is adequately able to examine the effects of nonshared environmental factors, and, further, that acknowledges that nonshared environments might also have nonshared *effects*.

Third, genetic factors explained a statistically significant amount of variance in all of the risk and protective factors as well as in the cumulative risk index and the cumulative protective factor index. These findings may come as a surprise to criminologists in that they indicate that factors that are routinely examined in quantitative criminological studies are under some level of genetic influence. This is true even of "environmental" factors, such as family risk, delinquent peers, social support, school attachment, and religiosity. Against this backdrop, we echo the sentiment of Vinkhuyzen et al. when they noted, "what we think of as measures of 'environment' are better described as *external factors* that might be partly under genetic control."[25] No longer is it tenable for criminologists to continue to think of certain variables as purely social; rather, a more realistic view, which comports with the empirical research, is that all variables are biosocial in that they are produced by a complex arrangement of biological and environmental factors. Newer paradigms of a "cell-to-society" approach where biology and environment are viewed as intertwined across multiple levels of analysis are likely to produce theories that are more in keeping with contemporary scientific trends.[26] Accepting such a view, however, also means that criminologists need to reevaluate the interpretation of previous research. For example, studies indicating links between criminological risk factors such as exposure to delinquent peers and delinquent involvement are likely capturing genetic effects as well as social processes. And, perhaps the reason why some risk factors are such strong predictors of antisocial behaviors is because they are "picking up" a substantial amount of genetic variance. Although criminologists have limited access to genetically sensitive data, future research needs to test this possibility.

Resiliency research also often examines the interaction between risk factors and protective factors in the prediction of delinquency. Typically, if there is an interaction between the risk and protective factors, the conclusion is that the protective factor is fostering resiliency from delinquency. There is another interpretation: Perhaps these interactions are really measuring gene–environment interactions. An impressive amount of research indicating that not all adolescents respond to the same environments in the same way because of the differences in genetic material exists. Perhaps the most well known example comes from research conducted by Caspi and colleagues, who examined whether a genetic risk factor, monoamine oxidase A (MAO-A), interacted with childhood maltreatment to predict variation in antisocial behaviors.[27] The results of their study revealed that the effect of childhood maltreatment was much stronger for males who were genetically at risk for antisocial behaviors. In other words, a genetic factor moderated the effect of the risk factor (i.e., childhood maltreatment) on antisocial behavior. It is quite possible that the measures of risk factors and/or protective factors in prior resiliency research are simply markers for genetic predispositions to antisocial behaviors. For now, that remains an open empirical question awaiting future investigation. The substantial amount of variance explained by nonshared environmental effects in the present study, for example, probably encompasses not only the effects of environments unique to each sibling, but also differential reactions by siblings to the same environments; in other words, person– (or gene–) environment interactions.

Studies examining risk and protective factors have pushed the field of criminology forward by generating a great deal of evidence indicating which risk and protective factors are salient to the development of antisocial behavior. In his presidential address to the American Society of Criminology, Farrington noted: "Improving the risk factor prevention paradigm is not merely an academic exercise designed to advance knowledge about explaining and preventing crime. It is also an intensely practical exercise designed to reduce crime and to improve people's lives."[28] And improving knowledge about risk and protective factors is tied in large part to being able to identify the factors that contribute to differential exposure to them. As the results of our study indicate, most risk and protective factors are multifactorial and produced by an assortment of genetic and environmental factors. Being able to identify the precise genetic and environmental factors that are involved will likely go a long way toward achieving the two major goals of the field—namely, improving our understanding of the causes of antisocial behavior and increasing the effectiveness of crime and delinquency prevention and intervention programs.

GLOSSARY

Heritability—the proportion of variance in a trait or behavior that is attributable to genetic factors

Nonshared environment—the proportion of variance in a trait or behavior that is attributable to unique factors between siblings, such as outside peer associations

Protective factor—a construct that reduces the likelihood of antisocial behavior

Risk factor—a construct that increases the likelihood of antisocial behavior

Shared environment—the proportion of variance in a trait or behavior that is attributable to common factors between siblings, such as family environment

NOTES

1. Rutter, M. (1979). Protective factors in children's response to stress and disadvantage. In M. W. Kent & J. E. Rolf (Eds.), *Primary prevention of psychopathology* (Vol. 3, pp. 49–74). Hanover, NH: University Press of New England.

2. Jessor, R., Van Den Bos, J., Vanderryn, J., Costa, F. E., & Turbin, M. S. (1995). Protective factors in adolescent problem behavior: Moderator effects and developmental change. *Developmental Psychology, 31,* 923–933; Rutter, M. (1987). Psychosocial resilience and protective mechanisms. *American Journal of Orthopsychiatry, 26,* 316–331.

3. Turner, M. G., Hartman, J. L., Exum, M. L., & Cullen, F. T. (2007). Examining the cumulative effects of protective factors: Resiliency among a national sample of high-risk youths. *Journal of Offender Rehabilitation, 46,* 81–111.

4. Farrington, D. P., & Welsh, B. C. (2007). *Saving children from a life of crime: Early risk factors and effective interventions.* New York: Oxford University Press; Lipsey, M. W., & Wilson, D. B. (1998). Effective intervention for serious juvenile offenders: A synthesis of research. In R. Loeber & D. P. Farrington (Eds.), *Serious and violent juvenile offenders: Risk factors and successful interventions* (pp. 313–345). Thousand Oaks, CA: Sage.

5. Cleveland, H. H., & Wiebe, R. P. (2008). Understanding the progression from adolescent marijuana use to young adult serious drug use: Gateway effect or developmental trajectory? *Development and Psychopathology, 20,* 615–632.

6. Turner et al. (2007), see Note 3; Farrington, & Welsh (2007), see Note 4.

7. Plomin, R., DeFries, J. C., McClearn, G., & McGuffin, P. (2008). *Behavioral genetics* (5th ed.). New York: Worth Publishers.

8. Ferguson, C. J. (2010). Genetic contributions to antisocial personality and behavior: A meta-analytic review from an evolutionary perspective. *Journal of Social Psychology, 150,* 1–21; Mason, D. A., & Frick, P. J. (1994). The heritability of antisocial behavior: A meta-analysis of twin and adoption studies. *Journal of Psychopathology and Behavioral Assessment, 16,* 301–323; Miles, D. R., & Carey, G. (1997). Genetic and environmental architecture of human aggression. *Journal of Personality and Social Psychology, 72,* 207–217; Rhee, S. H., & Waldman, I. D. (2002). Genetic and environmental influences on antisocial behavior: A meta-analysis of twin and adoption studies. *Psychological Bulletin, 128,* 490–529.

9. Boardman, J. D., Blalock, C. L., & Button, T. M. (2008). Sex differences in the heritability of resilience. *Twin Research and Human Genetics, 11,* 12–27.

10. Gottesman, I., & Gould, T. D. (2003). The endophenotype concept in psychiatry: Etymology and strategic intentions. *American Journal of Psychiatry, 160,* 636–645.

11. Beaver, K. M., Wright, J. P., DeLisi, M., & Vaughn, M. G. (2008). Genetic influences on the stability of low self-control: Results from a longitudinal sample of twins. *Journal of Criminal Justice, 36,* 478–485.

12. Rutter, M. (2006). *Genes and behavior: Nature-nurture interplay explained.* Malden, MA: Blackwell.

13. Plomin, R., & Bergeman, C. S. (1991). The nature of nurture: Genetic influence on "environmental" measures. *Behavioral and Brain Sciences, 14,* 373–427; Plomin, R., DeFries, J. C., & Loehlin, J. C. (1977). Genotype-environment interaction and correlation in the analysis of human behavior. *Psychological Bulletin, 84,* 309–322; Scarr, S. (1992). Developmental theories for the 1990s: Development and individual differences. *Child Development, 63,* 1–19; Scarr, S., & McCartney, K. (1983). How people make their own environments: A theory of genotype → environment effects. *Child Development, 54,* 424–435.

14. Kendler, K. S., & Baker, J. H. (2006). Genetic influences on measures of the environment: A systematic review. *Psychological Medicine, 37,* 615–626, p. 615.

15. Udry, J. R. (2003). *The National Longitudinal Study of Adolescent Health (Add Health), Waves I and II, 1994–1996; Wave III, 2001–2002* [machine-readable data file and documentation]. Chapel Hill, NC: Carolina Population Center, University of North Carolina at Chapel Hill; Harris, K. M., Florey, F., Tabor, J., Bearman, P. S., Jones, J., & Udry, J. R. (2003). The National Longitudinal Study of Adolescent Health: Research design. Retrieved from www.cpc.unc.edu/projects/addhealth/design.

16. Harris, K. M., Tucker Halpern, C., Smolen, A., & Haberstick, B. C. (2006). The National Longitudinal Study of Adolescent Health (Add Health) twin data. *Twin Research and Human Genetics, 9*, 988–997.

17. Beaver, K. M., Sak, A., Vaske, J., & Nilsson, J. (2010). Genetic risk, parent-child relations, and antisocial phenotypes in a sample of African-American males. *Psychiatry Research, 175*, 160–164.

18. Bellair, P. E., Roscigno, V. J., & McNulty, T. L. (2003). Linking local labor market opportunity to violent adolescent delinquency. *Journal of Research in Crime and Delinquency, 40*, 6–33.

19. Lynam, D., Moffitt, T., & Stouthamer-Loeber, M. (1993). Explaining the relation between IQ and delinquency: Class, race, test motivation, school failure, or self-control? *Journal of Abnormal Psychology, 102*, 187–196; Moffitt, T. E. (1990). The neuropsychology of juvenile delinquency: A critical review. In M. Tonry & N. Morris (Eds.), *Crime and justice: An annual review of research* (Vol. 12, pp. 99–169). Chicago: University of Chicago Press; Moffitt, T. E., Lynam, D. R., & Silva, P. A. (1994). Neuropsychological tests predicting persistent male delinquency. *Criminology, 32*, 277–300.

20. Radloff, L. S. (1977). The CES-D Scale: A self-report depression scale for research in the general population. *Applied Psychological Measurement, 1*, 385–401.

21. McNeely, C. A., Nonnemaker, J. M., & Blum, R. W. (2002). Promoting school connectedness: Evidence from the National Longitudinal Study of Adolescent Health. *Journal of School Health, 72*, 138–146.

22. Beaver, K. M., Gibson, C. L., Jennings, W. G., & Ward, J. T. (2009). A gene × environment interaction between DRD2 and religiosity in the prediction of adolescent delinquent involvement in a sample of males. *Biodemography and Social Biology, 55*, 71–87.

23. DeFries, J. C., & Fulker, D. W. (1985). Multiple regression analysis of twin data. *Behavior Genetics, 15*, 467–473.

24. Harris, J. R. (1998). *The nurture assumption: Why children turn out the way they do.* New York: The Free Press; Rowe, D. C. (1994). *The limits of family influence: Genes, experience, and behavior.* New York: Guilford.

25. Vinkhuyzen, A. A. E., van der Sluis, S., de Geus, E., Boomsma, D., & Posthuma, D. (2010). Genetic influences on 'environmental' factors. *Genes, Brain and Behavior, 9*, 276–287, p. 276.

26. Vaughn, M. G., Beaver, K. M., & DeLisi, M. (2009). A general biosocial paradigm of antisocial behavior: A preliminary test in a sample of adolescents. *Youth Violence and Juvenile Justice, 7*, 279–298.

27. Caspi, A., McClay, J., Moffitt, T. E., Mill, J., Martin, J., Craig, I. W., Taylor, A., & Poulton, R. (2002). Role of genotype in the cycle of violence in maltreated children. *Science, 297*, 851–854.

28. Farrington, D. P. (2000). Explaining and preventing crime: The globalization of knowledge—The American Society of Criminology 1999 Presidential Address. *Criminology, 38*, 1–24, p. 28.

PART TWO

Delinquency (Adolescence and Early Adulthood)

CHAPTER EIGHT

General Strain Theory and Offending over the Life Course

Stephen W. Baron and Robert Agnew

KEY TERMS

Failure to achieve positively valued goals
General strain theory
Presentation of negative stimuli
Removal of positively valued stimuli

INTRODUCTION

Criminologists have become increasingly aware that offenders exhibit different patterns of offending over the course of their lives. They begin offending at different ages, commit different amounts and types of crime, and stop offending at different ages. Most offenders begin their illegal activities as they enter adolescence, commit small to moderate amounts of largely minor crime, and desist from crime as they enter adulthood. Researchers have coined the term *adolescence limited* to describe this offending pattern.[1] Research also suggests that there is a smaller group of individuals who become involved in crime at an earlier age, offend at high rates, commit serious and minor crimes, and continue offending well into their adult years. Individuals in this group, known as the *life-course-persistent* offenders, encompass perhaps 5 to 10 percent of all offenders, yet account for the majority of all serious crime. Their frequent and serious offending makes them a special concern for law enforcement officials, policymakers, and academics, alike. Still other patterns of offending have been identified, but most research focuses on the adolescence-limited and life-course-persistent patterns. This chapter describes how General Strain Theory (GST) explains these two patterns of offending. It then draws on GST to explain patterns of offending committed by street youth, an important but neglected population. We begin by providing an overview of GST, then discuss how it can be used to explain adolescence-limited and life-course-persistent offending, and conclude by explaining patterns of offending among street youth.

AN OVERVIEW OF GENERAL STRAIN THEORY

General strain theory states that a range of strains or negative events and conditions increase crime. These strains lead to negative emotions, such as anger, which create much pressure for corrective action. Crime is one way to cope with strains and negative emotions. Individuals may engage in crime to escape from or reduce their strain (e.g., run away from abusive parents, steal the money they need), seek revenge against the source of strain or related targets (e.g., assault the peers who are bullying them), or alleviate negative emotions (e.g., use illicit drugs to feel better). Whether the strain and negative emotions lead to crime, however, is influenced by several factors—including coping skills and level of social support.[2]

GST focuses on three general types of strain. The first is the actual or anticipated **failure to achieve positively valued goals**, such as monetary goals, status or respect, and thrills/excitement. The second is the actual or threatened **removal of positively valued stimuli**, such as the loss of one's job, homelessness, and the breakup of relationships. The third is the actual or threatened **presentation of negative stimuli**, such as criminal victimization, child abuse, negative school and work experiences, and discrimination.

GST distinguishes between "objective" and "subjective" strains. Objective strains are events and conditions that most people in a given group dislike. Subjective strains are events and conditions disliked by the people experiencing them. Individuals may differ in their subjective reaction to the same objective strain. For example, some people may be devastated by their divorce and others may view their divorce as a cause for celebration. Both objective and subjective strains may impact crime, but we would expect subjective strains to have a greater effect.

Strains most likely to lead to crime have certain characteristics. They are high in magnitude, meaning that they are severe, frequent, long in duration, recent, and expected to continue into the future. They also affect the central values, goals, needs, identities, and/or activities of individuals. Strains conducive to crime are also seen as unjust (e.g., a deliberate versus accidental bump), which contributes to anger. Further, these strains are associated with low social control. For example, the strain of parental rejection is associated with weak bonds to parents, while the strain of unemployment is associated with weak bonds to society. Finally, the strains most likely to result in crime create some pressure or incentive for criminal coping. For example, certain strains (e.g., peer abuse) involve exposure to criminal models and foster beliefs conducive to crime (e.g., crime is a way to right a wrong). Also, crime is sometimes an expedient way to cope with certain strains (e.g., theft is an easy way to obtain money).[3]

Strains may lead to a range of negative emotions, such as anger, frustration, depression, and humiliation. However, the critical emotion in GST is anger. Anger is conducive to crime because it energizes individuals for action, creates a desire for revenge, reduces concern for the consequences of one's behavior, provides a justification for crime (righting a wrong), and undermines certain non-criminal forms of coping, such as negotiation. Most individuals who experience strain and anger, however, do not cope by committing crime. Several factors influence whether people engage in criminal coping. These factors include coping skills and resources, such as social and problem-solving skills, self-efficacy, self-esteem, and financial resources. They include the availability and quality of conventional social supports. Social support may come in the form of financial assistance, guidance, and emotional support—all of which reduce the likelihood of criminal coping. Criminal coping is also more likely when the benefits of crime are high and the costs are low. For example, criminal coping is more likely among those who have little to lose by engaging in crime, such as those with weak family ties, little interest in school, and no jobs. Further, criminal coping is more likely among those who are disposed to crime, including those with criminal peers, those with beliefs favorable to crime, and those with traits such as negative emotionality and low constraint. Criminal peers increase the likelihood of criminal coping by reinforcing criminal behavior, modeling criminal acts, defining certain illegal activities as appropriate responses to strain, and serving as instigators to crime. Individuals with negative emotionality are easily upset and tend to have aggressive interactional styles. Those with low constraint tend to be impulsive risk takers, with little empathy for others.

THE GST EXPLANATION OF ADOLESCENCE-LIMITED OFFENDING

As noted earlier, the most common pattern of offending is known as *adolescence limited* and involves an increase in mostly minor offending during the adolescent years. According to GST, offending peaks during the adolescent years because adolescents are more likely than children and/or adults to: (1) experience certain strains conducive to crime, (2) perceive these strains as highly aversive, and (3) cope with these strains through crime.

An Increase in Strain

Adolescents are more likely to experience strains conducive to crime for several reasons. Children live in a small, closely supervised world dominated by parents and teachers; as such, they are more often protected from strains. Parents, for example, are quick to intervene if anyone threatens their child. Adolescents, however, are given more freedom than children, and thus they often interact with others away from the watchful eye of parents and teachers. Many adolescents, for example, spend afternoons and evenings interacting with peers in settings outside home and school. Further, adolescents are increasingly expected to cope with problems on their own. As such, adolescents are much less likely to be protected from strains than are children.[4]

Further, adolescents live in a larger, more diverse, and more demanding social world than children and adults. Adolescents enter secondary schools that are typically much larger and more diverse than the elementary schools they attended. They change classes several times a day, interacting with large groups of peers, including many they do not know well. They also spend much time with others in public settings outside of school. Further, their social world becomes more demanding. Schoolwork becomes more challenging and grading more difficult. Peer relations become paramount, including romantic relationships, but they often lack the skills to effectively navigate complex social relationships. As a result of such changes, adolescents are exposed to more people (e.g., peers) who can treat them in a negative manner. The increased demands they face increase the likelihood of negative treatment, such as failure at school and rejection by romantic partners. But as adolescents enter adulthood, their social world shrinks, they have more control over it (e.g., control over the peers they interact with, where they live), and they develop the skills necessary to manage complex social relationships.

In addition, adolescents have more trouble achieving their goals than children and adults. The goals of children are limited, and adults provide much assistance in achieving them. Adolescents, however, come to desire an array of privileges that are usually reserved for adults, such as status, autonomy, and money (often necessary to achieve popularity with peers). Adolescents, however, are often prevented from achieving these goals through legal channels. Parents and teachers, for example, often deny them the autonomy and status they desire. And adolescents are often unable to secure the money they need from parents or work. Adolescents in these situations may resort to delinquent behavior as a method of asserting autonomy or adult status (e.g., sexual intercourse, alcohol consumption, disorderly behavior), obtaining money (e.g., theft), or venting frustration against those who block their achievement of goals (e.g., assaulting parents or teachers). Adults, by contrast, are better able than adolescents to achieve goals such as autonomy, status, and money through legal channels.

GST recognizes that not all adolescents experience increased levels of negative treatment or strain. Nor does it suggest that all individuals moving from adolescence into adulthood will experience a decline in strain. Some adolescents are protected from strain by attending small, well supervised schools. Further, some adolescents are less inclined to spend time in social environments where negative treatment is likely, perhaps because they are intensely focused on academics or are socially awkward. This allows us to understand why a small percentage of adolescents does not engage in crime. In a similar vein, there are adults who do not enter into positive employment, stable marriages, or other environments that reduce exposure to negative treatment. This helps us to partially understand the small percentage of people who continue their criminal participation as they enter adulthood.

An Increased Tendency to View Negative Relations as Aversive

Not only are adolescents more likely to experience strains, they are also more likely to view them as aversive. Adolescents begin to become aware of and upset by events and experiences that escape younger children's consideration. Adolescents are also more inclined to blame others for the strains they experience, a tendency that declines in adulthood. Further, many of the strains experienced by adolescents involve their peers or become known to their peers. These peers often remind adolescents of the strains they have experienced, thus exacerbating their effect. For example, peers may regularly remind one another of the insults they have experienced. The strains experienced by adults, however, are less likely to become public knowledge.

An Increased Likelihood of Responding to Strain with Crime

Finally, adolescents are more likely to cope with strains through crime. Adolescents lack experience at coping. Parents cope on behalf of their children, but adolescents are increasingly expected to cope on their own. However, adolescents are not as experienced at coping as adults, and they also lack many of the resources that facilitate coping, such as power and money. For example, adolescents "are compelled to live with their family in a certain neighborhood; to go to a certain school; and, within limits, to interact with the same age group of peers and neighbors."[5] As such, there is little they can do to legally avoid or escape from these people if they are mistreated. Adults, however, have the ability to end relationships, change schools or jobs, and move to new neighborhoods when they experience negative treatment.

Adolescents are also more likely to engage in criminal coping because they tend to lack access to conventional social supports. Parents and teachers often expect adolescents to cope on their own. Adolescents are often reluctant to turn to parents and teachers when they have problems, in part because of the cultural expectation that they should cope on their own. And peers are often unable to provide effective assistance at coping, because they too lack coping experience and resources.

Adolescents are also lower in many forms of social control. In particular, they are not as well supervised as children, their bonds to family members are weaker, they often lack strong commitments to conventional institutions such as school and work, and they are not yet subject to the harsher sanctions of the adult criminal justice system. This too increases the likelihood of criminal coping, since adolescents have less to lose through crime. Further, adolescents are more likely to associate with criminal peers who reinforce criminal coping, model such coping, and teach beliefs favorable to such coping. Criminal peers, for example, may foster the belief that violence is an appropriate response if others treat one in a disrespectful manner. Finally, adolescents are less able than adults to exercise self-control if they are tempted or provoked, again increasing the likelihood of criminal coping.

In sum, GST explains the peak in offending during adolescence by arguing that adolescents are more likely than children and adults to experience strains conducive to crime, to view these strains as averse, and to cope with them through crime.

THE GST EXPLANATION OF LIFE-COURSE-PERSISTENT OFFENDING

As noted earlier, there is a small group of offenders who engage in high rates of crime, including some serious crime, over much of their lives. According to GST, this is the case because such offenders are more likely to experience strain, interpret such strain as highly aversive, and engage in criminal coping *over much of their lives.* This is partly the case because such offenders possess the traits of negative emotionality and low constraint, traits that tend to be relatively stable over the life course. While these traits can undermine levels of social control and contribute to the social learning of crime, they also increase the likelihood that individuals will "a) experience objective strains conducive to crime, b) interpret these strains as high in magnitude and unjust, and c) cope with them in a criminal manner."[6]

Individuals with negative emotionality/low constraint experience more strain because they more often engage in behaviors that provoke negative responses from parents, peers, teachers, spouses, employers, and others. Such individuals might be described as "quick to anger" and "out-of-control"; as such, they often exasperate and provoke others, eliciting negative responses. For example, they are harshly punished by parents, rejected by conventional peers, frequently disciplined by teachers, and fired by employers. People with these traits are also more likely to select themselves into environments where the probability of negative treatment is high. For example, they are more likely to associate with delinquent peers because they are rejected by conventional peers and attracted by the risky behavior of delinquent peers. Delinquent peers, however, are more likely to verbally and physically abuse one another, as well as get into conflicts with others. To give another example, individuals with these traits are more likely to end up in undesirable jobs because of their poor school performance and bad work habits. Such jobs are characterized by low pay, few benefits, unpleasant working conditions, and coercive forms of control (e.g., threats, yelling). Further, the negative experiences encountered by people with these traits serve to reinforce the traits. Verbal and physical abuse by others, for example, increases negative emotionality. A vicious cycle is thereby set into motion: These traits increase negative treatment or strain, and strain fosters these traits. This cycle helps maintain high levels of offending over the life course.

Individuals with the traits of negative emotionality and low constraint are not only more likely to experience objective strains, but also to interpret these strains as highly aversive and unjust. In particular, such individuals are very sensitive to strains and tend to blame their problems on others. Further, they are more likely to cope in an aggressive or criminal manner, reflecting their higher anger, tendency to act without thinking about the consequences of their behavior, and attraction to risky activities. In brief, the traits of negative emotionality and low constraint contribute to life-course-persistent offending partly by increasing objective strain, subjective strain, and the tendency to cope in a criminal manner.

GST also posits that individuals without the traits of negative emotionality/low constraint sometimes engage in high rates of offending over their life course. Individuals who follow this pattern are most likely to be poor and live in poor communities. Such individuals are often more exposed to a range of strains related to their poverty, including problems with family members, peers, school, and neighborhood residents. The stresses associated with poverty, for example, increase the likelihood of abuse and harsh/erratic parental discipline, although this does not imply that most poor individuals experience these problems. To give another example, individuals in poor neighborhoods are more likely to be the victims of violence and to witness violence against others. Further, such individuals are more likely to cope with strains through crime. Among other things, they lack certain coping skills and resources, such as money; are lower in conventional social supports; and are more often exposed to peers who encourage criminal coping. The strains and the criminal coping of these individuals frequently contribute to further strains. For example, the criminal acts committed by these individuals often provoke negative treatment by parents, teachers, neighbors, and police. The result, once more, is a vicious cycle that contributes to offending over the life course. It is important to note, however, that most poor individuals do not become life-course-persistent offenders. A range of factors, such as coping skills, support from family, positive experiences at school, and mentoring by adults in the community helps prevent this.

Finally, life-course-persistent offending may result from what is known as the proliferation of strain.[7] Current experiences of strain may lead to or create other forms of strain. This may come about in two different ways. First, this may occur through *primary proliferation*. Here, the initial strains come to impact areas of one's life that are closely associated with the primary strain. Thus, job dissatisfaction could lead to conflict with colleagues. Second, there may also be situations where the primary strain comes to influence other aspects of the individual's life. This is referred to as *secondary proliferation*. For example, the loss of one's job may lead to conflict with one's spouse. Strain proliferation suggests then that strains can begin to cluster and accumulate over time, increasing the likelihood that offending will persist across the life course. It should be noted that the strains most likely to proliferate tend to be experienced more by disadvantaged groups and minorities, with these strains including "incarceration, illness, unemployment, and chronic financial strain."[8]

While the GST offers an explanation of life-course-persistent offending it also offers an explanation for those adult offenders who desist in their behavior.[9] Some offenders desist because they experience reductions in strain; for example, they obtain decent jobs or become involved in good relationships. Related to this, some desist because they become less likely to respond to strains with crime. Perhaps they experience an increase in social control (e.g., by getting a decent job), receive support from others, develop their coping skills, or leave their criminal peers.

PATTERNS OF OFFENDING AMONG STREET YOUTH

The central argument advanced in this chapter is individuals' offending at any given time is a function of their levels of objective and subjective strain, as well as their tendency to cope with strains in a criminal manner. These points are next illustrated in the following discussion of patterns of offending among street youth.

Background Strains

The term *street youth* usually refers to youths who have run away or been expelled from their homes and/or who spend some or most of their time in public locations. Their path to the street begins with their exposure to strain in their family backgrounds. Research suggests that many of these youth begin their lives in homes suffering economic difficulties where parents are unemployed or working poorly paying jobs. The economic pressures in these homes often lead to stresses that undermine family relations and break up the family. The conflict within the family, the parenting stresses associated with disruption, and the economic pressures all leave these families more at risk to utilize volatile and inconsistent methods of discipline in the home. As a result, the youths who flee to the streets from these homes tend to have experienced high rates of various forms of abuse.[10]

Research shows that many street youths have experienced high rates of emotional abuse where family members regularly say hurtful or insulting things. At the same time, these youths have experienced emotional neglect where they feel little affection, attachment, or support from the members of their family. Further, they experience high rates of physical abuse where members of their family strike them so violently that the attacks leaves bruises or marks, and the injuries are often so severe that the youths require medical attention. In addition, consistent with the economic problems of the parental home, these youths often suffer from physical neglect where there is not enough food in the home for them to gain regular meals, clothes are not washed, and parents are so preoccupied with alcohol and/or drugs that they are unable to provide their children with proper care. Finally, these youths are often exposed to sexual abuse in which others have attempted to make them engage in, or watch, unwanted sexual activities.[11]

A recent study showed that nearly all (98 percent) of the 400 street youths interviewed met the criteria for at least one form of maltreatment and more than one-third (34.5 percent) had experienced all five types. Prevalence in the individual forms of maltreatment ranged from 45.0 percent for sexual abuse to 90.0 percent for emotional neglect. The most common co-occurring maltreatment types were emotional neglect and emotional abuse (83.0 percent) and emotional abuse and physical neglect (73.7 percent).[12] Thus, we see that these youth experience a range of serious negative experiences. The following passage, drawn from separate interview data gathered as part of Baron's project on strain, illustrates these experiences:[13]

> *Because my mom, when I was growing up, this is when I was an only child, 'cause my sisters are young still, she was always doing everything. They were always doing dope. They used to do needles. Always drinkin', and I always had all kinds of babysitters, so I was everywhere.*

Within GST, these are serious strains that are likely to be linked to criminal coping behavior. Child abuse threatens a child's goals, values, needs, activities, and/or identities and is likely to be seen as unjust. Harsh physical punishment, verbal abuse, emotional and physical neglect, as well as sexual abuse can all be seen as strains that are high in magnitude. Parental violence, hostility, detachment, and sexual victimization undermine the attachments between parent and child, leading to low social control and crime. Further, evidence suggests that these forms of abuse can have long-term effects. Emotionally abused children often become hostile and aggressive and come to perceive the world as a hostile environment. Children from violent homes may model the violent behavior of their parents and adopt aggressive interpersonal strategies as a means of problem solving that can evolve to become favorable to violence.[14] Abusive experiences may also diminish the ability to cope with stress, inhibit the development of empathy, and can lead youths to seek out peers who support or encourage violence.[15] Similarly, there is evidence that sexual abuse can lead to feelings of betrayal, causing hostility and anger; the powerlessness associated with the abuse can impair judgments and decrease coping skills while increasing insecurities, anxieties, and the desire to protect oneself. Shame, guilt, and stigmatization also leave victims at risk for association with stigmatized others, including deviant peers.[16]

Many of the characteristics that evolve in response to abuse are similar to those that are encompassed in the development of negative emotionality/low constraint.[17] The development of negative emotionality in response to abuse leads to these youth becoming more easily upset and angered and developing impulsivity and risk-taking behaviors. Youth with these characteristics tend to select into negative environments like the street, unemployment, and peer groups that expose them to additional negative treatment. Further, this negative emotionality leaves the youth more likely to be sensitive to future strains, to interpret a broader array of situations as strain, and to view various conditions as unjust.[18] In sum, the strains experienced by youths prior to coming to the streets have a long-term impact on their utilization of criminal coping in reaction to future strains.

These negative experiences in the home also serve to undermine educational activities and achievements. Hagan and McCarthy note that these conflictual and unsupportive backgrounds serve not only to restrict success at school, but also lead to increased difficulties with educational authorities.[19] Conflict with school personnel and school failure serves as yet another form of strain and further undermines commitments to conventional lines of action and attachments to conventional others.

Upon entry to the street, these background strains continue to have an impact on criminal behavior. Research shows that street youths who have suffered either physical or sexual abuse are more likely to become involved in crime, generally once they land on the streets. Physical abuse, in particular, is strongly linked to violent offending on the street and appears to be linked to serious outcomes where the victims of these youths are more likely to suffer significant injuries. Backgrounds of physical abuse also appear to be drawn upon to increase participation in robberies where force, or the threat of force, is utilized to gain the possessions of others on the street.[20] Finally, youths tend to cope with backgrounds of abuse with hard drugs. For example, physical abuse is directly associated with harder drug use.[21] The relationship between abuse and hard drug use appears to be strengthened when youths have adopted values that support the use of crime. Street youths with backgrounds of physical abuse, sexual abuse, and emotional abuse and who hold strong values regarding the validity of breaking the law are more likely to use hard drugs. Further, those with backgrounds of sexual abuse tend to be more likely to engage in violence if they hold these values. In addition, contact with deviant peers, who may also have suffered sexual victimization and who can help contextualize the experience of this strain, increases the likelihood that youths will respond to their sexual abuse with violence once on the street. Lastly, it appears that street youths who somehow manage to retain their self-esteem despite these negative experiences are more likely to respond to their physical abuse with violence and their emotional abuse with property offending. Self-esteem appears to be a resource that allows these youths to channel their reactions to strain into criminal sources of coping.[22]

Homelessness

For youths who leave their homes because of adverse circumstances, the street itself becomes a new source of strain and also exposes them to an array of factors that increase their disposition toward offending. These youths enter an environment where they lack shelter, adequate sources of food, and access to other resources. GST maintains that homelessness is an extremely powerful source of strain for a number of reasons. Homelessness challenges a broad range of identities, needs, values, and goals, and therefore can be seen as a strain that is high in magnitude. Homelessness is often perceived as unfair, and, because it decreases people's contact with and attachment to conventional society, it is associated with low levels of social control. Homelessness provides opportunities for crime since youths lacking shelter may wander the streets at all hours, increasing the likelihood that they will encounter human and property targets for victimization and theft, respectively. Homelessness is strongly related to the social learning of crime because it exposes youth to other offenders who can offer tutelage in and support for crime.[23] Further, there is a culture on the street or "street lifestyle" that encourages offending, providing values that support all forms of offending, including violence and drug use. Lastly, homelessness can contribute to the development of negative emotionality/low constraint.[24] Those with negative emotionality and low constraint are drawn to others on the street with similar characteristics, who then shape the way negative emotionality is channelled behaviorally. For those who come to the street *without* negative emotionality/ low constraint, the negative economic circumstances, the often dangerous environment, and the characteristics of those who are already there, often lead to the development over time of negative emotionality/low constraint.

Evidence suggests that the objective strain of homelessness tends to lead to criminal coping as the time without a permanent shelter lengthens. Thus, as the duration of the strain grows, the impact of the strain deepens, and it is more likely to lead to an array of offending options. Street youths begin to engage in property offenses to obtain food and shelter. They turn to the distribution of illegal drugs to meet these same needs. Further, drug distribution facilitates gaining access to funds to help finance their own use of these substances as a coping mechanism. The use of these substances, particularly harder forms of drugs, tends to increase as time on the street becomes more extended. Finally, violence can be used as a method to gain funds (robbery), but can also be viewed as a way to cope with the stress created by being homeless.[25] Participants in Baron's study of strain were asked in separate qualitative interviews what they did with the proceeds from their criminal activities. Their responses illustrate how their illegal activities helped them cope with their adverse circumstances.[26]

> *I buy cigarettes, I buy food, I buy pot, and if there's money left over, we may go out and buy a little bit of crystal now and then. But other than that, we use it for things that we need before things that we want.*

> *'Cause I'm sober and I don't have a harsh drug addiction. So all my money goes towards food and housing instead of drugs.*

The street itself is a dangerous environment, which means that the more time one spends on the street increases the likelihood that one will participate in violent events. The culture on the street that supports the use of crime as a method for coping with strain also contributes to the relationship between homelessness and crime. Youths who come to the street encounter a world where crime is not viewed negatively and those youth who come to adopt these views tend to react to their lengthening homelessness with a range of criminal activities.[27]

Unemployment

Many youths are on the street and unable to support themselves because they lack viable employment. They often come to the street with incomplete education, which disqualifies them from all but the most menial of jobs. These youths' limited employment experiences tend to be characterized by a series of poorly paying jobs with few benefits and little opportunity for advancement. This means that most youths on the street have decreased attachments to the labor market and little to keep them involved in conventional society.[28]

The failure to acquire employment fits into all three types of strain outlined in GST, including goal blockage, the failure to achieve positively valued goals, and the presentation of negative stimuli. GST asserts that the desire for money is a major goal for many in our society and may be particularly important for those in lower socioeconomic locations.[29] Therefore, unemployment as an objective strain can be seen to increase individuals' motivation to commit crime in an effort to overcome economic problems. Unemployment may also leave individuals with fewer stakes to jeopardize through crime and can reduce beliefs regarding the legitimacy of rules or norms. Further, the unemployed may pass time together, facilitating the learning and undertaking of crime. Research suggests, however, that the link between unemployment and crime for street youths is complex, and it appears that criminal coping is more likely under a certain range of circumstances, and that subjective interpretations of economic situations, or subjective strain, may be as or more important to understanding the link between economics and crime in this population.[30]

First, as would be expected from GST, unemployment generates anger in street youths. It does not do so directly, however. Instead, street youths become angry over their unemployment, blaming others for their circumstances without seeing themselves as blameworthy. When this blaming of others is accompanied by peers who help shape the manner in which the strain is interpreted, it reinforces their anger. Youths on the street are also likely to become angry over their unemployment when they begin to compare themselves to those around them in the broader society and come to view themselves as unfairly deprived. This dissatisfaction with their meager financial situations feeds into their anger.

For those street youths who want to work in legitimate jobs to get off the street and who are willing to work hard, the anger over unemployment is increased. This anger over unemployment in turn leads to greater violent offending and drug distribution. Unemployment is also more likely to lead to coping through the use of violence and drug distribution when the youths report that they are not happy with their monetary circumstances. Without this interpretive state of affairs of their objective economic situation, lengthy unemployment does not lead to criminal coping. Thus, subjective strain helps to link objective strain with criminal coping. Additional data gathered as part of Baron and Hartnagel's study outlines some of the issues regarding attributions, anger, and perceptions of unfairness:[31]

> *Well, you try and get yourself out of it trying to get yourself off the street. But society is holding you back. They're telling the younger generation we want you to live on welfare the rest of your life.*

Unemployment is also more likely to lead to criminal coping when the street youth has adopted values that support the use of crime. As the strength of these values increases and the length of unemployment increases, so does the likelihood of engaging in violence as a coping mechanism.

One of the key arguments of GST is that the people may view their situations as unjust. One of the key tenets of the "American Dream" is that those who work hard will have a chance to succeed. For street youths experiencing unemployment, this cultural imperative may leave them feeling that they are failures. However, strain theorists also argue that the failure to achieve positively valued goals can lead to alienation and a withdrawal of their belief in the legitimacy of dominant norms. Research suggests that street youths' economic experiences make them likely to reject the dominant meritocratic ideology, which increases the likelihood that their poor employment situations will lead to criminal coping. The impact of this frustration is strengthened when they blame others for their situation, when they have adopted values supportive of the use of crime to cope, and when they associate with criminal peers. Unemployment and homelessness may lead similarly situated youths into the company of one another. These peers may help to contextualize the unemployment experience, and facilitate and encourage criminal behaviors as well as teaching, supporting, and rewarding beliefs that justify and rationalize criminal conduct.[32]

Subjective Economic Strains

Street youths' subjective interpretations of negative economic circumstances and perceptions of relative deprivation are extremely important contributors to strains. According to GST, this type of strain is a function of the attainments of those in one's comparative reference group(s) and one's own failure to achieve. Strain is more likely when individuals believe that they are worse off monetarily than those to whom they compare themselves.[33] In North American society, with the dominant meritocratic ideology, people are encouraged to compare themselves with a broad range of people. Street youths lacking shelter, employment, finances, and regular sustenance who adopt this frame of reference are very likely to feel strained. There is evidence that this perception of deprivation leads to a range of offenses, including property crime and violence. The impact of this perceived deprivation, however, becomes even stronger when objective economic strains are more severe. For example, when relative deprivation is coupled with long-term homelessness, it leads street youths to engage in high rates of offending. Further, peers appear to be important in contextualizing how one interprets one's degree of deprivation. In the case of street youths, those who have more associations with criminal peers are more likely to react to their relative deprivation with violence and drug dealing. The following data gathered for Baron's work provides evidence of these issues of comparison and deprivation:[34]

> They, they, they, they make, they make, they belittle you. They make you feel like less of a person like, because you're lower end of the social class, you're, you're shit. You know?

Another important factor in whether street youths react to their negative financial situations with crime appears to be how unhappy they are with their economic circumstances. This dissatisfaction pressures or compels people to commit crime.[35] Thus, we find that street youths who are unhappy with their current economic situation are more likely to engage in crime. Further, subjective evaluations of financial happiness appear to be influenced by adverse economic situations. This dissatisfaction with the current economic situation is more likely to lead street youths to crime as the length of homelessness increases and as time since previous employment increases.

Finally, there is the goal of financial success itself. According to GST, having this sort of goal and perceiving that this goal will not be reached can create the need for criminal coping strategies. Street youths who have the goal of monetary success but who believe they will not reach this goal are more likely to engage in drug dealing and violent means to reach this goal. The importance of this goal in creating criminal coping is exacerbated by unemployment and homelessness. Having this ambition for financial success under extremely adverse economic circumstances in a culture that encourages criminal coping also leads to increased offending behavior. Street youths who have lofty financial goals, and who hold values supportive of breaking the law, tend to engage in instrumental offenses to achieve these goals as their bouts of homelessness and unemployment lengthen and as their perception of financial dissatisfaction grows.[36]

Street Victimization

Being homeless also exposes these youths to a great deal of victimization. According to GST, victimization is typically seen as unjust and high in magnitude. Victimization is also most likely to occur in settings with low social control—such as unsupervised settings where young people gather. Further, people may learn from their victimization that physical aggression is necessary to deter future victimization and ensure the safety of themselves and their property. Offending behavior may also increase because prior victimization leads people to legitimate the use of crime. Because victimization is more common in delinquent peer groups and involves exposure to a criminal model, it is often associated with the social learning of crime and subcultural expectations that condone retaliation. Interviews conducted as part of Baron's study show the seriousness of some of this victimization:[37]

> Well, it's kinda weird. I don't know why, but about twenty guys came, like I was in ah, like, Burnaby or somewhere. About twenty guys came and rushed me and beat me up.

> Um, me and a girlfriend were walking down the street and a gang of punks decided to jump us and wanted us, wanted our shit and I said no. So I ended up fighting five guys.

Street youths experience not only direct victimization in terms of physical attacks and property loss, but their dangerous lifestyle also leaves them exposed to "vicarious" and "anticipated" strains. On the street these youth eye witness the negative experiences of friends and associates (e.g., assaults), hear the negative events taking place (e.g., screams), or learn about the negative events from others (i.e., from victims). These experiences are viewed as unjust, are associated with low social control, are great in magnitude, and increase street youths' perceptions that they will suffer similar strains, leading to a greater likelihood of criminal coping.[38] Street youths exposed to these forms of vicarious victimization undertake criminal behavior as a method to avert future harm to themselves and to those around them. Criminal behavior can also be an avenue for revenge or reprisal against the individuals judged to be accountable for the harm.

Research shows that street youths who encounter more direct physical victimization are more likely to engage in violent offending. They are more likely to engage in assaults of both minor and serious forms, get involved in group fights, and engage in robbery.[39] Further, this form of victimization is more likely to lead to violent offenses when the youth have attitudes that support the use of violence and have greater sense of self-esteem and self-efficacy that allows them to use violence as a coping mechanism. Physical victimization is also likely to lead to violent coping when street youths have high levels of negative emotionality and low levels of constraint.[40] Their experience with robbery also invites violent coping, particularly for those who have been able to retain high levels of self-esteem. Property victimization promotes a violent response when youths hold attitudes that support the use of criminal coping. Street youth are more likely to respond to violent and property victimization through the use of hard drugs when they hold values supportive of criminal coping and when they associate with peers who engage in criminal coping.[41] Research also shows that vicarious forms of violent victimization where peers have experienced violent events lead street youths to utilize criminal coping, particularly when they also have negative emotionality/low constraint. Further, the anticipation of physical victimization leads to a greater likelihood of violent coping when youths have acquired negative emotionality/low constraint.

While homelessness and the objective and subjective strains that accompany it embed street youths in situations that continually lead to criminal coping, some of these youths do manage to exit from the street and crime. Research suggests that participation in employment assists in decreasing criminal participation. The time commitment required of employment, the access to nonstreet contacts, the acquisition of job skills, and the establishment of employment histories create experiences for street youths that are dissonant from their street and illegal activities. Youth able to secure even marginal, low-wage employment spend less time interacting with street friends, panhandling, searching for food and shelter, using drugs, and engaging in crime. The "embeddedness" in the street crime networks begins to decline and employment serves as a turning point in changing these youths' trajectories toward a life off of the street.[42]

CONCLUSION

General strain theory offers a valuable avenue from which to explore and understand issues of stability and change in offending behavior. Going beyond its utilization as an explanation of behavior between offenders, GST can be used to explain the behavior of individuals and employed to comprehend adolescent-limited and life-course offending patterns. It shows that adolescent-limited offending can be understood in the context of a life-course period in which there is an increased exposure to strain. Life-course-persistent offending can be explained by a stable trait of negative emotionality/low constraint and chronic exposure to strain. As applied to street youths, we can see how negative experiences can lead to increased levels of criminal coping, how decreases in strain can lead to a desistance from offending, and why continued experiences with strain can lead to persistence in criminal coping.

GLOSSARY

Failure to achieve positively valued goals—actual or anticipated failure to achieve monetary goals, status or respect, and thrills/excitement

General strain theory—Agnew's theory that a range of strains or negative events and conditions increase crime; these strains lead to negative emotions, such as anger, which create pressure for corrective action

Presentation of negative stimuli—actual or threatened presence of negativity, such as criminal victimization, child abuse, negative school and work experiences, and discrimination

Removal of positively valued stimuli—actual or threatened removal of one's job, home, relationships, or other valued thing

NOTES

1. Agnew, R. (1997). Stability and change in crime over the life course: A strain theory explanation. In P. T. Thornberry (Ed.), *Developmental theories of crime and delinquency: Advances in criminological theory.* (Vol. 7). New Brunswick, NJ: Transaction Publishers; Agnew, R. (2006). *Pressured into crime: An overview of general strain theory.* Los Angeles: Roxbury; Moffitt, T. E. (1993). Adolescence-limited and life-course-persistent antisocial behavior: A developmental taxonomy. *Psychological Review, 100,* 674–701.

2. Agnew, R. (1992). Foundation for a general strain theory of crime and delinquency. *Criminology, 30,* 47–66; Agnew, R. (2001). Building on the foundation of general strain theory: Specifying the types of strain most likely to lead to crime and delinquency. *Journal of Research in Crime and Delinquency, 38,* 319–361.
3. Agnew (1992, 2001, 2006), see Notes 1 and 2.
4. Agnew, R. (2003). An integrated theory of the adolescent peak in offending. *Youth & Society, 34,* 263–299.
5. Agnew, R. (1985). A revised strain theory of delinquency. *Social Forces, 64,* 151–167, p. 156.
6. Agnew (2006), p. 122, see Note 1.
7. Slocum, L. (2010a). General strain theory and the development of stressors and substance use over time: An empirical examination. *Journal of Criminal Justice, 38,* 1100–1112; Slocum, L. (2010b). General strain theory and continuity in offending over time: Assessing and extending GST explanations of persistence. *Journal of Contemporary Criminal Justice, 26,* 204–223.
8. Slocum (2010b), p. 215, see Note 7.
9. Eitle, D. (2010). General strain theory, persistence, and desistance among adult males. *Journal of Criminal Justice, 38,* 1113–1121.
10. Hagan, J., & McCarthy, J. (1997). *Mean streets: Youth crime and homelessness.* New York: Cambridge University Press.
11. Forde, D. R., Baron, S. W., Scher, C., & Stein, M. B. (2012). Factor structure and reliability of the Childhood Trauma Questionnaire and prevalence estimates of trauma for male and female street youth. *Journal of Interpersonal Violence, 27*(2), 364–379.
12. Forde et al. (2012), see Note 11.
13. Baron, S. W. (2004). General strain, street youth and crime: A test of Agnew's revised theory. *Criminology, 42,* 457–483.
14. Baron, S. W., Kennedy, L. W., & Forde, D. R. (2001). Male street youths' conflict: The role of background, subcultural and situational factors. *Justice Quarterly, 18,* 759–789.
15. Baron, S. W., & Hartnagel, T. F. (1998). Street youth and criminal violence. *Journal of Research in Crime and Delinquency, 35,* 166–192.
16. Baron (2004), see Note 13.
17. Baron, S. W. (2009a). Differential coercion, street youth, and violent crime. *Criminology, 47,* 239–268.
18. Baron, S. W., Forde, D. R., & Kay, F. M. (2007). Self control, risky lifestyles, and situation: The role of opportunity and context in the general theory. *Journal of Criminal Justice, 35,* 119–136.
19. Hagan, & McCarthy (1997), see Note 10.
20. Baron (2004), see Note 13; Baron, & Hartnagel (1998), see Note 15.
21. Baron, S. W. (1999). Street youth and substance use: The role of background, subcultural, and economic factors. *Youth & Society, 31,* 3–26; Gallupe, O., & Baron, S. W. (2009). Street youth, relational strain, and drug use. *Journal of Drug Issues, 39,* 523–545.
22. Baron (2004), see Note 13.
23. Baron S. W., & Hartnagel, T. F (1997). Attributions, affect and crime: Street youths' reactions to unemployment. *Criminology, 35,* 409–434.
24. Baron, S. W. (2003). Self control, social consequences, and criminal behavior: Street youth and the general theory of crime. *Journal of Research in Crime and Delinquency, 40,* 403–425; Baron, S. W. (2009b). Street youths' violent responses to violent personal, vicarious, and anticipated strain. *Journal of Criminal Justice, 35,* 119–136.
25. Baron, & Hartnagel (1998), see Note 15.
26. Baron (2004), see Note 13.
27. Baron (2004), see Note 13.
28. Baron, S. W. (2001). Street youth labour market experiences and crime. *Canadian Review of Sociology and Anthropology, 38,* 189–215; Baron, S. W. (2008). Street youth unemployment and crime: Is it that simple? Using general strain theory to untangle the relationship. *Canadian Journal of Criminology and Criminal Justice, 50,* 399–434.
29. Agnew (2001), see Note 2.

30. Baron (2008), see Note 28; Baron, & Hartnagel (1997), see Note 23; Baron, S. W., & Hartnagel, T. F. (2002). Street youth and labor market strain. *Journal of Criminal Justice, 30*, 519–533.
31. Baron, & Hartnagel (1997), see Note 23.
32. Baron, & Hartnagel (2002), see Note 30.
33. Baron, S. W. (2006). Street youth, strain theory, and crime. *Journal of Criminal Justice, 34*, 209–223.
34. Baron (2004), see Note 13.
35. Agnew, R., Brezina, T., Wright, J. P., & Cullen, F. T. (2002). Strain, personality traits, and delinquency: Extending general strain theory. *Criminology, 40*, 43–71; Baron (2006), see Note 33; Baron (2008), see Note 28.
36. Baron, S. W. (2011). Street youths and the proximate and contingent causes of instrumental crime: Untangling anomie theory. *Justice Quarterly, 28*, 413–436.
37. Baron (2004), see Note 13.
38. Baron (2009b), see Note 24.
39. Baron, & Hartnagel (1998), see Note 15.
40. Baron (2004), see Note 13; Baron (2009b), see Note 24.
41. Gallupe, & Baron (2009), see Note 21.
42. Hagan, & McCarthy (1997), see Note 10.

CHAPTER NINE

Race, Ethnicity, Juvenile Offending, and Justice Court Responses

Michael J. Leiber and Jennifer H. Peck

KEY TERMS

Differential offending
Disproportionate minority confinement/contact
Race/ethnic selection bias

INTRODUCTION

Race/ethnicity has generated an enormous amount of research and debate in criminology.[1] One factor that underlies much of this activity is the overrepresentation of minority youth, particularly African Americans, in the juvenile justice system.[2] Between 1985 and 2007, for example, the average total arrest rate per 100,000 youth ages 10 to 17 in each racial group was as follows: Black youth (6083) were twice the rate of White youth (3130) and Native American youth (3137) and four times the rate of Asian youth (1411).[3] In 2008, White youth, who made up 78 percent of the at-risk population ages 10–17, accounted for 67 percent of arrests. African Americans were greatly over-represented: They comprised 17 percent of the youth population, but accounted for 31 percent of arrests.[4] Minority youth are not only overrepresented in arrests, but also throughout the juvenile justice system. At secure detention, the rate per 100 for White youths is 18.6 while for African American youths it is 25.4. For youth taken out of the home and placed in secure corrections (i.e., state training school), the rate per 100 youth for Whites is 25.9 and for African American, it is 31.6. In short, racial disparities, especially for African American youth, are quite prominent throughout the juvenile justice system, though it is most prominent at the front end of the system (e.g., arrest, court referrals, secure detention).[5]

This chapter discusses the explanations for the overrepresentation of minority youth in the juvenile justice system. The first part of the chapter focuses on the contention that minorities commit more crime and/or more serious crime, or what is known as the **differential offending** explanation for the overrepresentation. The second section discusses the theoretical perspectives for understanding why minorities commit more crime or more serious crime. Next, the differential selection argument is provided; it details explanations for minority overrepresentation in the juvenile justice system that center on racism or bias.

RACE DIFFERENCES IN OFFENDING

As noted in the introduction to this chapter, African Americans are more likely to be arrested than Whites. This seems to be true for almost every offense category. In addition, between 1985 and 2007 the African American arrest rate for property crimes annually remained near twice the White arrest rate. Over the same time period for violent crimes, the African American arrest rate averaged five times the White rate.[6] The data that arrest information is based on come from the *Uniform Crime Reports* (published by the Federal Bureau of Investigation [FBI]). Because the decision to stop, release, refer, and arrest youth is contingent upon a variety of factors beyond the offense and its severity—patrolling patterns, the organizational style and goals of a police department, the socioeconomic makeup of a community, racial profiling—questions emerge concerning whether arrest data reflect bias in police decision making.[7] Still, research exists that provides results confirming a differential offending argument.

Hindelang, for example, compared results from the *Uniform Crime Reports (UCR)* to those by the National Crime Victimization Survey (NCVS) to assess the differential involvement and selection bias hypotheses.[8] Victimization surveys ask household residents to report personal victimizations, regardless of whether they reported these crimes to the police. Thus, if the data from the NCVS reveal race differences in crime that are in basic agreement with arrest data, there is greater confidence that race differences in arrest reflect real differences in offending. If they do not, then the system bias hypothesis becomes more plausible. Using rape, robbery, and assault data from the *UCR* (arrest statistics) and NCVS (victimization statistics), Hindelang discovered that both measurements showed Blacks to be overrepresented compared to their representation in the general population for rape, robbery, and assault. Still, Hindelang found that for the crimes of rape and assault there was some evidence of unexplained disparity.[9] The Black arrest rate was higher than the Black NCVS rate. Overall, although he found some evidence of differential selection bias, most of the racial overrepresentation in the arrest data was shown in the victimization data. Therefore, Hindelang concluded that African Americans commit more crime, particularly more violent crime.

Victims can only tell researchers about characteristics of offenders where there has been face-to-face contact. For "victimless" crimes (e.g., drug offenses), property crimes, and homicides, victim descriptions of offenders are unavailable. Furthermore, even in crimes involving face-to-face contact, victims may not always be able to accurately identify the offender's age and race. Still, victim reports obtained from the annual NCVS indicate that minorities are overrepresented among offenders who commit serious violent crimes.[10] For robbery, the proportion of juvenile offenders whom victims identified as African American is only slightly lower than the proportion shown in police data. For sexual and aggravated assaults, African Americans are also overrepresented, although not nearly to the extent indicated in arrest data.

Using the FBI's 1997 and 1998 National Incident-Based Reporting System (NIBRS) involving data from 17 states, Pope and Snyder examined the most serious offenses and victim accounts to study whether race bias appeared to play a role in the decision to arrest.[11] Through NIBRS, the FBI asks law enforcement agencies to record a substantial amount of information on each reported crime and each arrest. For example, the agency is asked for the following information: age, sex, and race of the victim(s); offense(s) involved; date and time of the incident; type of place where the incident occurred; each victim's level of injury; weapon(s) used, if any; victim's perception of the demographics of the offender(s), including age, sex, and race; victim–offender relationship(s); and the demographics of arrestee(s). Thus, researchers can examine NIBRS data on the types of incidents likely to involve victim–offender interaction, determine the victim's perception of the offender in each incident, study which incidents resulted in arrests, and then compare, for example, the arrest probabilities of White and non-White juvenile offenders for similar crimes. Taking these factors into account, Pope and Snyder did not find direct evidence that racial bias exists in arrest decisions.[12] White juvenile offenders were more likely to be arrested than their non-White counterparts, especially for violent crimes. It is important to note that the study was based on data from only 17 states and that different juvenile arrest patterns may emerge if other states were included in the analysis. Furthermore, the data indicated evidence of an indirect bias effect in the arrest of non-White juveniles in that they are more likely to be arrested when the victim is White than when the victim is non-White.

The self-report survey is an alternative method of measuring crime independent of the police and victims.[13] Self-reports ask high school and other samples of youth to anonymously report any offenses they have committed and whether or not they were apprehended. But self-reports may not be equally valid for all racial groups. Some researchers have suggested that African Americans tend to underreport serious misconduct, while others have found no differences in the accuracy of reporting across racial groups.[14]

The National Youth Survey is a self-report administered to a nationally representative sample of teens. Using these data, Elliott found that African American youth admitted greater involvement in violent behavior than Hispanic youth, who in turn reported greater involvement than Whites.[15] These findings are consistent with those from the Denver, Pittsburgh, and Rochester Youth Studies, where White youth reported involvement in violent crimes at lower rates than Hispanic youth, and African American youth reported the highest levels of involvement. Although the self-reported race differences in violent offending across all these studies are substantial, they are not nearly as great as those found in police arrest data.[16]

In summary, victimization data and results from self-report surveys suggest that African American youth are more likely to commit more crime and more violent crimes than Whites. The disparities, however, are not as great as those reported in official arrest data. For example, for property and drug crimes (for which victimization data are not available), self-reports indicate minimal race differences in offending. With respect to drug offenses, Monitoring the Future—an annual national survey of high school students begun in 1975—has consistently shown that the highest proportions of drug use of all kinds are found among Whites, followed by Hispanics, with African Americans reporting the lowest levels of illicit drug use. These race differences in self-reported crime are not at all consistent with police arrest data, which show substantial overrepresentation of minorities for vandalism, theft, weapons, and drug offenses.[17]

Comparisons of arrest data with victimization data and self-report data provide some clarity to race being a correlate of crime. That is, in most instances minority youth, particularly African American youth, evidence greater frequency of offending and more serious offending, which supports the differential offending explanation of their overrepresentation in the juvenile justice system. But, minority overrepresentation is also accounted for by differences in the way justice officials (i.e., police, courts) respond to White and minority juveniles who engage in similar behavior. Researchers generally acknowledge that both of these explanations may account for the overrepresentation of minority youth in the juvenile justice system. The following section describes theories that attempt to explain why minorities commit more crime throughout the life course compared to Whites.

THEORETICAL EXPLANATIONS FOR DIFFERENTIAL OFFENDING

For the most part, general or traditional criminological theories have lacked in their attention to the relationship between race and crime, and more specifically racial and ethnic differences in offending. Some have argued that most mainstream theories are not designed to explain why minorities commit more frequent and serious crimes than Whites, therefore race differences in delinquency and criminal offending are indirectly related to individual differences in offenders, family processes, or structural and community explanations.[18] Others have argued that criminological theories can universally apply to all racial and ethnic groups, regardless of any differences in individual offenders, while others have stated that no one theory had adequately addressed why some racial and ethnic groups have higher levels of offending than others.[19] However, there are some race-specific theories and perspectives for understanding why minorities are overrepresented as offenders. This section describes how some general and race-specific theories have explained racial differences in offending.

Beginning with individual-level theories, Gottfredson's and Hirschi's general theory of crime argues that differences in offending between minorities (more specifically African Americans) and Whites are due to African Americans having lower levels of self-control, which results from inept and erratic socialization from parents.[20] In other words, Gottfredson and Hirschi suggest that minority families are less able to monitor, recognize, and punish deviant behavior in children, and are ineffective at childrearing. This process results in minority youth having lower levels of self-control, and in turn committing more frequent and serious acts. They also argue that minority families in general are more likely to include parental criminality, larger family sizes, more single-parent families, and mothers working outside of the home, which are additional examples of why minority youth have lower levels of self-control. It is important to note, however, that Gottfredson and Hirschi did not say that African Americans inherently have lower levels of self-control, but it is inept parenting that resulted in this assumption. Also, while Gottfredson and Hirschi did not test this specific aspect of the theory, and previous research has not examined if race disparities in offending can be attributed to differences in levels of self-control, research has found conflicting results on the universal applicability of the general theory of crime. Some research has found that levels of self-control do not differ between Whites and non-Whites, suggesting that the general theory of crime is race-neutral, while others suggest the opposite.[21]

Moffitt's developmental theory or taxonomy of antisocial behavior argues that African Americans are more likely to be overrepresented as both life-course-persistent (LCP) offenders (those who offend throughout the life course) as well as adolescent-limited (AL) offenders (those who offend only during adolescent years).[22] Moffitt's justification for this argument is that African American offenders are a product of living in structurally disadvantaged communities that hinder their life chances at success. From this, African American life-course-persistent offenders have experienced inadequate prenatal care, exposure to hazardous toxins, lack of familial attachment and bonds, and disadvantaged school environments. These developmental situations have resulted in poor educational success and limited employment opportunities. Thus, African Americans will be more likely to continue offending as adults because they as a group experience higher levels of risk factors (e.g., unemployment, single parents, drinking, drug use) than do Whites.

Concerning African Americans as adolescent-limited offenders, Moffitt argues that this type of African American offender is stuck within a maturity gap that White youth do not experience for a long duration of time. In other words, African American youth spend more years within a maturity gap because they do not necessarily shift to an "adult status" with valued adult roles (e.g., employment) and privileges. This delayed entrance into adulthood and conventional adult roles results in African American youth becoming overrepresented as adolescent-limited offenders, who mimic offending patterns of delinquent peers. These types of offenses symbolize adult social status, such as theft, drinking, and drug use.

Empirical research has found support for Moffitt's explanation of racial differences in offending. For example, some results have indicated that African Americans, compared to Whites, are more likely to persist in violent offending due to reduced opportunities for economic wellbeing and employment when in adolescence, supporting an economic maturity gap perspective.[23] Research has also found that for African Americans certain LCP characteristics of offenders (developmental risk and adverse family situations) were intensified by community disadvantage. However, this result was not applicable to Whites.

General strain theory (GST) is an individual-level perspective that has been applied to examine racial and ethnic differences in offending. Researchers have argued that GST can explain racial differences in offending because African Americans and minorities generally experience a disproportionate amount of and qualitatively different types of strain that lead to increased levels of offending. Kaufman, Rebellon, Thaxton, and Agnew argued that African Americans experience different types of strains more than other groups, including economic strain (poverty, unemployment), family strain (poor parenting practices), educational strain (problems with students and teachers, poor grades), criminal victimization, racial discrimination, and community strain (living in structurally disadvantaged neighborhoods).[24] For example, Simons, Chen, Stewart, and Broidy found that African American youth who experience discrimination are more likely to cope through delinquency; Jang and Johnson found that GST can be applied to offending in African Americans; and results from Eitle and Turner indicated that being African American (race) is a marker for increased risk of stress exposure, later resulting in delinquency coping.[25] In other words, a possible reason that African Americans are more likely to be involved in crime compared to Hispanics and Whites is because they are exposed to a greater level of strain and stress.

As mentioned earlier, researchers who attempt to understand why minorities differentially offend compared to Whites have examined theories at the individual and community levels. For example, Peeples and Loeber examined individual and neighborhood characteristics in relation to juvenile offending and found that, while African American youth were more frequent and serious delinquents, when African American youth did not live in disadvantaged neighborhoods, their delinquent offending was similar to that of Whites.[26] This research shows that when considering racial differences in offending, both micro- and macro-level explanations—especially those that place an emphasis on structural factors (e.g., rates of community inequality)—should be considered.

One of the most utilized macro-level, community, or structural theories that attempts to understand why minorities offend more than Whites is *social disorganization theory*. Social disorganization theories contend that the reason why African Americans disproportionately offend is because they are more likely to reside in disadvantaged communities with little collective efficacy.[27] *Collective efficacy* is a fairly recent concept formulated by Sampson, Raudenbush, and Earls, in which neighborhoods that have collective efficacy are identified by a social cohesion among neighbors that includes a willingness to intervene on behalf of the community.[28] In other words, African Americans are more likely to live in disadvantaged communities with low levels of collective efficacy, and this results in their higher crime rates.

Social disorganization theory has also been applied to Hispanic offending, where research has indicated that Hispanics do not commit homicides at as high levels as African Americans or as low levels of Whites. Hispanics (Latinos) are involved in a "Latino paradox," indicating that even though

Hispanics are more likely to live in disadvantaged communities comparable to African Americans, they have not been found to have higher crime rates. A possible reason why Hispanics have lower rates of homicide compared to African Americans is that they are more socially integrated, especially in the labor market.[29] Overall, social disorganization has been examined more specifically concerning differential offending of African American and Hispanic offending, not just minorities in general compared to Whites.

Shifting the focus away from general criminal theories, subcultural theories are race-specific perspectives aimed at examining crime in regard to racial and ethnic minorities. Certain subcultural perspectives consist of race-related themes, which attempt to explain why minorities differentially offend compared to Whites; however, it could be argued that some components of subcultural theories measure class as a proxy for race. The most popular subcultural theory is Wolfgang and Ferracuti's subculture of violence theory.[30] Wolfgang examined the relationship between race and homicides, specifically the influence of victim-precipitated homicides.[31] He argued that there is a subculture of violence that does not consider assaults as wrong or antisocial, where physical aggression is socially approved, and where violence is considered normal but often ends in death. Later with Ferracuti, the theory articulated seven propositions that encompass the subculture of violence.[32] Some of these propositions include that the subculture is found within all ages of society, however most prominently in late-adolescent and middle age. It does not consider violence as an illicit occurrence and there are no feelings of guilt, but those in the subculture are not violent all of the time. Throughout the 1970s and 1980s support for this theory was mixed, with some studies finding a subculture of violence among African Americans, while others failed to find support where it was argued that poverty was the most consistent predictor of homicide, not race.[33]

Hawkins suggested numerous critiques about the theory, stating that the economic, political, and social disadvantages of African Americans can produce high rates of homicides as well, regardless of subcultural values.[34] These examples can be considered additional reasons why minorities offend more than Whites. He also provided three additions to the theory, first arguing that African American lives are valued less than White lives within criminal law, therefore African Americans can murder each other without a fear of being punished. Second, little attention is paid to violence in African American neighborhoods, which leads to a poor relationship between community members and police. Third, there is an association between structural disadvantage and violence, which until then was not included in the original theory. However, no support has been found for the theory throughout more recent examinations.[35]

Wilson and Bernard formed a more structural/cultural perspective (compared to a purely subcultural theory) that centered on the "truly disadvantaged."[36] This perspective focused on explaining aggression and an "intent to harm" among residents who live in disadvantaged communities, and centered on issues including the urban environment, low social position or class, racial discrimination, and social isolation as explanations for why disadvantaged communities are highly aggressive and violent compared to other types of neighborhoods. Sampson and Wilson extended this argument to propose that residential inequality between African Americans and Whites has lead to social isolation and concentration of the "truly disadvantaged."[37] From these structural barriers, cultural adaptations emerge. In other words, in order to survive, the "truly disadvantaged" adapt and form a subculture. Support for this perspective has been found in the prison setting, where Blacks were more than twice as likely to be found guilty of violent misconduct while incarcerated, compared to Whites.[38]

A more recent structural/cultural theory supported by empirical research that can also be considered race specific is Anderson's "code of the streets" thesis, where residents in disadvantaged communities adopt a "street code" in order to survive. This perspective has found considerable support. For example, disadvantaged neighborhood characteristics, living in a "street" family, and racial discrimination have been found to relate to minorities adopting the "street code."[39] Research suggests that this perspective is applicable to African Americans, Hispanics, and even rap music.[40] The theory has also been applied to victimization in that residents who adopt a "street code" are at a heightened risk for victimization—above and beyond living in disorganized and violent communities.[41]

In sum, there are a number of theoretical reasons why minority youth commit more frequent and serious offenses than Whites throughout the life course. Research results have generally found four factors related to the continued occurrence of minorities as offenders and those processed in the juvenile justice system. First, educational deficits in the form of poor school attendance, lack of emphasis on the value of education, disciplinary problems, and learning and cognitive abilities have been related to more delinquency among minorities. Second, certain types of family dysfunctions have also been shown to be related to involvement in delinquent behavior, especially for minorities who live in single-parent homes, reside in high-poverty neighborhoods, experience parental abuse, and have insufficient supervision. Third, studies have found that many youth referred to juvenile court suffer from mental and emotional disabilities, often lacking the necessary services to address these issues due to family and financial problems. Finally, structural characteristics of juvenile courts may lead to minority overrepresentation. That is, juvenile courts in urban areas are characterized by racial inequality, underclass poverty, high crime rates, and adaptations to values tolerant or conducive to involvement in delinquent behavior that results in a greater likelihood of being arrested and processed through the system.

The following section focuses on **race/ethnic selection bias** as a second explanation to explain minority overrepresentation in arrests and presence in the juvenile justice system. Both differential offending and differential selection/bias help explain the relationship between race and juvenile justice system proceedings.

RACE/ETHNIC SELECTION BIAS

According to the consensus tradition, law, punishment, and treatment derive from a broad consensus of societal norms and values. State intervention into individual liberty and the incarceration of individuals result primarily from the occurrence, distribution, and severity of delinquent behavior. Due to the *parens patriae* foundation of the juvenile court, extralegal factors such as age (younger youth may be viewed differently than older youth), school (conduct problems or not attending), and the family (assessments about the ability of the youths' family to effectively supervise and socialize their child) also enter into decision-making outcomes.[42] Under a consensus perspective, differences between Whites and minorities in case outcomes are attributed to differential involvement in crime (i.e., more crime, more serious crime) and/or problematic school or family situations. Accordingly, when these factors are considered, race/ethnicity will not be a significant predictor of police and court decisions.

A significant number of studies have examined the extent to which legal criteria, race, and other extralegal factors influence juvenile justice decision making.[43] Research has shown that legal factors, such as crime severity, rather than race predict decision making and lend support for a consensus interpretation for differential offending among minorities to explain the overrepresentation of minorities in arrests and the juvenile justice system.[44] Some research has also made claims of non-race effects, even when race is predictive of case outcomes, on the basis that legal factors either were more often statistically significant and/or were stronger in their influence on decision making than race.[45]

According to the race/ethnic selection perspective, laws and procedures as well as legal and extralegal criteria may be racially/ethnically tainted and often indirectly work to the disadvantage of minorities but not Whites. While overt or intentional discrimination may still exist, effects of race/ethnicity on police decisions to arrest and on court case outcomes appear to be more implicit, subtle, or subconscious, but no less harmful than overt bias. For example, assessments made about the youth's family, his or her progress in school, and even the maturity level of the youth (most often captured by the youth's age) may be more likely to negatively impact minority youth than White youth. Prior research indicates that White adolescents, for example, are viewed as more immature, impressionable, and amenable to treatment than African American youth who commit the same crimes.[46] Likewise, legitimate criteria, such as assessments about the family or age, have also been discovered to negatively impact case outcomes for males and African Americans compared to females or Whites, respectively.[47]

Overall, at least seven comprehensive reviews of the existing literature report that legal and extralegal factors alone are unable to account for race differences in involvement in the juvenile justice system and lend support for the base premises of the race/ethnic selection bias perspective. Race/ethnicity still matters.[48] For example, Pope and Feyerherm discovered that roughly two-thirds of the studies conducted during the years 1970 through 1988 found minority youth, primarily African Americans, received more severe outcomes relative to similarly situated White youth.[49] From the pervasiveness of these results and through their own review of more than 150 studies on race and juvenile justice decision making, Bishop and Leiber concluded, "The issue is no longer simply whether Whites and youths of color are treated differently. Instead, the preeminent question for scholars is to explain *how* these differences come about."[50] Next, theories of differential selection bias are presented to provide some insights into why and how differences exist in the social control of Whites and minority youth.

THEORIES OF DIFFERENTIAL SELECTION

The conflict approach has been considered a more traditional theory of differential selection to understand why minorities are overrepresented throughout the life course in the justice system. According to conflict theory, majority groups (i.e., Whites) exercise social control over a powerless group (i.e., minorities) in order to maintain the status quo and protect their own interests. This is done by the majority group's ability to control the law, law enforcement, and court system. This results in minorities becoming labeled as outsiders, deviants, delinquents, and criminals, and therefore subject to harsher punishment in the criminal justice system. Refinements to the conflict perspective and in general the differential selection bias explanation have centered on the contexts or conditions as to when race/ethnicity matters.[51]

Within these approaches minorities are believed to be stereotyped based on negative labels and the inability to acquire necessary resources, which is why they are differentially treated in the court system in comparison to Whites. Juvenile justice decision makers, for example, may possess negative images and stereotypes of minorities (e.g., dangerous, drug users) when deciding culpability, blameworthiness, or treatment possibilities of minority youth.[52] Accordingly, minority youth may not be provided the same opportunities to participate in diversion programs in general, and specifically as an alternative to outcomes involving secure detention or corrections. Adding to this is the lack of bicultural and bilingual staff and the use of English-only information materials that foster further stereotyping and lead to an inability to effectively communicate with non-English-speaking youth and families—resulting in youth and their family being unable to successfully navigate the juvenile justice system. Last, laws and policies may also work to the disadvantage of minority youth more so than White youth. Examples of these include zero-tolerance and transfer/waiver adult certification of youths.

Macro-Level Theories

As mentioned earlier, some theories have taken into consideration contextual and structural factors of youth (e.g., social disorganization theory) as to why minority youth commit more frequent and serious crime compared to Whites. Macro-level theories have also been constructed to understand the salience of racial and ethnic biases when examining decision-making outcomes within the juvenile justice system.[53] Community- and neighborhood-level theories expand beyond conflict and labeling perspectives to take into account structural characteristics that may also influence decision making. For example, levels of impoverishment, minority representation, crime rates, and political orientation are some of the factors that research has found to influence decision making within the juvenile and adult systems.

A historical example of a macro-level theory of differential selection is Blalock's minority group power threat thesis, which argues that as the minority group population grows within a community, the White majority population becomes threatened.[54] This perceived threat comes in the form of competition for economic resources (e.g., jobs, money, property), and increases in income and wealth by the minority group. From this, prejudicial and discriminatory attitudes and practices are employed by Whites in order to diffuse this threat, which results in increased social control and more harsh outcomes for minorities in the juvenile and adult court systems. However, results have not always been supported regarding the accuracy of the power threat thesis when applied to the juvenile justice system. Some research has found prejudicial and harsh police arresting practices in regard to minority youth within communities with a higher proportion of minorities, yet these disparities were corrected in later stages of court proceedings.[55]

A more recent contextual and structural theory is Sampson and Laub's macro-level perspective, which argues that the poor, underclass, and minorities are perceived by juvenile justice decision makers as threatening and in need of increased social control.[56] This is due to the amount of minorities who reside in communities with high levels of economic and racial inequality. Sampson and Laub propose that juvenile justice decision makers hold beliefs that minority youth are aggressive, sexual, lack discipline, and suggest that they pose symbolic threats to public safety and middle-class standards. Also important to this perspective is the interrelationship between the War on Drugs of the 1980s, stereotyping, and prevalence of disadvantaged communities that lead to the typecasting of minority males as drug users and dealers. Sampson and Laub argued that as a result of these three situations, poor Black men (especially those who were involved with drugs) would be subject to increased social control by the juvenile justice system.

Sampson and Laub found support for their theory when examining more than 200 counties across the United States. Measures of underclass poverty and racial inequality were significantly related to harsher treatment of juvenile offenders, especially during the stages of predisposition detention, and judicial disposition.[57] Rodriguez also indicated that racial and ethnic biases occurred throughout numerous juvenile court stages.[58] Black and Hispanic youth were more likely to be detained than White youth, while youth in general who lived in disadvantaged communities were more likely to be detained. They were also indirectly treated more severely in later stages of juvenile justice proceedings. In other words, decisions made at earlier stages (e.g., detention) affect outcomes at later stages (especially judicial disposition), which can result in a cumulative disadvantage for Black and Hispanic youth compared to Whites.

Micro-Level Theories

Micro-level theories in comparison to macro-level perspectives focus more on the individual and social psychological characteristics of juvenile justice decision makers. For example, the roles of stereotypes and discretion concerning court outcomes for minority youth play an important part in decision making. An example of this is Tittle and Curran's symbolic threat thesis, which argues that juvenile justice decision makers may view minority youth as aggressive, sexual, and lacking discipline, which makes juvenile court officers feel uneasy, uncomfortable, and unable to identify with the youth.[59] This leads to greater social control and harsher sanctioning of minority youth. These examples of symbolic threats are connected to structural characteristics, minority threat, inequality, and public perceptions of gang involvement.[60] While this threat is more symbolic (threat to middle-class standards and public safety) than real, it can still impact how minority youth are perceived within juvenile justice proceedings. Tittle and Curran found that a larger minority and youth population lead to more harsh case outcomes for minority youth who were charged with drug and/or sexual offenses. More recent research also supports this finding: Minority youth who were drug offenders were differentially treated compared to other youth and were more likely to be referred farther into the juvenile court system.[61]

An additional micro-level perspective to understand why minority youth are overrepresented as offenders in the juvenile justice system is shown in the research by Bridges and Steen.[62] They incorporated attribution theory and prior research on racial discrimination to argue that decision makers develop and rely on past experiences and prior cases to influence case outcomes. They found that probation officers used different types of attributions to assess delinquent behavior between Black and White youth. Delinquent acts committed by African American youth were seen by decision makers as internal or dispositional attributions (e.g., defects of character, lack of responsibility), while delinquency committed by White youth was related to external causes (e.g., disadvantaged neighborhoods, delinquent peers, poor school performance). In turn, decision makers perceived Black youth to be more at risk to re-offend, less amenable to treatment, and more dangerous. This resulted in longer sentences for Black youth than for White youth. In sum, Bridges and Steen argued that certain values and beliefs of juvenile probation officers lead to a stereotyped image of juvenile offenders that resulted in differential treatment.[63] Also supporting this theory, Gaarder, Rodriguez, and Zatz found that attributions related to delinquency and victimization assigned to Hispanic females by juvenile justice decision makers were linked to racial and gender stereotypes.[64] In another study, Hispanic females were treated more severely than their White counterparts, whether they were from economically advantaged and disadvantaged communities.[65]

Concerning the use of discretion by juvenile justice decision makers, the *liberation hypothesis* identifies certain conditions that relate to decision makers' ability to "liberate" certain legal criteria and uniformity when deciding court outcomes. In research that examined the adult court system, jurors were found to be more likely to exercise discretion when evidence against the defendant was weak and a less serious crime occurred. Some studies found that judges were less likely to exercise discretion involving cases of murder, robbery, or rape, but in less serious cases, discretion was enhanced. Blacks who were convicted of less serious crimes were treated harsher at sentencing compared to Whites, while Blacks who were convicted of more serious crimes were no more likely than Whites to receive a harsher sentence.

The liberation hypothesis has also been examined within the juvenile justice system, which suggests that in less serious cases, court outcomes will be influenced by race. This is due to the ability of court officers to make decisions based on their own discretion. Research has found some support for the liberation hypothesis at the juvenile level; however, this depends on which stages are examined.[66] Overall, it seems that there is some support for the applicability of the liberation hypothesis throughout the life course of minorities' involvement within the juvenile and adult court systems.

The focal concerns perspective has also been applied to why minorities are disproportionately involved in the juvenile and adult justice system. Steffensmeier argues that the underlying premise of this perspective is that judges have a limited amount of time and information about defendants, so they may rely on three focal concerns or attributions involving race, gender, and class stereotypes when making decisions: (1) the defendants' blameworthiness and culpability, (2) society's concern to protect the community, and (3) organizational considerations involving available correctional resources.[67] While the three focal concerns are interrelated, it is argued that judges do not have complete information about each component and end up exercising their own discretion by relying on the offender's age, race, and social class to make decisions. In other words, judges develop a "perceptual shorthand," where they rely on both legal factors (crime severity, prior record) and racial stereotypes to determine case outcomes.

The focal concerns approach has been modified in order to be applicable to the juvenile justice system. Compared to the adult system, the juvenile court system has a dual focus of social control and a concern for social welfare of all youth, with an emphasis on treatment and rehabilitation. However, juvenile court officers may rely on stereotypes involving a youth's family, peers, and school situations. For example, juvenile court officers may believe the stereotype that single-parent families are unable to adequately provide for their children, which has been shown to influence decision making.[68]

In an important development to this approach, Bishop and colleagues integrated the focal concerns perspective with the "loose coupling" perspective.[69] They argued that race effects may vary by each stage in juvenile justice proceedings due to numerous decision makers, responsibilities, and concerns. They also propose that the influence of race is more likely to exist at the loosely coupled stages of intake and judicial disposition where there are multiple actors and multiple goals. These goals are often tied to racial stereotypes since there is more opportunity to utilize discretion. At the tightly coupled stages of petition and adjudication (where there are fewer decision makers involved), legal factors and the severity of the crime are more likely to influence decision making. In these two stages, discretion is constrained, and race effects are predicted to have less of a role in decision making. Consistent with the focal concerns and loose coupling perspective, results supported their arguments.

POLICY IMPLICATIONS

In 1989, the **disproportionate minority confinement/contact** (DMC) mandate was passed by Congress as part of the reauthorization of the Juvenile Justice and Delinquency Prevention Act (JJDP Act) of 1974. In 2002, the JJDP Act was modified, shifting the emphasis from "disproportionate minority confinement" to "disproportionate minority contact," requiring the examination of possible minority youth overrepresentation throughout all decision points in the juvenile justice system. All states receive funds from the federal government in an attempt to comply with the JJDP Act, including DMC. The underlying goal of the DMC mandate is the equitable treatment of all youth within the juvenile justice system.[70]

In addition to the federal mandate, multiple-prong strategies and interventions are needed that address both delinquent offending and system issues for a reduction of minority youth overrepresentation in the juvenile justice system. Interventions that have been used to address minority youth overrepresentation include direct services, training and education, and system change.[71] Direct services center on prevention and intervention programs that address the various needs of at-risk youth, including skill development, educational attainment, and positive relationships with family and peers. Beyond the focus on prevention and intervention programs, direct services also include diversion programs/services, advocacy, and alternatives to secure detention. Another intervention strategy is cultural diversity training that exposes and educates individuals about racial/ethnic biases in the juvenile justice system. The training of law enforcement and juvenile court personnel has centered on the reduction of racial/ethnic stereotypes and how such biases influence the processing of youth within the juvenile justice system.[72]

Significant system change has been slow to occur. When system change has occurred, it has come in the form of legislative reform, administrative changes for diversification, and structural and procedural changes that affect decision making.[73] The most popular effort has focused on changing detention procedures with an emphasis on the use of alternatives to secure detention. Spearheading this effort has been the Annie E. Casey Foundation's Juvenile Detention Alternatives Initiative (JDAI). This initiative seeks to improve detention systems and the overall efficiency of the juvenile justice system (i.e., reduce the number of youth in secure detention, improve public safety).

CONCLUSION

Minority youth overrepresentation in arrests and presence in the juvenile justice system is a significant issue. While the differential offending perspective and the race/ethnic selection bias perspective have been used to explain this occurrence, these perspectives have all too often have been pitted against one another. Research has shown that both can explain the disproportionate presence in arrests and the juvenile justice system. These findings have led some to argue that there is a need to more adequately address or link the factors associated with differential offending (i.e., inequality, subcultural values as adaptations to structural inequities) with the contexts of race and social control in terms of police contact and processing within juvenile court.[74]

Thus, theory and policy efforts need to focus on the conditions that contribute to involvement in crime and foster the reproduction by justice officials of cultural stereotypes of minority offenders, their families, and the neighborhoods in which they live when deciding what to do with youth. This point is underscored by the fact that unnecessary intervention may have implications that further disadvantage youth and "may become an endogenous form of inequality that is difficult to escape."[75] Having a prior record and being incarcerated at a young age, even for a relatively short period of time, may increase the risk of recidivism, future incarceration, and diminish future life chances in terms of the pursuit of education, employment, and marriage.[76]

GLOSSARY

Differential offending—minorities commit more crime and/or more serious crime

Disproportionate minority confinement/contact—also known as DMC and is a federal mandate that is part of the Juvenile Justice and Juvenile Delinquency Prevention Act

Race/ethnic selection bias—policies, procedures, and practices result in minorities receiving more disadvantaged outcomes than Whites

NOTES

1. Leiber, M. (2008). Theories of racial and ethnic bias. In M. Lynch, E. B. Patterson, & K. Childs (Eds.), *Racial divide: Race, ethnicity and criminal justice theories of racial and ethnic bias* (pp. 15–38). Monsey, NY: Criminal Justice Press; Leiber, M. (2002). Disproportionate minority confinement (DMC) of youth: An analysis of state and federal efforts to address the issue. *Crime & Delinquency, 48*, 3–45; Sampson, R. J., & Lauritsen, J. L. (1997). Racial and ethnic disparities in crime and criminal justice in the United States. *Crime and Justice, 21*, 311–374.

2. Leiber, M., Bishop, D., & Chamlin, M. (2011). Juvenile justice decision-making before and after the implementation of the disproportionate minority contact (DMC) mandate. *Justice Quarterly, 28*(3), 460–492; Pope, C. E., Lovell, R., & Hsia, H. (2002). *Disproportionate minority confinement: A review of the research literature from 1989 to 2002.* Washington, DC: U.S. Department of Justice, Office of Juvenile Justice and Delinquency Prevention; Rodriguez, N. (2010). The cumulative effect of race and ethnicity in juvenile court outcomes and why preadjudication detention matters. *Journal of Research in Crime & Delinquency, 47*(3), 391–413.

3. Snyder, H. (2012). Juvenile delinquents and juvenile justice clientele: Trends and patterns in crime and justice system responses. In D. Bishop & B. Feld (Eds.), *Juvenile Justice* (pp. 1–30). Cambridge, UK: Oxford University Press.

4. Puzzanchera, C., & Adams, B. (2011). *National disproportionate minority contact Databook.* Developed by the National Center for Juvenile Justice for the Office of Juvenile Justice and Delinquency Prevention. Retrieved from www.ojjdp.gov/ojstatbb/dmcdb/index.html.

5. Bishop, D., & Leiber, M. (2012). Race, ethnicity, and juvenile justice: Racial and ethnic differences in delinquency and justice system responses. In D. Bishop & B. Felds (Eds.), *Juvenile Justice* (pp. 445–484). Oxford University Press; Puzzanchera, & Adams (2011), see Note 4.

6. Snyder (2012), see Note 3.

7. Huizinga, D., Thornberry, T., Knight, K., & Lovegrove, P (2007). *Disproportionate minority contact in the juvenile justice system: A study of differential minority arrest/referral to court in three cities.* Washington, DC: U.S. Department of Justice; Sampson, R. (1986). Effects of socioeconomic context on official reaction to juvenile delinquency. *America Sociological Review, 51*, 876–885; Terrill, W., & Reisig, M. (2003). Neighborhood context and police use of force. *Journal of Research in Crime and Delinquency, 40*, 291–321.

8. Hindelang, M. (1978). Race and involvement in common personal crimes. *American Sociological Review, 43*, 93–109.

9. Hindelang (1978), see Note 8.

10. Sampson, & Lauritsen (1997), see Note 1.

11. Pope, C., & Snyder, H. (2003). Race as a factor in juvenile arrests. *Juvenile Justice Bulletin.* Washington, DC: U.S. Department of Justice, Office of Justice Programs, Office of Juvenile Justice and Delinquency Prevention.

12. Pope, & Snyder (2003), see Note 11.

13. Krohn, M., Thornberry, T., Gibson, C., & Baldwin, J. (2010). The development and impact of self-report measures of crime and delinquency. *Journal of Quantitative Criminology, 26*, 509–525.

14. Farrington, D. P., Loeber, R., Stouthamer-Loeber, M., Van Kammen, W. B., & Schmidt, L. (1996). Self-reported delinquency and a combined delinquency seriousness scale based on boys, mothers, and teachers: Concurrent and

predictive validity for African-American and Caucasians. *Criminology, 34,* 493–517; Hindelang M., Hirschi, T., & Weis, G. (1981). *Measuring delinquency.* Beverly Hills, CA: Sage Publications; Huizinga, D., & Elliott, D. S. (1986). Reassessing the reliability and validity of self-report delinquent measures. *Journal of Quantitative Criminology, 2,* 293–327; Thornberry, T. P., & Krohn, M. D. (2000). The self-report method for measuring delinquency and crime. In D. Duffee, R. Crutchfield, S. Mastrofski, L. Mazerolle, & D. McDowall (Eds.), *Criminal justice 2000, Vol. 4: Measurement and analysis of crime and justice.* Washington, DC: National Institute of Justice, pp. 33–83.

15. Elliott, D. (1994). Serious violent offenders: Onset, developmental course, and termination: The American Society of Criminology 1993 presidential address. *Criminology, 32*(1), 1–21.

16. Bishop, & Leiber (2012), see Note 5; Huizinga et al. (2007), see Note 7.

17. Bishop, & Leiber (2012), see Note 5; Snyder (2012), see Note 3.

18. Sampson, & Lauritsen (1997), see Note 1.

19. Hawkins, D. F. (1995). *Ethnicity, race, and crime: Perspectives across time and place.* Albany, NY: State University of New York Press; Leiber, M. J., Mack, K. Y., & Featherstone, R. A. (2009). Family structure, family processes, economic factors, and delinquency: Similarities and differences by race and ethnicity. *Youth Violence and Juvenile Justice, 7*(2), 79–99.

20. Gottfredson, M. R., & Hirschi, T. (1990). *A general theory of crime.* Stanford, CA: Stanford University Press.

21. Higgins, G. E., & Ricketts, M. L. (2008). Self-control theory, race, and delinquency. *Journal of Ethnicity in Criminal Justice, 3*(3), 5–22; Leiber, M. J., Brubaker, S. J. & Fox, K. C. (2009). A closer look at the individual and joint effects of gender and race on juvenile court decision making. *Feminist Criminology, 4*(4), 333–358; Pratt, T. C., Turner, M. G., & Piquero, A. R. (2004). Parental socialization and context: A longitudinal analysis of the structural sources of low-self control. *Journal of Research in Crime and Delinquency, 41,* 219–243.

22. Moffitt, T. E. (1993). Adolescence-limited and life-course-persistent antisocial behavior: A developmental taxonomy. *Psychological Review, 100*(4), 674–701.

23. Haynie, D. L., Weiss, H. E., & Piquero, A. (2008). Race, the economic maturity gap, and criminal offending in young adulthood. *Justice Quarterly, 25*(4), 595–622.

24. Kaufman, J. M., Rebellon, C. J., Thaxton, S., & Agnew, R. (2008). A general strain theory of racial differences in criminal offending. *The Australian and New Zealand Journal of Criminology, 41*(3), 421–437.

25. Eitle, D., & Turner, R. J. (2003). Stress exposure, race, and young adult male crime. *The Sociological Quarterly, 44*(2), 243–269; Jang, S. J., & Johnson, B. R. (2003). Strain, negative emotions, and deviant coping among African Americans: A test of general strain theory. *Journal of Quantitative Criminology, 19*(1), 79–105; Simons, R. L., Chen, Y. F., Stewart, E. A., & Brody, G. H. (2003). Incidents of discrimination and risk for delinquency: A longitudinal test of strain theory with an African American sample. *Justice Quarterly, 20*(4), 827–854.

26. Peeples, F., & Loeber, R. (1994). Do individual factors and neighborhood context explain ethnic differences in juvenile delinquency? *Journal of Quantitative Criminology, 10*(2), 141.

27. Unnever, J. D., & Gabbidon, S. L. (2011). *A theory of African American offending: Race, racism, and crime.* New York: Routledge.

28. Sampson, R. J., Raudenbush, S. W., & Earls, F. (1997). Neighborhoods and violent crime: A multilevel study of collective efficacy. *Science, 5328,* 918–924.

29. Martinez, R. (2002). *Latino homicide: Immigration, violence, and community.* New York: Routledge.

30. Wolfgang, M. E., & Ferracuti, F. (1967). *The subculture of violence. Towards an integrated through in criminology.* London: Tavistock Publications.

31. Wolfgang, M. E. (1958). *Patterns in criminal homicide.* Philadelphia, PA: University of Pennsylvania Press.

32. Wolfgang, & Ferracuti (1967), see Note 30.

33. Curtis, L. A. (1975). *Violence, race, and culture.* Lexington, MA: Heath; Messner, S. F. (1983). Regional and racial effects on the urban homicide rate: The subculture of violence revisited. *American Journal of Sociology, 88,* 997–1007; Parker, R. N. (1989). Poverty, subculture of violence, and type of homicide. *Social Forces, 67,* 983–1007; Silberman, C. E. (1978). *Criminal violence, criminal justice.* New York: Random House.

34. Hawkins, D. F. (1983). Black and white homicide differentials: Alternatives to an inadequate theory. *Criminal Justice and Behavior, 10,* 407–440.

35. Cheatwood, D. (1990). Black homicides in Baltimore 1974–1986: Age, gender, and weapon use changes. *Criminal Justice Review, 15,* 192–207.

36. Bernard, T. J. (1984). Control criticisms of strain theories: An assessment of theoretical and empirical adequacy. *Journal of Research in Crime and Delinquency, 21,* 353–372; Wilson, W. J. (1987). *The truly disadvantaged.* Chicago: University of Chicago Press.

37. Sampson, R. J., & Wilson, W. J. (1995). Toward a theory of race, crime, and urban inequality. In J. Hagan & R. D. Peterson (Eds.), *Crime and inequality* (pp. 37–54). Stanford, CA: Stanford University Press.

38. Harer, M. D., & Steffensmeier, D. (1996). Race and prison violence. *Criminology, 34,* 323–355.

39. Stewart, E. A., & Simons, R. L. (2003). Structure and culture in African American adolescent violence: A partial test of "the code of the street" thesis. *Justice Quarterly, 23*(1), 1–33.

40. Baumer, E., Horney, J., Felson, R., & Lauristen, J. (2003). Neighborhood disadvantage and the nature of violence. *Criminology, 41,* 39–71; Kubrin, C. E. (2005). Gangstas, thugs, and hustlas: Identity and the code of the streets in rap music. *Social Problems, 52,* 360–378; Lopez, R., Roose, M. W., Tein, J. T., & Dinh, K. T. (2004). Accounting for Anglo-Hispanic differences in school misbehavior. *Journal of Ethnicity in Criminal Justice, 2,* 27–46.

41. Stewart, E. A., Schreck, C. J., & Simons, R. L. (2006). "I ain't gonna let no one disrespect me": Does the code of the street reduce or increase violent victimization among African American adolescents? *Journal of Research in Crime and Delinquency, 43*(4), 427–458.

42. Rodriguez, N., Smith, H., & Zatz, M. (2009). Youth is enmeshed in a highly dysfunctional family system: Exploring the relationship among dysfunctional families, parental incarceration, and juvenile court decision making. *Criminology, 47,* 177–208.

43. Bishop, D., Leiber, M., & Johnson, J. (2010). Contexts of decision making in the juvenile justice system: An organizational approach to understanding minority overrepresentation. *Journal of Youth Violence & Juvenile Justice, 8*(3), 213–233; Freiburger, T. L., & Burke, A. S. (2010). Adjudication decisions for black, white, hispanic, and native american youth in juvenile court. *Journal of Ethnicity in Criminal Justice, 8*(4), 231–247; Leiber, M., & Johnson, J. (2008). Being young and Black: What are their effects on juvenile justice decision making? *Crime & Delinquency, 54,* 560–581; Rodriguez (2010), see Note 2.

44. Pope, & Snyder (2003), see Note 11.

45. Harris, R., Huenke, C., Rodriguez-Labarca, J., & O'Connell, J. (1998). *Disproportionate representation of minority juveniles at arrest: An examination of 1994 charging patterns by race.* Dover, DE: Delaware Statistical Analysis Center.

46. Bridges, G., & Steen, S. (1998). Racial disparities in official assessments of juvenile offenders: Attribution stereotypes as mediating mechanisms. *American Sociological Review, 63,* 554–570; Graham, S., & Lowery, B. S. (2004). Priming unconscious racial stereotypes about adolescent offenders. *Law and Human Behavior, 28*(5), 483–504.

47. Leiber et al. (2009), see Note 19; Leiber, & Johnson (2008), see Note 43; Rodriguez et al. (2009), see Note 42.

48. Bishop, D. M. (2005). The role of race and ethnicity in juvenile justice processing. In D. Bishop & B. Felds (Eds.), *Our children, their children: Confronting racial and ethnic differences in American juvenile justice* (pp. 23–82). Chicago: The John D. and Catherine T. MacArthur Foundation, Research Network on Adolescent Development and Juvenile Justice, University of Chicago Press; Bishop & Leiber (2012), see Note 5; Engen, R., Steen, S., & Bridges, G. (2002). Racial disparities in the punishment of youth: A theoretical and empirical assessment of the literature. *Social Problems, 49*(2), 194–220; Leiber (2002), see Note 1; Pope, C., & Feyerherm, W. (1992). *Minorities and the juvenile justice system: Full report.* Washington, DC: Office of Juvenile Justice and Delinquency Prevention; Pope et al. (2002), see Note 2; Pope, C., & Leiber, M. (2005). Disproportionate minority contact (DMC): The federal initiative. In D. Bishop & B. Felds (Eds.), *Our children, their children: Confronting racial and ethnic differences in American juvenile justice* (pp. 351–389). Chicago: The John D. and Catherine T. MacArthur Foundation, Research Network on Adolescent Development and Juvenile Justice, University of Chicago Press.

49. Pope, & Feyerherm (1992), see Note 48.

50. Bishop, & Leiber (2012), see Note 5.

51. Leiber, M. (2003). *The contexts of juvenile justice decision making: When race matters*. Albany, NY: State University of New York Press; Sampson, R. J., & Laub, J. (1993). *Crime in the making*. Cambridge, MA: Harvard University Press.

52. Graham, & Lowery (2004), see Note 46.

53. Leiber (2003), see Note 51; Rodriguez (2010), see Note 2; Sampson, & Laub (1993), see Note 51.

54. Blalock, H. M. (1967). *Toward a theory of minority group relations*. New York: Wiley.

55. Bishop, Leiber, & Johnson (2010), see Note 43.

56. Sampson, & Laub (1993), see Note 51.

57. Sampson, & Laub (1993), see Note 51.

58. Rodriguez (2010), see Note 2.

59. Tittle, C. R., & Curran, D. A. (1988). Contingencies for dispositional disparities in juvenile justice. *Social Forces*, *67*(1), 23–58.

60. Fagan, J. (2010). The contradictions of juvenile crime & punishment. *Dædalus, 139*(3), 43–61.

61. Bishop, Leiber, & Johnson (2010), see Note 43; Leiber, & Johnson (2008), see Note 43.

62. Bridges, & Steen (1998), see Note 46.

63. Bridges, & Steen (1998), see Note 46.

64. Gaarder, E., Rodriguez, N., & Zatz, M. (2004). Criers, liars, manipulators: Probation officers' views of girls. *Justice Quarterly, 21*(3), 547–578.

65. Rodriguez, N. (2007). Juvenile court context and detention decisions: Reconsidering the role of race, ethnicity, and community characteristics in juvenile court processes. *Justice Quarterly, 24*, 629–656.

66. Guevara, L., Boyd, L. M., Taylor, A. P., & Brown, R. A. (2011). Racial disparities in juvenile court outcomes: A test of the liberation hypothesis. *Journal of Ethnicity in Criminal Justice, 9*(3), 200–217.

67. Steffensmeier, D., Ulmer, J., & Kramer, J. (1998). The interaction of race, gender, and age in criminal sentencing: The punishment cost of being young, black, and male. *Criminology, 36*, 763–797.

68. Bishop, D., & Frazier, C. (1996). Race effects in juvenile justice decision-making: Findings of a statewide analysis. *Journal of Criminal Law & Criminology, 86*, 392–414; Pope, & Leiber (2005), see Note 48.

69. Bishop, Leiber, & Johnson (2010), see Note 43.

70. Leiber, M., & Rodriguez, N. (2011). The implementation of the disproportionate minority confinement/contact (DMC) mandate: A failure or success? *Race and Justice, 1*(1), 103–124.

71. Leonard-Kempf, K. (2007). Minority youths and juvenile justice: Disproportionate minority contact after nearly 20 years of reform efforts. *Youth Violence and Juvenile Justice, 5*, 71.

72. Leiber, & Rodriguez (2011), see Note 70.

73. Leiber, & Rodriguez (2011), see Note 70; Pope, & Leiber (2005), see Note 48.

74. Bishop, & Leiber (2012), see Note 5; Leiber, & Rodriguez (2011), see Note 70; Leonard-Kempf (2007), see Note 71; Piquero, A. (2008). Disproportionate minority contact. *The Future of Children, 18*(2), 59–79.

75. Fagan (2010), see Note 60.

76. Fagan (2010), see Note 60; Laub, J., & Sampson, R. (2003). *Shared beginnings, divergent lives: Delinquent boys to age 70*. Cambridge, MA: Harvard University Press; Pager, D. (2007). *Marked: Race, crime, and finding work in an era of mass incarceration*. Chicago: University of Chicago Press.

CHAPTER TEN

Media Violence and the Development of Aggressive Behavior

Edward L. Swing and Craig A. Anderson

KEY TERMS

Aggression
Cross-sectional correlational studies
Experimental studies
Longitudinal studies
Media violence
Violence

INTRODUCTION

Americans' history of exposure to violent media has spanned more than a century, from the film *The Great Train Robbery* to the television show *Dragnet* to the video game *Far Cry 2*. Through films, television shows, music, and video games, people in all modern societies consume violent entertainment media on a regular basis. There is also a long history of concern that exposure to violent media might have negative consequences for viewers. In 1954, Senator Estes Kefauver's judiciary subcommittee investigated the role of violent television shows in juvenile delinquency and issued a warning about such shows. In 1969, the National Commission on the Causes and Prevention of Violence identified television violence as a contributor to societal violence. A 1972 report by Surgeon General Jesse Steinfeld recognized an overwhelming consensus among the Scientific Advisory Committee members that screen violence (films and television) contributed to aggressive behavior.[1] A more recent Surgeon General's report again identified television violence as an important risk factor for youth violence and discussed similar findings for violent music and video games.[2] A panel brought together at the request of the Surgeon General concluded that "Research on violent television and films, video games, and music reveals unequivocal evidence that **media violence** increases the likelihood of aggressive and violent behavior in both immediate and long-term contexts."[3] A number of scientific and health organizations have reached similar conclusions. In 2000, six organizations (the American Academy of Child and Adolescent Psychiatry, American Academy of Family Physicians, American Academy of Pediatrics, American Medical Association, American Psychiatric Association, and American Psychological Association) issued a joint statement that the research evidence "point[s] overwhelmingly to a causal connection between media violence and aggressive behavior in some children."[4] The conclusions of these scientific organizations and governmental agencies justify the examination of media violence as a causal risk factor for antisocial behavior.

AGGRESSION AND ANTISOCIAL BEHAVIOR

This chapter focuses primarily on aggression and violent behavior. Within the social psychological research tradition in which most of the discussed research and theories are based, **aggression** usually is defined as behavior that is primarily intended to cause harm and that is carried out with the expectation of causing harm to an individual who is motivated to avoid that harm. This means that trying to shoot someone with a gun but missing still constitutes aggression. Breaking an inanimate object does not constitute aggression unless the intent in breaking the object was to cause harm to the owner of that object. Neither behavior that unintentionally results in harm (e.g., a car accident) nor behavior for which harm is an unintentional byproduct (e.g., a dental operation) constitutes aggression.

Similarly, playing soccer assertively and with confidence is not aggression, even though many coaches and athletes use the term to describe such assertive and confident play; the primary goal usually is to win the match within the rules of the game, not to harm the opposing players. Unfortunately, some coaches and players do occasionally intend to physically harm opposing players; in such cases the behavior crosses a fine line and becomes aggressive.

Aggression is defined broadly enough to include verbal aggression (e.g., insulting someone, thereby causing emotional harm) and relational aggression (e.g., spreading rumors about someone, thereby harming their social relationships). Nevertheless, most aggression research on the effects of media violence has focused on physical aggression (e.g., pinching, hitting, slapping, pushing, punching, choking). **Violence** refers to extreme forms of physical aggression, such as assault and murder. All violence is aggression, but not all aggression is violence. Neither definition makes distinctions based on the criminality of the act. Physical punishment of children may be legal, but still constitutes aggression. Even acts of violence may be legal, such as a justifiable shooting by a police officer. Nonetheless, most of the violent behaviors studied in the context of media violence are at least potentially criminal.

Other research has focused on variables believed to mediate the effects of media violence on aggression—variables such as aggressive beliefs and thoughts, aggression-related emotions (i.e., anger), physiological arousal, and desensitization to violence. That is, media violence has its effects on later aggressive behavior by changing one of these cognitive, affective, or physiological mediating variables. These variables are discussed where relevant; however, the term *aggression* refers only to aggressive behavior.

Antisocial behavior is broader than aggression. Certain antisocial behaviors, such as lying, cheating, stealing, or destroying property, do not always fit our definition of aggression. For example, a person may steal without the primary goal of intending to harm another individual, in which case it would not be considered aggression. Such behaviors have not been thoroughly studied as a potential outcome of media exposure. Nonetheless, the theoretical framework and mechanisms discussed in this chapter may be helpful in predicting and explaining these behaviors as well.

This chapter begins with a review of past media violence studies, including studies of violent television shows, films, music, and video games, along with evidence that exposure to such media can increase aggression. It also summarizes some studies that use various research methods and populations. The general aggression model (GAM) is presented to provide a theoretical framework for understanding how media violence can lead to specific episodes of aggressive behavior, as well as how these episodes create long-term individual changes in a person's general aggressiveness. The possible role of executive control as a mechanism in media violence effects on aggression is also discussed.

PAST MEDIA VIOLENCE EFFECTS RESEARCH

Research on the effects of media violence can be divided into three broad types of research designs: experimental studies (laboratory or field), cross-sectional correlational studies, and longitudinal studies. Each design has strengths that can complement the weaknesses of the other designs. **Experimental studies** provide the strongest evidence of causality (i.e., media violence causes an increase in aggression). Because participants in experimental studies are randomly assigned to different conditions (e.g., viewing a violent film or viewing a nonviolent film), these groups tend to be equivalent on any preexisting individual difference variables. This allows researchers to eliminate almost all alternative explanations for the results, which greatly strengthens the causal argument. The primary limitation of experimental research designs is that, for ethical reasons, researchers cannot study the most serious forms of aggression (i.e., violence) in these studies.

Cross-sectional correlational studies allow researchers to measure participants' past behavior, including severe forms of aggression (violence) that cannot be used in experimental studies. A well designed cross-sectional study also can test, and therefore eliminate, some alternative explanations (alternatives to the causal hypothesis under investigation), for example, by including covariates in analyses of the relationship of interest. Nonetheless, researchers cannot anticipate and address all potential alternative explanations through a cross-sectional study, so this research is weaker in terms of drawing causal conclusions.

Longitudinal studies allow researchers to examine how a causal candidate variable measured at one point in time predicts changes in an individual's aggression measured at a later time, controlling for level of aggression at the earlier time. A well designed longitudinal study allows stronger causal conclusions than a cross-sectional study because it controls for all alternative explanatory variables that impinged upon the earlier aggression measure. Furthermore, the longitudinal design has the potential to examine serious forms of aggression. The primary limitations of longitudinal studies are that they are expensive and time consuming, limiting the extent to which they are used.

Each research design has its place in the study of the effects of media violence on aggressive behavior, and strong causal conclusions depend on consistent results across each of these designs. This consistency across designs provides a compelling case that the causal effects observed among milder forms of aggression are also true for more severe forms of aggression. Because hundreds of studies have been conducted on media violence effects, we cannot here report a comprehensive review. Instead, the following sections provide representative examples of each of the different types of research designs for each type of media.

Film and Television

Violent television and film effects have been studied the longest and, consequently, constitute the largest area within the media violence research literature. More studies have been conducted on such "passive" screen media than on all other types of media violence combined.

Experimental Studies

An experimental study of 5- and 6-year-old children in Finland randomly assigned half of these children to watch a violent film and the other half to watch a nonviolent film.[5] The children were then observed by raters as they played together in another room. These raters were unaware of which type of film the children had viewed. Those children who had viewed the violent film were more likely to engage in physical aggression (e.g., hitting other children, wrestling) than those who viewed the nonviolent film. In another study, 7- to 9-year-old boys were randomly assigned to watch a violent film or a nonviolent film.[6] Next, the boys played floor hockey. Half of the boys in each film condition were assigned to have hockey referees who carried walkie-talkies (half had no walkie-talkies). This was intended to serve as a cue to the violent film's content, as the violent film also depicted walkie-talkies. The children's behavior as they played floor hockey was rated for aggression (e.g., hitting, tripping, elbowing, kneeing, pulling hair, and insulting) by observers unaware of the type of film the boys had watched. Boys who had initially tested as above average in aggressiveness, watched the violent film, and had the walkie-talkies present behaved more aggressively in the hockey game than boys in any other condition of the study.

Experiments also demonstrated film violence effects among teenagers. In one study, a treatment center for delinquent boys in Belgium was randomly divided into three cottages.[7] The boys in one cottage viewed violent films every night for 5 straight nights, whereas the boys in the other two cottages viewed nonviolent films every night for the same period of time. During this time observers coded the boys' behavior for hitting, slapping, choking, or kicking the other boys. The boys who viewed violent films engaged in more physical aggression. A similar study among American juvenile offenders found that boys assigned to watch violent films engaged in more physical and verbal aggression than those who viewed nonviolent films.[8]

Experiments have demonstrated a violent film effect on aggression among individuals of college age as well. In one study, college men were randomly assigned to view either a film of a boxing match (violent) or a nonviolent track meet.[9] The participants then had an opportunity to deliver painful noise blasts to another student who had previously provoked them. Those who had viewed the violent film delivered more intense noise blasts than those who had viewed the nonviolent film. Another study demonstrated that combining violent and sexual content was especially effective in producing aggression.[10] In this study, male university students were randomly assigned to view a film containing sexual and violent content, a nonviolent film containing sexual content, or a nonviolent, nonsexual film. They were then able to deliver electric shocks to a woman who had provoked them. Those who had viewed the violent, sexual film delivered stronger shocks than those in either of the other two film conditions.

These studies provide examples of some of the experimental designs used to examine television and film violence effects. As with any area of scientific research, the results of experimental studies of television and film violence vary somewhat from study to study. Meta-analytic reviews provide researchers with an effective way to determine whether an association exists and the strength of that association based on all available studies. One of the most comprehensive meta-analyses of television and film violence effects on aggression found a moderate to strong overall effect, although this effect was smaller, but still significant, when only criminal violence was assessed.[11] In other words, the available experimental research demonstrates that exposure to violent television and films increases aggression.

Cross-Sectional Studies

Violent television and film effects were also examined in a variety of cross-sectional correlational studies. These studies have examined both mild and serious forms of physical aggression. One study found television violence viewing among 8-year-old boys to be associated with greater physical aggression. Another study showed a significant correlation between violent television viewing among 6- to 10-year-old boys and concurrent aggressive behavior.[12] In both studies, the association between television violence and concurrent aggression was nonsignificant among the same sample in later adolescence and early adulthood. However, other cross-sectional studies found television and film violence exposure to be associated with aggressive behavior among adolescents and young adults. A study of junior high and high school students in Wisconsin and Maryland found small to moderate correlations between television violence viewing and aggressive delinquency (e.g., hitting, fighting) for both males and females.[13] A study of 12- to 17-year-old English boys found that high television violence viewers engaged in 49 percent more acts of violence in the previous 6 months than low television violence viewers.[14] Overall, the meta-analytic findings for cross-sectional studies of television and film violence effects on physical aggression against another person are generally consistent with the findings of experimental studies, supporting the idea that the causal effects of television and film violence extend to serious forms of aggression as well.

Longitudinal Studies

Compared with the other research designs, relatively few longitudinal studies of the effects of television and film violence exposure have been conducted. One of these studies found that boys' exposure to media violence at age 8 was associated with their aggressive behavior, based on peer nominations, 10 years later.[15] This association remained significant (and moderately strong) when several potentially relevant variables, including childhood aggressiveness, child's IQ, family socioeconomic status, and parental punishment and nurturance, were statistically controlled. Another longitudinal study examined television violence effects in five countries, beginning in the late 1970s. This study assessed children's violent television viewing, aggression (based on peer nominations), and several control variables from ages 6 to 8 or ages 8 to 11. For girls, early violent television viewing was associated with later aggressive behavior even when early aggression, socioeconomic status, and educational achievement were statistically controlled. Television exposure alone did not predict later aggression in boys; however, boys who frequently watched violent television in early childhood and identified with aggressive television characters were rated by their peers to be the most aggressive children. A 15-year follow up to this study interviewed a subsample from the United States, now in their early 20s.[16] The results in this follow up were even more clear. Television violence exposure in childhood was associated with aggression in both men and women. Even when the outcome was restricted to more extreme forms of aggression (e.g., punching, beating, choking, threatening or attacking with a knife or gun), childhood exposure to violent television remained a significant predictor. For example, men who watched a lot of violent television as children were much more likely to report having pushed, grabbed, or shoved their spouse (42 percent) than those who watched very little violent television as children (22 percent). Though meta-analyses have not examined longitudinal studies of television and film violence separately, one meta-analysis examined longitudinal media violence studies, most of which were studies of television and film violence. This meta-analysis found a significant longitudinal association between media violence exposure and aggression. These findings provide further support that viewing violent films and television increases aggression and that this causal association extends from mild to serious forms of aggression.[17]

Music

Compared with other types of media, such as television, films, and video games, there is considerably less research on the effects of violent music lyrics and violent music videos on aggression. Nonetheless, the research conducted in this area can be compared with research on other types of media violence. Because relatively little research has examined this topic, some studies that measure only exposure to music in general (e.g., watching MTV, listening to rap or heavy metal music) are presented when such studies are the most representative studies available; as such, general exposure is almost certain to be positively associated with exposure to violent music in particular (e.g., a person who listens to a lot of rap music is likely to listen to more violent rap music than a person who does not listen to rap music at all). In general, using measures of total media exposure, rather than violent media exposure, tends to underestimate the effect of violent media on aggression. Additionally, studies using certain theoretically relevant outcomes other than aggressive behavior are also reported, because such results are informative of likely outcomes using actual behavioral measures.

Experimental Studies

Anderson, Carnagey, and Eubanks reported several experimental studies of violent music. University student participants in two of these studies were randomly assigned to listen to either a violent song or a nonviolent song before completing other measures. In the first study, after listening to the song, participants reported their state of hostility by indicating the extent to which a series of anger-related words described them. Those who had listened to a violent song reported greater hostility than those who listened to a nonviolent song. In the second study, after listening to the violent or nonviolent song, participants rated how similar a series of words were. Some of these words were related to aggression (choke, fight, gun) and others were neutral (bottle, night, stick). Those who had listened to a violent song rated the neutral words as more related to aggression words than those who listened to a nonviolent song, reflecting greater aggressive thinking after a listening to violent song. This pair of studies shows that violent music increases aggressive feelings and aggressive thoughts, both of which are theoretically related to aggressive behavior.[18]

Another study randomly assigned young adult African American men to listen to either a violent or a nonviolent rap song. Those who listened to the violent rap song were more likely to endorse a violent response to a potential conflict situation than were those who listened to the nonviolent rap song. This indicates greater aggressive thinking after violent music than nonviolent music. We found only one experimental study that examined aggressive music lyric effects on aggressive behavior.[19] Participants listened to either sexually aggressive or neutral music and later decided how much hot chili sauce another person would have to eat. Because that other person had previously indicated that he or she did not like spicy food, the amount of hot chili sauce served as a valid measure of aggression. Males who had heard the sexually aggressive song gave more hot chili sauce to a female confederate than did participants in any other condition. Similarly, researchers in a psychiatric inpatient ward noted that aggressive behavior decreased significantly when MTV was removed from the ward.[20] Because much MTV content was violent at the time of the study, it provides further evidence that the findings for violent music effects on aggressive affect and cognition may extend to aggressive behavior as well.

Cross-Sectional Studies

There are very few cross-sectional correlational studies of violent music. Research on music preferences revealed associations between preferences for certain types of music and aggression. For example, male college students who prefer heavy metal music tend to have more negative attitudes toward women, whereas college students who prefer rap music tend to be more distrustful.[21] These hostile attitudes could theoretically lead to greater aggression. Other researchers found associations between the preference for rap or heavy metal music and worse academic performance, behavioral problems in school, sexual activity, and criminal behavior.[22] Roberts, Christenson, and Gentile showed that time spent watching MTV among third-, fourth-, and fifth-grade students was associated with more physical fights. Peers and teachers rated those who watched more MTV as physically and relationally aggressive.[23]

In sum, the existing experimental and cross-sectional studies provide some evidence that violent music increases aggression and aggression-related internal states, but the evidence is quite limited. Furthermore, there aren't any longitudinal studies of violent music effects. Thus, conclusions about the effects of violent music must be more tentative than those made for other types of violent media. The tentativeness is not due to studies failing to find music lyric or music video effects; rather, there simply have been too few studies.

Video Games

Violent video games have a history of more than 30 years, dating back to games such as *Air-Sea Battle* for the Atari 2600. The violence depicted in video games became considerably more intense in graphics during the 1990s, when games such as *Mortal Kombat* and *Doom* were released. Compared with violent television and films, violent video games are relatively new and have received research attention only more recently. Consequently, fewer studies have examined the effects of violent video games. Nevertheless, research on the effects of violent video games has flourished in recent years and has yielded clear results.

Experimental Studies

In one study, 7- and 8-year-old boys were randomly assigned to play either a violent video game or a nonviolent video game.[24] Participants were told they would compete to be the first one to color a picture. The boy that the participant was competing with was actually a confederate (an individual working for the experimenter who behaves in a predetermined way) who disobeyed the adult and began coloring his picture before the real participant had a marker. In other words, the confederate cheated. This situation was intended to induce frustration in the participant. The boys' aggressive behaviors toward the confederate were rated. Those who had played the violent video game behaved more aggressively toward the confederate during the frustrating situation than did the boys who had played a nonviolent video game.

Another study tested violent video game effects in a Dutch sample of adolescent boys.[25] Participants were randomly assigned to play a violent or nonviolent video game. They also reported the extent to which they identified with their game character. After playing the game, participants completed a competitive reaction time task in which they were told they would be able to deliver harmful noise blasts to another participant (there was no actual participant). Those participants who played the violent video game selected more intense noise blasts than those who had played the nonviolent game. This was particularly true for those who highly identified with their video game character.

Experimental studies also examined violent video game effects in young adults. For example, one study assigned male and female university students to play one of three versions of a racing video game. In one version, participants were rewarded for violent behavior, such as running over pedestrians in their car. Another version punished violent behaviors. In the third version of the game, there were no pedestrians and violence was not possible. Participants then completed a competitive reaction time task, and those participants who had played the video game that rewarded violence (the first version) delivered louder noise blasts to their (fictitious) opponent than those participants who played the nonviolent version of the game and those who were punished for violent behavior.[26] Meta-analytic reviews of violent video game effects consistently find that experimental studies yield significant effects in the moderate range, with methodologically better experimental studies yielding larger effects than methodologically weaker studies. Consistent with the examples provided here, the meta-analyses find that playing violent video games causes an increase in aggressive behavior.[27]

Cross-Sectional Studies

A number of cross-sectional correlational studies examined the association between playing violent video games and various measures of aggression. One study examined the video game–playing habits and aggressive behaviors of eighth and ninth graders.[28] Those who played more violent video games were more likely to have been in arguments with teachers and in physical fights. Another cross-sectional study examined media exposure, aggressive behaviors, and a variety of other potentially relevant variables among a sample of 9th to 12th graders.[29] Violent video game exposure was assessed by asking participants to name their favorite video games, rate how violent each game is, and report how often they play each video game. The aggressive behaviors reported included serious violent behaviors, such as attacking another person with the idea of causing serious harm or using force to get money or other things from others. Those who played more violent video games reported more mild physical aggression and more severe violent behaviors than those who were less exposed to violent video games. The association between violent video game playing and physical aggression and violence remained significant even when gender, total time spent viewing any type of screen media, and aggressive beliefs and attitudes were statistically controlled. A similar cross-sectional study of university students yielded very similar results. Those who played more violent video games were also more violent and also engaged in more nonviolent forms of delinquency, such as stealing or cheating.[30] The results of these sample studies, as well as the meta-analytic findings, provide further evidence that the causal effect of violent video games on aggression exists outside of the laboratory and applies to serious forms of aggression as well.

Longitudinal Studies

The first longitudinal study (at least among those published in English) directly tested the effect of violent video game exposure on aggressive behavior among third, fourth, and fifth graders over a period of approximately 5 months. This study assessed violent video game exposure through the same measure as used by prior research (i.e., violent content of favorite video games and how often each is played). Violent video game playing at the beginning of the study was associated with later physical aggression. This association remained significant even when a variety of important variables were statistically controlled, including initial aggression, gender, race, and time spent with all types of screen media. A more recent longitudinal study compared longitudinal effects of video game exposure on physical aggression as assessed in two Japanese samples and one U.S. sample of school children. All three samples yielded significant longitudinal effects on aggression, despite differences in culture, age (12–15 and 13–18 in the Japanese samples, 9–12 in the U.S. sample), time lag (3–4 months in the Japanese samples, 5–6 months in the U.S. sample), and specific measures of video game violence exposure and of aggressive behavior.[31]

Another longitudinal study provides some evidence relevant to violent video game effects on aggression. This study examined the association between media violence exposure and aggression in sixth and seventh graders over a 2-year period.[32] Media violence, as assessed in this study, included how often the participant watched action movies, played video games that involved shooting a weapon, or visited Internet sites that described or recommended violence. Because the media violence measure is based in part on the frequency of playing video games that involve firing a weapon, the results of this study are relevant to the effects of violent video games. This study found that media violence was associated with later physical aggression, even when relevant variables, including earlier physical aggression, were statistically controlled. The reverse direction of causality (aggressive individuals preferring to consume more violent media) was not supported by the time-lag analyses. Though this study does not distinguish video game violence from other types of media violence, the study nevertheless provides longitudinal evidence that violent media exposure, based partially on video game violence, is associated with subsequent aggression. Though more longitudinal research is clearly needed on violent video game effects on behavior, these studies provide some early evidence that violent video game playing increases aggressive behavior across time outside of the laboratory.

Intervention Research

A small number of studies examined the effectiveness of interventions targeting media violence on children's subsequent aggressive behavior. One longitudinal study of 132 second and fourth graders who viewed large amounts of violent television assigned some children to an intervention intended to reduce the effect of television violence and the other children to a neutral control intervention. Children assigned to the treatment intervention were taught that most people do not behave as violently as those on television, and that filming techniques and special effects allow television shows to depict violence in unrealistic ways. These children learned the methods that normal people use to resolve situations of potential conflict. The children also made a video to teach other children about television violence. Those who took part in this intervention were rated as less aggressive by peers at a 4-month follow up than those in the control intervention, even when gender, grade, and initial aggression were controlled. Another longitudinal study of third- and fourth-grade students used an intervention intended to reduce time spent with television, films, and console video games.[33] Though this intervention reduced exposure only to types of screen media (violent and nonviolent), presumably this included a reduction in exposure to all types of violent media. Children from one school received this intervention, whereas children from another matched school did not. In a 7-month follow up, those who participated in the intervention were rated by their peers as less aggressive than those children who did not take part in the intervention. Not only do these studies provide further evidence that media violence exerts a causal effect on aggressive behavior, but they also provide evidence that interventions that reduce media violence exposure or teach children that media violence is unrealistic may be effective in reducing aggressive behavior.

Overall Findings

Though more research is needed to examine certain types of media using certain research designs, a consistent picture has emerged across the hundreds of media violence studies conducted to date. Exposure to violent media, including television, film, music, and video games, causes an increase in the probability of aggressive behavior. This association exists both inside and outside of the laboratory. Violent media can increase both mild forms of aggression and serious forms of violence. Interventions that reduce media violence exposure or teach that media violence is unrealistic can effectively reduce aggression. Alternative explanations for the association between media violence and aggression (e.g., aggressive individuals prefer violent media) cannot fully account for these findings.

AGGRESSION THEORY AND MEDIA VIOLENCE

A variety of theoretical explanations has been used to explain aggressive behavior. According to the social learning theory, people learn to behave aggressively by watching the aggressive behavior of others.[34] This is particularly true when those others are rewarded for their aggressive behavior. Several types of learning occur. One can learn specific ways to harm someone, including novel use of initially innocuous items. For example, aficionados of professional wrestling television programs are likely to view folding chairs as potential weapons because of their frequent use in such broadcasts. One can also learn what types of situations call for aggression. And, of course, one also learns a host of related attitudes and beliefs concerning aggression, many of which may be objectively incorrect but which influence the media violence consumer's perceptions and actions in the real world. Social learning theory can certainly explain some types of media violence effects, because media such as television, films, and video games all provide opportunities for people to view the behavior of others and often depict rewards for that behavior. For example, the hero of a film may be praised and rewarded for using violence against the villain.

Cognitive neoassociation theory posits that aversive events such as frustrations and provocations automatically produce negative affect. In turn, negative affect automatically stimulates or primes memories and thoughts, expressive motor reactions (e.g., angry face), and physiological reactions that are associated with both fight and flight tendencies. These fight and flight associations give rise to rudimentary feelings of anger and fear. One key feature of this theory is that stimuli present in an aversive situation, even originally neutral cues, can become associated with the aversive situation and with the aggressive thoughts and feelings that were stimulated by the situation. Indeed, even in nonaversive situations, originally innocuous stimuli can become associated with thoughts and feelings with which they are paired. If paired with thoughts about aggression, such stimuli themselves can cue aggressive thoughts or concepts.[35]

Cognitive neoassociation theory is helpful in explaining many media violence findings, especially those involving cognitive priming effects. For example, in the Josephson study described earlier, the walkie-talkies that were seen in the violent boxing film became associated with ideas about violence. Later, they cued (primed) aggressive ideas (scripts) during the game of floor hockey played by those boys who had previously watched the violent film when they saw referees carrying walkie-talkies.

Script theory posits that individuals develop sets of highly associated concepts (i.e., scripts) that guide perception of social events and enactment of social behavior.[36] As an aggressive concept becomes integrated into a particular script, activating one part of that script can lead the entire script (including the aggressive concept) to be activated, making aggression more likely. This is relevant to media violence exposure, in that media depictions can influence the content of our social scripts and the relative accessibility of various types of social scripts. If a person views television shows in which verbal insults are frequently followed by retaliation in the form of physical violence, this could become part of that person's script for how to deal with verbal insults. That person would subsequently be more likely to react violently when insulted, because the script involving violent retaliation is activated.

Excitation transfer theory states that when a person experiences physiological arousal and is subsequently in a situation that produces anger before the arousal can dissipate, the prior arousal can be misattributed to the source of the anger.[37] This misattribution can further increase the likelihood of aggression. This can happen in the context of media violence effects. Violent media can increase arousal (heart rate, blood pressure), as shown in the video game literature. Therefore, some violent media effects on aggression may be caused by the transfer (or misattribution) of arousal from violent media exposure to a subsequent provocation. This misattribution of arousal can increase the self-perception of anger, which in turn can make aggressive behavior a more likely response.

These past theories are capable of making contributions to the understanding of media violence effects on aggression, yet none of them is broad enough to explain all the research findings. For example, excitation transfer theory cannot explain why in Carnagey and Anderson's study the violent (rewarded) video game led to greater aggression than a nonviolent video game, even when these video games were equally arousing. Similarly, cognitive neoassociation cannot easily explain media violence effects in studies that show no difference in negative affect between those who were exposed to violent versus nonviolent media. This is not a failing of these theories; each works well within its specific domain. But a broader theoretical model is useful for considering the mechanisms and processes relevant to aggression. The general aggression model (GAM) integrates the various (affective, cognitive, and arousal-based) mechanisms described by many of these more specific theories of aggression.[38]

GENERAL AGGRESSION MODEL

The GAM represents an ongoing effort to integrate the older, more specific aggression theories, such as those described in the previous sections, into a cohesive, broad model. Note that the GAM, like its progenitors, is not specifically about media violence effects. Rather, it attempts to model a wide range of factors that influence human aggression. The GAM begins with a social episode with person and situational inputs that determine the person's present internal state (affect, cognition, and arousal). The present internal state guides appraisal and decision processes, which eventually lead to some type of action. The resulting behavior (potentially an aggressive behavior) exerts an influence on the current social encounter and on longer term changes in the individual and that individual's typical life situations.

Inputs

Person Inputs

People bring a variety of relatively stable individual characteristics with them to each episode of potential aggression that can influence their behavior in that episode. These inputs include some innate characteristics of the individual, such as gender and genetic predisposition. U.S. crime statistics generally show that men are much more likely to commit murder than are women, often by a margin of approximately 10 to 1. Person inputs also include personality traits, such as narcissism, which can lead to greater aggression when a narcissistic individual's unrealistic positive self-view is challenged.[39] Other person inputs include long-term goals, beliefs, attitudes, values, scripts, perceptual schemata, and expectation schemata. Furthermore, a person brings to each situation some less stable characteristics, such as his or her current mood state.

Situational Inputs

Whether aggressive behavior occurs or not depends not just on characteristics of the individual, but also on characteristics of the immediate situation. One situational characteristic, one of the most powerful instigating factors for aggressive behavior, is provocation. There are many forms of provocation, such as verbal aggression (e.g., insults), physical aggression, or interfering with a person's attainment of a goal. Another situational factor with the potential to increase aggression is frustration. Frustration occurs when there is an impediment to one's goal. This can lead to aggression when another person is identified as the source of that frustration. Frustration often leads to aggression when the source of frustration is fully justified in his or her behavior, or even aggression against someone who is not responsible for the initial frustration at all. Pain and discomfort also increase the likelihood of aggression. Unpleasant conditions such as immersing a hand in ice water or exposure to hot temperatures, loud noises, or even unpleasant odors can increase aggressive behavior. This effect appears to depend primarily on anger, though aggressive thoughts or arousal may also be involved.[40]

Another major situational influence on aggression is the presence of aggressive cues. For example, the presence of weapons in a situation can increase aggressive behavior through the priming of aggressive thoughts. Even pictures of weapons or words for weapons are sufficient to prime aggressive thoughts and aggressive behavior. Similarly, media violence probably increases aggression at least in part by acting as an aggressive cue. Research has demonstrated that violent television shows, films, and video games are capable of priming aggressive thoughts.[41]

The use of drugs is another situational input that can influence the expression of aggressive behavior. Drugs such as alcohol or caffeine can increase the probability of aggressive behavior. These drugs may indirectly increase aggression by making the effect of other aggression-enhancing inputs (e.g., provocation, frustration, or aggressive cues) much stronger while people are under the influence of the drugs. Person and situation variables often combine interactively in their influence on aggression.[42] For example, prior research found that pictures of assault guns increased the aggressive behavior of hunters more than did pictures of hunting guns, whereas the opposite pattern occurred for nonhunters. The explanation for this intriguing finding lies in the mediating processes described in later portions of the GAM.

Present Internal State

All the person and situational variables together determine an individual's present internal state. This present internal state consists of three components: affect, cognition, and arousal. Each component is influenced not only by the person and situational inputs, but by the other components of the present internal state as well. Also, more than one of these routes may be important in a particular situation.

Affect

Some input variables increase aggression primarily through their influence on affect. Pain increases aggression most directly through anger and hostility. There is also considerable evidence that violent media increases aggression-related affect, especially in short-term contexts. This suggests the possibility that violent media increases aggression at least in part by producing feelings of anger and hostility. There are also stable individual differences in how readily one becomes angry or hostile.[43]

Cognitions

Another route by which person and situational inputs can produce aggression is through cognitions. Input variables may combine to produce aggressive priming, a temporary increase in the accessibility of aggressive thoughts, concepts, and scripts. Media violence exposure can prime aggressive cognitions, indicating another potential route by which media violence may lead to aggressive behavior. The more frequently an aggressive thought is activated, the more easily it can be activated in the future.[44] The different effects of hunting versus assault guns on the aggressive behavior of hunters and nonhunters results from their differential effects on aggressive thoughts. For hunters, assault weapons produce more aggressive thoughts than do hunting guns, presumably because they associate hunting with enjoyable outdoor activities with family and friends, while they associate assault guns with war, terrorism, and violent crime. For nonhunters, however, hunting guns inspire greater aggressive thinking than assault guns, for reasons that are not entirely clear. In any case these differential effects on aggressive thoughts lead to the opposite patterns of gun cue effects on aggressive behavior.

Arousal

Person and situational inputs can heighten physiological arousal. Physiological arousal can lead to aggression in different ways. Arousal tends to increase the dominant response tendency, so individuals who have a history of aggression in the situation they are presently in tend to become aggressive when they experience heightened arousal, especially if other predisposing factors (e.g., provocation) are present. Arousal can also increase aggression if arousal from one source (e.g., exercise, exposure to media violence) is misattributed to another source, such as a person who gives some sort of provocation.

The affective, cognitive, and arousal present internal state variables combine to influence an individual's appraisal of his or her current situation. For example, an individual might decide that another person has pushed him or her too far. This immediate appraisal can produce a decision and behavior without the person even being aware that he or she has made a decision. This could be described as an impulsive action. In other cases, the individual may have the resources and the motivation to reappraise the current situation, searching for information that gives the situation a different meaning, possibly arriving at a different, more thoughtful decision. A reappraisal depends on both sufficient cognitive resources (time to think, ability to think) and sufficient motivation (the outcome of the situation must be sufficiently important). If either of these conditions is not met, the individual will reach a decision and behave based on his or her immediate appraisal.

More deliberative reappraisals do not always lead to less aggression than the immediate appraisal would have instigated. In some circumstances, reappraisal leads to a more conscious decision to behave aggressively. In most social situations, however, our society provides negative outcomes for aggressive behavior, so more thoughtful decisions probably tend to be less aggressive than impulsive decisions. There are exceptions to this, of course, as some environments (e.g., being with a gang) actively encourage aggression (e.g., against intruders).

Once these appraisal and decision processes are complete, the individual engages in some behavior (potentially an aggressive one) in the present situation. These components, from the person and situational inputs to the present internal state variables to the appraisal and decision processes and the resulting behavior, constitute a single aggressive episode. This episodic portion of the GAM explains how aggressive behavior is produced in the short term. As mentioned throughout the description of the GAM, the single episode portion of the GAM specifies several ways in which violent media exposure could lead to the short-term increases in aggression observed by researchers. It might prime an individual's aggressive scripts or other thoughts. It might increase an individual's anger or hostility. It might lead to aggression through arousal, either by increasing the dominant response tendency (the individual may not have the ability to engage in reappraisal of the situation when highly aroused) or by misattributing it to another source, heightening anger. Though such short-term increases in aggression are important, long-term increases in aggressive personality may be of even greater importance.

The GAM identifies a variety of knowledge structures that can be changed through repeated aggressive episodes, but also predicts aggression in the long term. These include aggression-related beliefs, attitudes, values, scripts, perceptual schemata, and expectation schemata. These knowledge structures can be regarded as the components of aggressive personality. Certain beliefs, attitudes, and values can be important person inputs to potentially aggressive social episodes. To the extent a person believes that aggression is likely to be effective, he or she will be more likely to use aggression. Various attitudes are also associated with the probability of certain aggressive behaviors. Attitudes toward violence against women are associated with both sexual and nonsexual aggression against women.[45] Other attitudes are associated with more aggressive behavior in a variety of contexts, including attitudes toward violence in war, corporal punishment of children, violence in intimate relationships, and violence in the penal code system. Some personal values are relevant to aggressive behavior. For example, individuals from certain cultures (e.g., the American South) place a high value on personal honor and answering insults to their honor. This cultural value is often associated with greater aggressive behavior.[46] Other aggression-related knowledge structures include scripts, perceptual schemata, and expectation schemata. Scripts, as described earlier, are a set of concepts that become associated through repeatedly being activated together. Scripts follow principles of cognitive-associative models of memory.[47]

Schemata are knowledge structures that organize knowledge about a subject in a way that guides expectations and perceptions. Perceptual schemata are those schemata that organize information about a perceptual target. One way in which perceptual schemata are relevant to media violence effects is through facial emotion perception. Individuals differ in how quickly they are able to identify happy and angry faces. Recognizing angry but not happy faces relatively quickly may be an indication that an individual tends to automatically perceive greater hostility in others. Researchers found that individuals who play more violent video games identify angry faces more quickly than those who do not play violent video games.[48] Expectation schemata organize information in a way that guides expectations for different situations. For example, some children are more likely to attribute hostile intentions to ambiguous behavior than others.[49] This hostile attribution bias is positively associated with violent video game playing and a greater likelihood of aggressive behavior.

A final component of aggressive personality is emotional desensitization to violence. Desensitization to violence refers to a reduction in the normal emotion-related physiological reaction to viewing violence. The normal human reaction to violence is negative emotion (fear, disgust) and increased arousal.[50] It is only through experience that this negative emotional reaction to violence is eliminated. Viewing violent media leads to desensitization to later violence.[51] Of course, desensitization is desirable in certain contexts. Systematic desensitization is used to treat conditions such as posttraumatic stress disorder.[52] Similarly, medical students must get desensitized to the sights, sounds, and smells of the operating room, and soldiers must get desensitized to the horrors of combat. In both cases, the blunting of normal negative emotional reactions is crucial to successful performance in those situations. However, desensitization to violence in the general population is a generally undesirable outcome because the GAM predicts that such desensitization leads to a greater likelihood of initiating aggression, the use of more severe forms of aggression, greater persistence in using aggression, and a decline in helping victims of violence.[53]

CONCLUSION

A considerable number of studies have examined the potential effects of violent media exposure on aggression. Overall, this research, whether conducted on violent television, films, music, or video games, consistently found that exposure to violent media is associated with aggressive behavior and with other theoretically relevant outcome measures. There is strong evidence that this media violence effect is causal and applies to laboratory measures of aggression, and both mild and serious forms of aggression outside of the laboratory. The GAM provides a framework for explaining these media violence effects on aggression. Existing individual differences and characteristics of the situation determine the person's thoughts, feelings, and level of arousal. These internal states serve as inputs to appraisal and decision processes that ultimately determine what type of behavior is enacted. Within this framework, media violence can produce short-term increases in aggression by increasing the accessibility of aggressive thoughts, making a person feel angry or hostile, or by increasing the person's physiological arousal. Long-term changes in aggressive personality are the result of repeated aggressive episodes that exert an influence on the individual's aggression-related attitudes, beliefs, and values, and that produce desensitization to violence. Violent media may also increase aggressive behavior by reducing an individual's ability to exert executive control in the short or long term, and by changing the situations and types of people that are routinely encountered.

GLOSSARY

Aggression—behavior that is primarily intended to cause harm and that is carried out with the expectation of causing harm to an individual who is motivated to avoid that harm

Cross-sectional correlational studies—conducted where the independent and dependent variables are measured at the same point in time

Experimental studies—subjects are randomly assigned to either a treatment or control group to determine whether a particular treatment causes a particular outcome

Longitudinal studies—conducted where the independent variable is measured before the dependent variable

Media violence—the portrayal and depiction of violent behaviors in different media outlets, such as movies and video games

Violence—extreme forms of physical aggression, such as assault and murder

NOTES

1. Steinfeld, J. (1972). *Statement in hearings before Subcommittee on Communications of Committee on Commerce* (U.S. Senate, Serial No. 92-52, pp. 25–27). Washington, DC: U.S. Government Printing Office.
2. Satcher, D. (2001). *Youth violence: A report of the Surgeon General.* Rockville, MD: Office of the Surgeon General.
3. Anderson, C. A., Berkowitz, L., Donnerstein, E., Huesmann, L. R., Johnson, J., Linz, D., Malamuth, N., & Wartella, E. (2003). The influence of media violence on youth. *Psychological Science in the Public Interest, 4,* 81–110.
4. American Academy of Pediatrics. (2000). Joint statement on the impact of entertainment violence on children. Retrieved from www.aap.org/advocacy/releases/jstmtevc.htm.
5. Bjorkqvist, K. (1985). *Violent films, anxiety, and aggression.* Helsinki: Finnish Society of Sciences and Letters.
6. Josephson, W. L. (1987). Television violence and children's aggression: Testing the priming, social script, and disinhibition predictions. *Journal of Personality and Social Psychology, 53,* 882–890.
7. Leyens, J. P., Camino, L., Parke, R. D., & Berkowitz, L. (1975). Effects of movie violence on aggression in a field setting as a function of group dominance and cohesion. *Journal of Personality and Social Psychology, 32,* 346–360.
8. Parke, R. D., Berkowitz, L., Leyens, J. P., West, S. G., & Sebastian, R. J. (1977). Some effects of violent and nonviolent movies on the behavior of juvenile delinquents. In L. Berkowitz (Ed.), *Advances in experimental social psychology* (Vol. 10, pp. 135–172). New York: Academic Press.
9. Geen, R. G., & O'Neal, E. C. (1969). Activation of cue-elicited aggression by general arousal. *Journal of Personality and Social Psychology, 11,* 289–292.
10. Donnerstein, E., & Berkowitz, L. (1981). Victim reactions in aggressive erotic films as a factor in violence against women. *Journal of Personality and Social Psychology, 41,* 710–724.
11. Paik, H., & Comstock, G. (1994). The effects of television violence on antisocial behavior: A meta-analysis. *Communication Research, 21,* 516–546.
12. Eron, L. D., Huesmann, L. R., Lefkowitz, M. M., & Walder, L. O. (1972). Does television violence cause aggression? *American Psychologist, 27,* 253–263; Huesmann, L. R., Moise-Titus, J., Podolski, C. L., & Eron, L. (2003). Longitudinal relations between children's exposure to TV violence and their aggressive and violent behavior in young adulthood: 1977–1992. *Developmental Psychology, 39,* 201–221.
13. McLeod, J. M., Atkin, C. K., & Chaffee, S. H. (1972). Adolescents, parents, and television use: Adolescent self-report measures from Maryland and Wisconsin samples. In G. A. Comstock & E. A. Rubinstein (Eds.), *Television and social behavior: A technical report to the Surgeon General's Scientific Advisory Committee on Television and Social Behavior: Vol. 3. Television and adolescent aggressiveness* (DHEW Publication No. HSM 72-9058, pp. 173–238). Washington, DC: U.S. Government Printing Office.
14. Belson, W. A. (1978). *Television violence and the adolescent boy.* Hampshire, England: Saxon House, Teakfield.
15. Lefkowitz, M. M., Eron, L. D., Walder, L. O., & Huesmann, L. R. (1977). *Growing up to be violent: A longitudinal study of the development of aggression.* New York: Pergamon Press.
16. Huesmann, L. R., & Eron, L. D. (Eds.). (1986). *Television and the aggressive child: A cross-national comparison.* Hillsdale, NJ: Erlbaum; Huesmann, L. R., Eron, L. D., Klein, R., Brice, P., & Fischer, P. (1983). Mitigating the imitation of aggressive behaviors by changing children's attitudes about media violence. *Journal of Personality and Social Psychology, 44,* 899–910; Huesmann, L. R., Lagerspetz, K., & Eron, L. D. (1984). Intervening variables in the TV violence-aggression relation: Evidence from two countries. *Developmental Psychology, 20,* 746–775.
17. Anderson, C. A., & Bushman, B. J. (2002a). The effects of media violence on society. *Science, 295,* 2377–2378.
18. Anderson, C. A., Carnagey, N. L., & Eubanks, J. (2003). Exposure to violent media: The effects of songs with violent lyrics on aggressive thoughts and feelings. *Journal of Personality and Social Psychology, 84,* 960–971.

19. Fischer, P., & Greitemeyer, T. (2006). Music and aggression: The impact of sexual-aggressive song lyrics on aggression-related thoughts, emotions, and behavior toward the same and the opposite sex. *Personality and Social Psychology Bulletin, 32,* 1165–1176.
20. Waite, B. M., Hillbrand, M., & Foster, H. G. (1992). Reduction of aggressive behavior after removal of music television. *Hospital and Community Psychiatry, 43,* 173–175.
21. Rubin, A. M., West, D. V., & Mitchell, W. S. (2001). Differences in aggression, attitudes towards women, and distrust as reflected in popular music preferences. *Media Psychology, 3,* 25–42.
22. Took, K. J., & Weiss, D. S. (1994). The relationship between heavy metal and rap music and adolescent turmoil: Real or artifact. *Adolescence, 29,* 613–621.
23. Roberts, D. F., Christenson, P. G., & Gentile, D. A. (2003). The effects of violent music on children and adolescents. In D. A. Gentile (Ed.), *Media violence and children* (pp. 153–170). Westport, CT: Praeger.
24. Irwin, A. R., & Gross, A. M. (1995). Cognitive tempo, violent video games, and aggressive behavior in young boys. *Journal of Family Violence, 10,* 337–350.
25. Konijn, E. A., Bijnank, M. N., & Bushman, B. J. (2007). I wish I were a warrior: The role of wishful identification in the effects of violent video games on aggression in adolescent boys. *Developmental Psychology, 43,* 1038–1044.
26. Carnagey, N. L., & Anderson, C. A. (2005). The effects of reward and punishment in violent video games on aggressive affect, cognition, and behavior. *Psychological Science, 16,* 882–889.
27. Anderson, C. A. (2004). An update on the effects of violent video games. *Journal of Adolescence, 27,* 133–122; Anderson, C. A., Carnagey, N. L., Flanagan, M., Benjamin, A. J., Eubanks, J., & Valentine, J. C. (2004). Violent video games: Specific effects of violent content on aggressive thoughts and behavior. *Advances in Experimental Social Psychology, 36,* 199–249.
28. Gentile, D. A., Lynch, P. J., Linder, J. R., & Walsh, D. A. (2004). The effects of violent video game habits on adolescent aggressive attitudes and behaviors. *Journal of Adolescence, 27,* 5–22.
29. Anderson, C. A., Gentile, D. A., & Buckley, K. E. (2007). *Violent video game effects on children and adolescents: Theory, research, and public policy.* New York: Oxford University Press.
30. Anderson, C. A., & Dill, K. E. (2000). Video games and aggressive thoughts, feelings, and behavior in the laboratory and in life. *Journal of Personality and Social Psychology, 78,* 772–790.
31. Anderson, & Dill (2000), see Note 30; Anderson, C. A., Sakamoto, A., Gentile, D. A., Ihori, N., & Shibuya, A. (2008). Longitudinal effects of violent video games aggression in Japan and the United States. *Pediatrics, 122,* e1067–e1072.
32. Slater, M. D., Henry, K. L., Swaim, R. C., & Anderson, L. L. (2003). Violent media content and aggressiveness in adolescents: A downward spiral model. *Communication Research, 30,* 713–736.
33. Robinson, T. N., Wilde, M. L., Navracruz, L. C., Haydel, K. F., & Varady, A. (2001). Effects of reducing children's television and video game use on aggressive behavior: A randomized controlled trial. *Archives of Pediatric Adolescent Medicine, 155,* 17–23.
34. Bandura, A. (1973). *Aggression: A social learning theory analysis.* Englewood Cliffs, NJ: Prentice Hall; Bandura, A. (1977). *Social learning theory.* New York: Prentice Hall; Bandura, A. (1983). Psychological mechanism of aggression. In R. G. Geen & E. I. Donnerstein (Eds.), *Aggression: Theoretical and empirical reviews* (Vol. 1, pp. 1–40). New York: Academic Press.
35. Berkowitz, L. (1989). Frustration-aggression hypothesis: Examination and reformulation. *Psychological Bulletin, 106,* 59–73; Berkowitz, L. (1993). Pain and aggression: Some findings and implications. *Motivation and Emotion, 17,* 277–293.
36. Huesmann, L. R. (1986). Psychological processes promoting the relation between exposure to media violence and aggressive behavior by the viewer. *Journal of Social Issues, 42,* 125–139; Huesmann, L. R. (1998). The role of social information processing and cognitive schema in the acquisition and maintenance of habitual aggressive behavior. In R. Geen & E. Donnerstein (Eds.), *Human aggression: Theories, research, and implications for policy* (pp. 73–109). New York: Academic Press.
37. Zillmann, D. (1983). Arousal and aggression: In R. Geen & E. Donnerstein (Eds.), *Aggression: Theoretical and empirical reviews* (Vol. 1, pp. 75–102). New York: Academic Press; Zillmann, D. (1988). Cognition-excitation interdependencies in aggressive behavior. *Aggressive Behavior, 14,* 51–64.

38. Anderson, C. A., & Bushman, B. J. (2002b). Human aggression. *Annual Review of Psychology, 53,* 27–51.

39. Bushman, B. J., & Baumeister, R. F. (1998). Threatened egotism, narcissism, self-esteem, and direct and displaced aggression: Does self-love or self-hate lead to violence? *Journal of Personality and Social Psychology, 75,* 219–229.

40. Dill, J., & Anderson, C. A. (1995). Effects of justified and unjustified frustration on aggression. *Aggressive Behavior, 21,* 359–369; Geen, R. G. (1968). Effects of frustration, attack, and prior training in aggressiveness upon aggressive behavior. *Journal of Personality and Social Psychology, 9,* 316–321; Geen, R. G. (2001). *Human aggression.* Philadelphia: Open University Press; Miller, N., Pedersen, W. C., Earleywine, M., & Pollock, V. E. (2003). A theoretical model of triggered displaced aggression. *Personality and Social Psychology Review, 7,* 75–97.

41. Anderson, C. A., Benjamin, A. J., & Bartholow, B. D. (1998). Does the gun pull the trigger? Automatic priming effects of weapon pictures and weapon names. *Psychological Science, 9,* 308–314; Bartholow, B. D., Anderson, C. A., Carnagey, N. L., & Benjamin, A. J. (2005). Interactive effects of life experience and situational cues on aggression: The weapons priming effect in hunters and nonhunters. *Journal of Experimental Social Psychology, 41,* 48–60; Berkowitz, L., & LePage, A. (1967). Weapons as aggression eliciting stimuli. *Journal of Personality and Social Psychology, 7,* 202–207; Bushman, B. J. (1998). Priming effects of violent media on the accessibility of aggressive constructs in memory. *Personality and Social Psychology Bulletin, 24,* 537–545; Carlson, M., Marcus-Newhall, A., & Miller, N. (1990). Effects of situational aggression cues: A quantitative review. *Journal of Personality and Social Psychology, 58,* 622–633.

42. Bushman, B. J. (1993). Human aggression while under the influence of alcohol and other drugs: An integrative research review. *Current Directions in Psychological Science, 2,* 148–152; Bushman, B. J. (1997). Effects of alcohol on human aggression: Validity of proposed explanations. In D. Fuller, R. Dietrich, & E. Gottheil (Eds.), *Recent developments in alcoholism: Alcohol and violence* (pp. 227–243). New York: Plenum Press.

43. Anderson, K. B., Anderson, C. A., Dill, K. E., & Deuser, W. E. (1998). The interactive relations between trait hostility, pain, and aggressive thoughts. *Aggressive Behavior, 24,* 161–171.

44. Sedikides, C., & Skowronski, J. J. (1990). Towards reconciling personality and social psychology: A construct accessibility approach. *Journal of Social Behavior and Personality, 5,* 531–546.

45. Anderson, C. A., & Anderson, K. B. (2008). Men who target women: Specificity of target, generality of aggressive behavior. *Aggressive Behavior, 34,* 605–622; Malamuth, N. M., Linz, D., Heavey, C. L., Barnes, G., & Acker, M. (1995). Using the confluence model of sexual aggression to predict men's conflict with women: A 10-year follow-up study. *Journal of Personality and Social Psychology, 69,* 353–369.

46. Nisbett, R. E., & Cohen, D. (1996). *Culture of honor: The psychology of violence in the south.* Boulder, CO: Westview.

47. Collins, A. M., & Loftus, E. F. (1975). A spreading activation theory of semantic processing. *Psychological Review, 82,* 407–428.

48. Kirsh, S. J., Olczak, P. V., & Mounts, J. R. (2005). Violent video games induce affect processing bias. *Media Psychology, 7,* 239–250.

49. Crick, N. R., & Dodge, K. A. (1994). A review and reformulation of social information processing mechanisms in children's adjustment. *Psychological Bulletin, 115,* 74–101; Dodge, K. A., & Coie, J. D. (1987). Social information processing factors in reactive and proactive aggression in children's peer groups. *Journal of Personality and Social Psychology, 53,* 1146–1158.

50. Cantor, J. (1994). Fright reactions to mass media. In J. Bryant & D. Zillmann (Eds.), *Media effects: Advances in theory and research* (pp. 213–245). Hillsdale, NJ: Erlbaum; Cantor, J. (1998). *"Mommy, I'm scared": How TV and movies frighten children and what we can do to protect them.* San Diego, CA: Harvest/Harcourt.

51. Carnagey, N. L., Anderson, C. A., & Bushman, B. J. (2007). The effect of video game violence on physiological desensitization to real-life violence. *Journal of Experimental Social Psychology, 43,* 489–496; Linz, D., Donnerstein, E., & Adams, S. M. (1989). Physiological desensitization and judgments about female victims of violence. *Human Communication Research, 15,* 509–522.

52. Pantalon, M. V., & Motta, R. W. (1998). Effectiveness of anxiety management training in the treatment of posttraumatic stress disorder: A preliminary report. *Journal of Behavior Therapy and Experimental Psychiatry, 29,* 21–29.

53. Bushman, B. J., & Anderson, C. A. (2009). Comfortably numb: Desensitizing effects of violent media on helping others. *Psychological Science, 20,* 273–277.

CHAPTER ELEVEN

Substance Use Careers and Antisocial Behavior: A Biosocial Life-Course Perspective

Michael G. Vaughn and Brian E. Perron

KEY TERMS

Heritable
Latent trait
Psychopathy
Substance dependence
Substance use

INTRODUCTION

Data from a variety of sources over time and in different societies demonstrate a highly stable association between substance abuse and crime. The use and abuse of, and dependence on, psychoactive intoxicants are present in a large proportion of antisocial behaviors, including offenses against person and property. Indeed, a survey of the American Society of Criminology professional membership regarding the causes of serious and persistent offending found that alcohol and "hard drugs" are perceived to be just as important as family factors, lack of educational opportunities, and impulsiveness, and more important than labeling, low IQ, and biological factors. A survey by the Bureau of Justice Statistics indicated that approximately 33 percent of adult inmates committed their offense while under the influence of drugs, and more than half of the adult inmates were using drugs in the month prior to their offense. Approximately 17 percent of state inmates and 18 percent of federal inmates reported that they had committed an offense to obtain money for drugs.[1]

Substance abuse is also predictive of recidivism and decreases the likelihood of cessation from offending. Studies of juvenile offenders consistently demonstrate that problematic **substance use** and delinquent acts are intertwined. For example, in a study of a statewide population of juvenile offenders, Vaughn and colleagues found that high rates of past drug abuse and property and violent offending clustered together at both the low and high ends of the use spectrum; youth who used the most drugs also possessed extensive offending histories. In a landmark study of 1829 youth in detention facilities in Chicago, Teplin and associates found that approximately half of males and females had a diagnosable substance use disorder.[2]

In addition to the linkages between drug use and crime, there are enormous social costs, with the majority of the economic burden attributable to crime and criminal justice. These costs include government crime control (e.g., law enforcement and police protection), incarceration, social services, and loss of productivity of victims. The range of services across these domains is great. Also, there are costs related to emergency room care and ongoing medical care for injuries sustained during violent encounters where alcohol or drug intoxication was a precipitating factor. Although estimates vary, the economic costs are clearly well over $100 billion annually.[3]

In general, the relationship between substance use and crime is reciprocal.[4] This is because the context of crime and substance use is shared or overlapping. This is not to say that all criminals abuse drugs or that drug users inherently possess a propensity for crime, although many do. From a comorbidity perspective, this may be a spurious relationship due to an unmeasured or unobserved "third" variable, such as a **latent trait** like psychopathic personality or low self-control that increases the risk for both substance use and crime. Other researchers claim that both drug abuse and crime are part of a general deviance pattern or problem behavior spectrum.[5] Although research consistently shows a link between substance use careers and crime, a more focused theoretical understanding of the causal mechanisms is warranted.

The best way to define patterns of substance use varies. Some advocate that the terms *abuse*, *addicted*, or *dependence* be changed to "substance misuse" or to "the troublesome use of substances."[6] One contemporary definition that succinctly captures the essence of addiction was developed by George Koob, who defined drug addiction as follows:

> *Drug addiction, also known as substance dependence, is a chronically relapsing disorder characterized by (1) a compulsion to seek and take the drug, (2) loss of control in limiting intake, and (3) emergence of a negative emotional state (e.g., dysphoria, anxiety, irritability) when access to the drug is prevented (defined here as dependence).*[7]

One can readily see the implications for crime inherent in this definition, given the compulsion to seek out drugs and corresponding loss of self-control.

One of the most common conceptualizations is a continuum of severity beginning from use to abuse and finally to dependence. This conceptualization is used in the *Diagnostic and Statistical Manual of Mental Disorders* of the American Psychiatric Association for differentiating two hierarchical disorders: abuse and dependence. Both disorders are characterized by maladaptive patterns of use and clinical distress. To be diagnosed with abuse, the individual must meet at least one of the following four criteria in a 12-month period (and have never met criteria for dependence): (1) recurrent substance use resulting in a failure to fulfill major role obligations at work, school, or home; (2) recurrent substance use in physically hazardous situations; (3) recurrent substance-related legal problems; and (4) continued substance use despite having recurrent social or interpersonal problems attributable to substance use.[8]

Substance dependence, the more severe form of substance use disorders, requires three or more of the following criteria to be meet in the same 12-month period: (1) tolerance; (2) withdrawal; (3) taking the substance in larger amounts or over a longer period of time than intended; (4) persistent or unsuccessful effort to cut down or control use; (5) spending a great deal of time to obtain the substance, use the substance, or recover from its effect; (6) important social, occupational, or recreational activities are given up or reduced because of substance use; and (7) substance use is continued despite knowledge of having a persistent physical or psychological problem that is likely to have been caused or exacerbated by the substance. It should be noted that in this chapter we refer to problematic substance use generally, which includes both abuse and dependence. A substance refers to a wide range of products used to get high, including alcohol, drugs, and other types of medications.

SUBSTANCE CAREERS AND BIOSOCIAL LIFE-COURSE THEORY

The goal of this chapter is to highlight the relationship of substance use to crime across the life course by using a "career" perspective. Although it is statistically quite normal to drink alcohol and experiment with drugs at some point across the life course, usually during adolescence, the substance abuse–crime connection can be advanced by using a career approach similar to that advanced by career criminal researchers. The reason for this reflects the reality that substance abuse and crime are not evenly distributed across the population or across the life course. Birth cohort studies, longitudinal studies, and research from the biopsychological sciences have shown that a small subset of the population accounts for most antisocial behavior.[9] As such, this chapter examines those factors across the life course that initiate, amplify, and maintain a substance use career in relation to forms of antisocial behavior. But first, factors that serve as fundamentals for understanding the vulnerability to substance abuse and crime, such as genetics and neurological and personality facets, are examined. In addition, the reward pathway in the brain is discussed because of its pivotal role as a mechanism for compulsive substance seeking. Although there are many factors to consider, such as the political economy of substances that influences their availability, a discussion of distal governmental and market forces is beyond the scope of this chapter.

Are Some Persons More Vulnerable to a Substance Abuse Career?

Behavioral genetics and neuroscience, two complementary research areas, have together provided a foundation for the contemporary understanding of problematic substance use career vulnerability. Twin studies indicate that approximately 50 to 60 percent of the variation in substance use disorders is **heritable**. Several genes have also been found to be associated with behavior disinhibition and substance dependence, including dopamine (e.g., DRD4) and GABAergic (e.g., GABRA1) receptor genes.[10] Much of the recent understanding and conceptualization of addiction as a brain disease has been facilitated by the rise of neuroimaging procedures, such as functional magnetic resonance imaging (fMRI). These techniques allow direct comparisons of substance-using and -nonusing individuals. Findings indicate clear differences in biochemical, structural, and functional processes in the brain that influence decision making, control, and even craving. Persons with developmental deficits in critical areas of the prefrontal cortex are vulnerable to dysregulation in inhibitory control and are at higher risk for substance-related crime, perhaps even more persistent and severe forms of offending. It has long been recognized that deficits in and trauma to areas of the frontal lobe result in aberrant and aggressive behavior characterized by marked reductions in behavior monitoring. These systems, although not fully understood, act as "brakes" for physiological and psychological impulses in the face of reward or stimulus.[11]

Personality can be defined simply as characteristic ways of thinking and behaving. Neurological substrates (coded for by various genes) comprising inhibitory control provide the building blocks for personality. Given the close association between genes and neural substrates, approximately 40 to 50 percent of the variance in personality can be attributed to genes. It seems probable that individual propensity to violence and vulnerability to substance abuse share many biological substrates. Certain personality traits are important factors in their relation to substance abuse and criminality.

In fact, antisocial personality disorder—the psychiatric diagnosis found in the *Diagnostic and Statistical Manual of Mental Disorders, Fourth Edition (DSM-IV)* most associated with violence, criminality, and alcohol and drug use disorders—is associated with the personality trait of novelty seeking. Novelty seeking has much in common with other constructs shown to be related to crime and drug abuse, including sensation seeking and impulsivity. In a study using the five-factor model of personality (NEO-PI-R), Terracciano and colleagues found that low conscientiousness, which is also associated with psychopathic personality, was associated with multiple forms of illicit drug use.[12]

Two constructs that have received much research attention by criminologists and psychologists are low self-control and psychopathy. Self-control theory in criminology gained prominence with Gottfredson and Hirschi's *A General Theory of Crime*. Since then, the low self-control construct has garnered much research attention and empirical support.[13] However, the concept of self-control is not new. In its various guises (e.g., self-regulation, neurodisinhibition, impulsivity), self-control constructs are quite common across the behavioral sciences. Individuals low in self-control can be generally described as impulse driven, self-centered, insensitive to others, hot-tempered, irresponsible, and prone to risky behaviors. With respect to substance abuse and dependence, Baler and Volkow stated, "Importantly, there seems to be intimate relationships between the circuits disrupted by abused drugs and those that underlie self-control."[14]

Psychopathy, or psychopathic personality, may be one of the oldest psychiatric personality constructs shown to predict violence and other forms of antisocial behavior among adults and juveniles.[15] Psychopathy is also convergent with career criminality.[16] Psychopathy can be defined by its various behavioral, emotional, and cognitive dimensions that express themselves as an inability to form warm bonds with others, low empathy, high manipulativeness, deception, callousness, sensation seeking, and poor impulse control. Overall, antisocial personalities are highly convergent with one another and typically comorbid with substance use disorders. Thus, it is quite typical for antisocial persons to have extensive substance use careers spanning many years.

The Reward Pathway: A Key Mechanism for Understanding Substance Abuse and Crime

Research has demonstrated major differences between the brains of addicted persons and nonaddicted persons.[17] The reward pathway, also known as the mesolimbic reward system, is one of the major neural circuits that is profoundly altered in the brains of addicted persons. This system is pivotal due to its role in survival, providing positive reinforcement for eating, drinking, sex, and other functions that are basic to survival. The reward pathway therefore has its roots in our evolutionary past. In the course of substance misuse, chemicals flood the reward pathway to supernormal levels, well beyond the effects of aforementioned food and sex. Thus, this neural circuit becomes "hijacked" and compulsive substance seeking ensues. The usual inhibitory control mechanisms are overwhelmed, particularly in those persons with structural or functional deficits in this area of the brain. Substances of abuse and dependence are known to increase levels of dopamine within the brain. Dopamine facilitates communication between receptors in the brain involved with heightened states of joy and physiological arousal. Drugs such as cocaine, methamphetamine, nicotine, and opiates activate dopamine in the nucleus accumbens. Neuroimaging studies found low numbers of type 2 dopamine receptors (i.e., DRD2) to be associated with heightened vulnerability to substance dependence. Conversely, it has been hypothesized that higher levels may exert a protective benefit or shield from substance use careers.[18]

Overall, two frontal areas of the reward system (anterior cingulate and orbitofrontal cortices) are compromised in substance-dependent individuals. These findings are critical because these regions have been implicated in inhibitory control. Variations in the sensitivity of these systems due to polygenic factors and neural deficits influence the response to drugs of abuse and the vulnerability to an extended substance abuse career. These variations create profound impacts over the life course.

Prenatal and Early Developmental Factors

Research has shown that having an antisocial parent or substance-abusing parent heightens the risk transmission of substance dependence and delinquent involvement.[19] Although there is commonly discussion about whether genes or environment are the primary mechanism, both are certainly involved. How can drugs of abuse cause harm to the newborn? Although we know very little about maternal behavior with respect to substance abuse and behavior, there are several ways in which this can happen. Women who themselves have had extensive substance use careers or are currently abusing substances are more likely to smoke during pregnancy and place their infants in risky situations. An accidental or intentional blow to the head of a fetus or newborn can cause subtle damage, resulting in deficits in inhibitory control, which is critical to executive governance later in life. Several studies have also found that maternal cocaine exposure can interrupt caregiving behaviors and perhaps harm early attachments.[20]

Adrian Raine pioneered research on the early developmental pathways and risk factors leading to antisocial behavior in adulthood. Raine's numerous studies found low physiological arousal or autonomic functioning—typically measured by heart rate at rest and under stressful demand—to be predictive of antisocial behavior in childhood. Raine and colleagues found that birth complications, such as low birth weight, low Apgar score, and lack of oxygen, are predictive of later violence. These birth complications interact with family factors in heightening the vulnerability to antisocial behavior.[21] Research is needed to assess the strength of these factors in relation to the formation and maintenance of substance abuse careers.

Childhood

There has been a greater amount of research on the childhood manifestations of early conduct problems that have important implications for the origins of substance use careers across the life course. Because antisocial behavior is multifaceted, researchers often identify risk factors for antisocial behavior across levels of analysis such as the individual, family, peer, and community. At the individual level, these include early externalizing behaviors, low IQ, hyperactivity, and low behavioral inhibition. Family and peer risk factors include parental psychopathology, familial antisocial behaviors, maltreatment, peer rejection, and deviant peer affiliations. School and community risk factors are poor academic performance, concentrated disadvantage, and access to weapons. Determining how these risk factors interact to produce antisocial phenotypes, including the initiation of early substance misuse, is a major challenge.

It appears that early behavioral problems indicative of prefrontal disturbances are predictive of later dependence on psychoactive substances. In a long running investigation, the Dunedin birth cohort study, Caspi and colleagues found that observational reports of 3-year-old boys described as impulsive, easily distractible, and prone to negative emotions were three times more likely to be alcohol dependent in early adulthood.[22] Other studies found that externalizing problems in childhood was instrumental in setting a course toward alcohol use problems, including dependence in adulthood. Similarly, deficits in neurobehavioral control scores, at ages 10 to 12 among low- and high-risk boys in a longitudinal sample, were found to predict substance use disorder with high accuracy. This construct was based on a pool of affective, behavioral, and cognitive indicators. Item response theory methods pointed to a single latent trait termed "neurobehavioral disinhibition." Conversely, the decision to terminate a substance use career was predicted by low neurodisinhibition scores.[23]

Adolescence

Adolescence is a period characterized by greater independence and, as such, experimentation with substances. The start of most substance use careers occurs during this period of adjustment between childhood and adulthood norms and expectations. The adolescent brain is not fully formed, and thus decision making associated with risk appraisal and impulse drives are often compromised during this developmental transition period. Decisions typically involve satisfying immediate emotional needs rather than following intellectually based decision making that considers a range of possible consequences. Brain systems that subserve impulses and risk are further along in development than those that involve control. For those adolescents who already possess prefrontal disturbances and are in contact with environmental pathogens, such as deviant peers or substance-using family members, substance abuse careers become a greater possibility.[24]

The origins of delinquent behavior and substance use are linked in decades of research, thus indicating similar etiologies such as those previously discussed. Adolescent juvenile offenders also initiate their substance use career earlier than nondelinquents, and court-referred youth use and abuse substances at rates higher than nonreferred youth. Further, illicit substance abuse heightens the risk for future referrals. Finally, substance use careers and crime among adolescent juvenile offenders involve comorbid mental health disorders, including attention deficit hyperactivity disorder (ADHD), anxiety, and depression.[25]

There have been several investigations of psychopathy in adolescence and substance abuse. In a study of adolescents using the Psychopathy Checklist: Youth Version (PCL:YV), Mailloux and colleagues found that the total scores of this measure, indicating higher levels of psychopathic behavior, were significantly related to higher scores on the Michigan Alcoholism Screening Test, Drug Abuse Screening Test, age at initiation, and numbers of drugs tried.[26] Another study of substance abuse treatment by O'Neill, Lidz, and Heilbrun showed that higher PCL:YV scores were associated with a lower percentage of clear urine screens and fewer consecutively clean urine screens.[27] These findings, as with the data on violence, parallel the research on adult samples of antisocial and psychopathic offenders.

Adulthood

Most studies of substance abuse and crime are based on samples of persons diagnosed with antisocial personality disorder.[28] Fewer investigations focused more narrowly on the relationship of psychopathy to substance use disorders. Empirical findings indicate that problematic substance use tends to go hand in hand with antisocial personality and psychopathy in particular. This finding is unsurprising given that sensation seeking, impulsivity, risk taking, and failure to plan ahead characterize both syndromes. Problematic substance use, career offending, and psychopathy are important to study conjointly because their etiologies appear to be intertwined, and thus insight into one may lead to greater knowledge about the other. Also, because both are so highly associated with aggression, another reason to study this relationship is the hope that it may lead to more effective interventions in children and adolescents at the earliest stages possible.

A key question in the research on substance use careers is: What are the long-term consequences of substance abuse and dependence? In a 33-year study of a community sample of 581 male heroin-dependent persons, Hser, Hoffman, Grella, and Anglin found that nearly half (48.9 percent) had died prematurely, with the majority of death attributable to overdose or poisoning, chronic disease, liver disease, homicide, accidents, and suicide. A similar study of substance use careers among 5168 persons (approximately 40 percent arrestees) found that alcohol, tobacco, marijuana, and crack/powder cocaine were the most prevalent substances used, often in conjunction.[29] In both studies more than half of study participants had used substances throughout the follow-up periods.

CONCLUSION

Can understanding the biosocial foundations of substance use careers be reconciled with the popular Goldstein tripartite framework? The answer is yes. In 1985, Paul Goldstein published an influential article organizing the "drugs/violence nexus" occurring around three domains: psychopharmacological, economically compulsive, and systemic. This typology was developed inductively based on data collected in New York City on substance abuse and its behavioral effects. The first of these phenomena, *psychopharmacological*, is related to violence due to the direct effects that drugs have on the brain. The second major domain of the drugs/violence nexus is termed *economically compulsive*, which is simply the result of drug-dependent persons engaging in robberies to provide cash to buy more drugs. The final level, *systemic*, refers to violence perpetrated as a part of the operation of drug markets and the business of distributing illegal substances. Because substance abuse hijacks the reward system in the brain and there is a corresponding need to continue to reinforce this pathway, there is simultaneously a relationship between Goldstein's psychopharmacological and economic-compulsive domains. Thus, these two areas of the framework are intertwined. The systemic domain certainly provides the context for the availability of drugs of abuse, yet it is the neurobiology of addiction that provides the ongoing fuel for systemic violence associated with substance abuse.[30]

Substance use careers and crime are inextricably linked. The causal structure is inherently complex and dynamic. Their association involves biological processes that occur at different levels of organization and that are linked to an ecological habitat and developmentally sensitive periods. The vulnerabilities to a substance use career and crime are not fully understood. In general, the more extreme the substance use or criminal career the greater the likelihood for convergence between the two. Additional research on the convergence between career criminal and addiction careers is warranted. Considering the disproportionate involvement in crime among a subset of persons, illuminating the overlap would be useful for both clinical and policy arenas.

Historically, the evidence associated with the widespread use of psychoactive agents across cultures and time suggests that the human species seeks to alter consciousness. Seemingly, the ingestion of both natural and synthetic chemicals, and resulting behavioral changes among humans, is an ever-present danger. Although some psychoactive substances (e.g., marijuana) are less likely to lead to crime and some (e.g., alcohol and cocaine) are more likely, identifying effective management methods is important and would likely benefit from a foundation of sound empirical evidence that is interpretable within a biosocial framework.

A fairly recent conceptualization that facilitates the theorizing of the reciprocal relations between drug abuse and crime is a general biosocial liability model that denotes risk across a continuum. Vaughn reviewed research and theory and developed a synthesis that is useful for directing and organizing findings and concepts that range from the biological (dispositional) to the environmental (proximal and distal contextual). Because the phenomenon of substance abuse careers and crime extends across multiple disciplinary fields, a transdisciplinary theoretical synthesis is needed. Without such a synthesis, a lack of biological–environmental interplay, which can lead to isolated findings not linking together, is a potential consequence. Explaining substance use careers and crime strictly in terms of a singular disciplinary focus is folly. Future experimental and prospective studies that combine behavioral genetics, neuroscience, and psychosocial factors are necessary to begin to build causal-based intervention knowledge.[31]

GLOSSARY

Heritable—the proportion of variance in a measure that is accounted for by genetic factors

Latent trait—personality traits that can be measured on a continuum

Psychopathy—a personality disorder that characterizes people who are egotistical, self-centered, impulsive, exploitive, lacking in remorse, and emotionally callous

Substance dependence—a severe form of substance use disorder that is typified by tolerance for the substance, difficulty quitting, and withdrawal symptoms

Substance use—a continuum of severity beginning from use to abuse and finally to dependence

NOTES

1. Ellis, L., Cooper, J. A., & Walsh, A. (2008). Criminologists' opinions about causes and theories of crime and delinquency: A follow-up. *The Criminologist, 33*, 23–26.
2. Schroeder, R. D., Giordano, P. C., & Cernkovich, S. A. (2007). Drug use and desistance processes. *Criminology, 45*, 191–222; Stoolmiller, M., & Blechman, E. A. (2005). Substance use is a robust predictor of adolescent recidivism. *Criminal Justice and Behavior, 32*, 302–328; Teplin, L. A., Abram, K. M., McClelland, G. M., Dulcan, M. K., & Mericle, A. A. (2002). Psychiatric disorders in youth in juvenile detention. *Archives of General Psychiatry, 59*, 1133–1143; Vaughn, M. G., Beaver, K. M., & DeLisi, M. (2009). A general biosocial liability model of antisocial behavior: A preliminary test in a prospective community sample. *Youth Violence and Juvenile Justice, 7*, 279–298.
3. Office of National Drug Control Policy. (2001). *The economic costs of drug abuse in the United States, 1992–1998.* Washington, DC: Executive Office of the President.
4. Mason, W. A., & Windle, M. (2002). Reciprocal relations between adolescent substance use and delinquency: A longitudinal latent variable analysis. *Journal of Abnormal Psychology, 111*, 63–76; Menard, S., Mihalic, S., & Huizinga, D. (2001). Drugs and crime revisited. *Justice Quarterly, 18*, 269–297.
5. Jessor, R., Donovan, J. E., & Costa, F. M. (1991). *Beyond adolescence: Problem behavior and young adult development.* New York: Cambridge University Press.
6. Miller, W. R., & Carroll, K. (2006). *Rethinking substance abuse: What the science shows, and what we should do about it.* New York: Guilford Press.
7. Koob, G. E. (2006). The neurobiology of addiction: A hedonic Calvinist view. In W. R. Miller & K. Carroll (Eds.), *Rethinking substance abuse: What the science shows, and what we should do about it* (pp. 25–45). New York: Guilford Press, p. 25.

8. American Psychiatric Association. (2000). *Diagnostic and statistical manual of mental disorders* (4th ed., text revision). Washington, DC: Author.

9. Blumstein, A. Cohen, J., Roth, J. A., & Visher, C. A. (Eds.). (1986). *Criminal careers and "career criminals."* Washington, DC: National Academies Press; DeLisi, M. (2005). *Career criminals in society.* Thousand Oaks, CA: Sage; Piquero, A. R., Farrington, D. P., & Blumstein, A. (2003). The criminal career paradigm: Background and recent developments. In M. Tonry (Ed.), *Crime and justice: A review of research* (Vol. 30, pp. 359–506). Chicago: University of Chicago Press.

10. Hasin, D., Hatzenbuehler, M., & Waxman, R. (2006). Genetics of substance use disorders. In W. R. Miller & K. Carroll (Eds.), *Rethinking substance abuse: What the science shows, and what we should do about it* (pp. 61–77). New York: Guilford Press; Kreek, M. J., Nielsen, D. A., Butelman, E. R., & LaForge, S. K. (2005). Genetic influences on impulsivity, risk taking, stress responsivity and vulnerability to drug abuse and addiction. *Nature Neuroscience, 8,* 1450–1457.

11. Golden, C. J., Jackson, M. L., Peterson-Rohne, A., & Gontkovsky, S. T. (1996). Neuropsychological correlates of violence and aggression: A review of the clinical literature. *Aggressive and Violent Behavior, 1,* 3–25; Lubman, D. I., Yucel, M., & Pantelis, C. (2004). Addiction, a condition of compulsive behaviour? Neuroimaging and neuropsychological evidence of inhibitory dysregulation. *Addiction, 99,* 1491–1502; Niehoff, D. (1999). *The biology of violence: How understanding the brain, behavior, and environment can break the vicious circle of aggression.* New York: The Free Press.

12. Bouchard, T. J., & McGue, M. (2003). Genetic and environmental influences on human psychological differences. *Journal of Neurobiology, 54,* 4–45; Cloninger, C. R. (2005). Antisocial personality disorder: A review. In M. Maj, H. S. Akiskal, J. E. Mezzich, & A. Okasha (Eds.), *Personality disorders* (pp. 125–169). New York: John Wiley & Sons; Fishbein, D. (2000). Neuropsychological function, drug abuse, and violence: A conceptual framework. *Criminal Justice and Behavior, 27,* 139–159; Terracciano, A., Löckenhoff, C. E., Crum, R. M., Bienvenu, J., & Costa, P. T. (2008). Five-factor personality profiles of drug users. *BMC Psychiatry, 8,* 1–10.

13. Gottfredson, M. R., & Hirschi, T. (1990). *A general theory of crime.* Stanford, CA: Stanford University Press; Pratt, T. C., & Cullen, F. T. (2000). The empirical status of Gottfredson and Hirschi's general theory of crime: A meta-analysis. *Criminology, 38,* 931–964.

14. Baler, R. D., & Volkow, N. D. (2006). Drug addiction: The neurobiology of disrupted self-control. *Trends in Molecular Medicine, 12,* 559–566, p. 559.

15. Cleckley, H. (1941/1976). *The mask of sanity* (5th ed.). St. Louis, MO: Mosby; Hare, R. D. (1996). Psychopathy: A clinical construct whose time has come. *Criminal Justice and Behavior, 23,* 25–54; Vaughn, M. G., & Howard, M. O. (2005). The construct of psychopathy and its role in contributing to the study of serious, violent, and chronic youth offending. *Youth Violence and Juvenile Justice, 3,* 235–252.

16. DeLisi, M., & Vaughn, M. G. (2008). Still psychopathic after all these years. In M. DeLisi & P. J. Conis (Eds.), *Violent offenders: Theory, research, public policy, and practice* (pp. 155–168). Sudbury, MA: Jones & Bartlett; Vaughn, M. G., & DeLisi, M. (2008). Were Wolfgang's chronic offenders psychopaths? On the convergent validity of psychopathy and career criminality. *Journal of Criminal Justice, 36,* 33–42.

17. Leshner, A. I. (1997). Addiction is a brain disease, and it matters. *Science, 278,* 45–47.

18. Childress, A. R. (2006). What can human brain imaging tell us about vulnerability to addiction and to relapse. In W. R. Miller & K. Carroll (Eds.), *Rethinking substance abuse: What the science shows, and what we should do about it* (pp. 46–60). New York: Guilford Press.

19. Lipsey, M. W., & Derzon, J. H. (1998). Predictors of violent or serious delinquency in adolescence and early adulthood: A synthesis of longitudinal research. In R. Loeber & D. P. Farrington (Eds.), *Serious and violent juvenile offenders: Risk factors and successful interventions* (pp. 86–105). Thousand Oaks, CA: Sage; Merikangas, K. R., Stolar, M., Stevens, D. E., Goulet, J., Preisig, M. A., Fenton, B., et al. (1998). Familial transmission of substance use disorders. *Archives of General Psychiatry, 55,* 973–979.

20. LeGasse, L. L., Messinger, D., Lester, B. M., Seifer, R., Tronick, E. Z., Bauer, C. R., et al. (2003). Prenatal drug exposure and maternal and infant feeding behaviour. *Archives of Disease in Childhood: Fetal and Neonatal Edition, 88,* OF391–399; Mayes, L. C., Bornstein, M. H., Chawarska, K., & Granger, R. H. (1995). Information

processing and developmental assessments in 3-month-old infants exposed prenatally to cocaine. *Pediatrics, 9,* 778–783.

21. Raine, A., Brennan, P., & Mednick, S. A. (1997). Interaction between birth complications and early maternal rejection in predisposing individuals to adult violence: Specificity to serious, early-onset violence. *American Journal of Psychiatry, 154,* 1265–1271; Raine, A., Venables, P. H., & Mednick, S. A. (1997). Low resting heart rate at age 3 years predisposes to aggression at age 11 years: Findings from the Mauritius joint child health project. *Journal of the American Academy of Child and Adolescent Psychiatry, 36,* 1457–1464.

22. Caspi, A., Moffitt, T. E., Newman, D., & Silva, D. A. (1996). Behavioral observations at age 3 years predict adult psychiatric disorders: Longitudinal evidence from a birth cohort. *Archives of General Psychiatry, 53,* 1033–1039.

23. Englund, M. M., Egeland, B., Oliva, E. M., & Collins, A. (2008). Childhood and adolescent predictors of heavy drinking and alcohol use disorders in early adulthood: A longitudinal development analysis, *Addiction, 103,* 23–35; Kirisci, L., Tarter, R. E., Reynolds, M., & Vanyukov, M. (2005). Individual differences in childhood neurobehavioral disinhibition predict decision to desist substance use during adolescence and substance use disorder in young adulthood: A prospective study. *Addictive Behaviors, 31,* 686–696; Tarter, R. E., Kirisci, L., Mezzich, A., Cornelious, J. R., Pajer, K., Vanyukov, M., et al. (2003). Neurobehavioral disinhibition in childhood predicts early age at onset of substance use disorder. *American Journal of Psychiatry, 160,* 1078–1085.

24. Casey, B. J., Jones, R. M., & Hare, T. A. (2008). The adolescent brain. *Annals of the New York Academy of Sciences, 1124,* 111–126.

25. Dembo, R., Pacheco, K., Schmeidler, J., Fisher, L., & Cooper, S. (1997). Drug use and delinquent behavior among high risk youths. *Journal of Adolescent Substance Abuse, 6,* 1–25; Dembo, R., Williams, L., Fagan, J., & Schmeidler, J. (1993). The relationships of substance abuse and other delinquency over time in a sample of juvenile detainees. *Criminal Behaviour and Mental Health, 3,* 158–179; Rounds-Bryant, J. L., Kristiansen, P. L., Fairbank, J. A., & Hubbard, R. L. (1998). Substance use, mental disorders, abuse and crime: Gender comparisons among a national sample of adolescent drug treatment clients. *Journal of Child and Adolescent Substance Abuse, 7,* 19–34; Thompson, L., Riggs, P., Mukilich, S., & Crowley, T. (1996). Contribution of ADHD symptoms to substance problems and delinquency in conduct disordered adolescents. *Journal of Abnormal Child Psychology, 24,* 325–347; Van Kammen, W. B., Loeber, R., & Stouthamer-Loeber, M. (1991). Substance use and its relationship to conduct problems and delinquency in young boys. *Journal of Youth and Adolescence, 20,* 399–413; Wasserman, G. A., McReynolds, L. S., Lucas, C. P., Fisher, P., Santos, L. (2002). The voice DISC-IV with incarcerated male youths: Prevalence of disorder. *Journal of the American Academy of Child and Adolescent Psychiatry, 41,* 314–318.

26. Mailloux, D. L., Forth, A. E., & Kroner, D. G. (1997). Psychopathy and substance use in adolescent male offenders. *Psychological Reports, 81,* 529–530.

27. O'Neill, M. L., Lidz, V., & Heilbrun, K. (2003). Adolescents with psychopathic characteristics in a substance abusing cohort: Treatment process and outcomes. *Law and Human Behavior, 27,* 299–313.

28. Alterman, A. I., & Cacciola, J. S. (1991). The antisocial personality disorder diagnosis in substance abusers. *Journal of Nervous and Mental Disease, 179,* 401–409; Cadoret, R. J., O'Gorman, T. W., Troughton, E., & Heywood, E. (1985). Alcoholism and antisocial personality. *Archives of General Psychiatry, 42,* 161–167; Gerstley, L. J., Alterman, A. I., McClellan, A. T., & Woody, G. E. (1990). Antisocial personality disorder in substance abusers: A problematic diagnosis? *American Journal of Psychiatry, 147,* 173–178; Moeller, G. F., & Dougherty, D. M. (2001). Antisocial personality disorder, alcohol, and aggression. *Alcohol Research and Health, 25,* 5–11.

29. Hser, Y. I., Boyle, K., & Anglin, M. D. (1998). Drug use and correlates among sexually transmitted disease patients, emergency room patients, and arrestees. *Journal of Drug Issues, 28,* 437–454; Hser, Y. I., Hoffman, V., Grella, C. E., & Anglin, M. D. (2001). A 33-year follow-up of narcotics addicts. *Archives of General Psychiatry, 58,* 503–508.

30. Goldstein, P. J. (1985). The drugs/violence nexus: A tripartite conceptual framework. *Journal of Drug Issues, 15,* 493–506.

31. Vaughn, M. G. (2008). Biosocial dynamics: A transdisciplinary approach to violence. In M. DeLisi & P. J. Conis (Eds.), *Violent offenders: Theory, research, public policy, and practice* (pp. 63–77). Sudbury, MA: Jones & Bartlett.

CHAPTER TWELVE

Developmental Trajectories of Exposure to Violence

Daniel J. Flannery, Manfred H. M. van Dulmen, and Andrea D. Mata

KEY TERMS

Bullying
Comorbidity
Protective factors
Risk factors
Violence exposure

INTRODUCTION

Children and adolescents are exposed to violence at higher rates than adults. **Violence exposure** can include either being a victim of violence or witnessing violence. Rates of exposure to violence peak during middle childhood but then decline in adolescence. Despite declining rates of school violence, many children and adolescents are victims of violence, with almost half (44 percent) of all middle school students in one community sample reporting that they have been the victim of violence. Violence exposure occurs across different developmental contexts, including schools, neighborhoods, and at home. Many children are exposed to violence at school. Within the school context, children experience violence ranging from bullying all the way to becoming victims of fatal school violence. These high prevalence rates, the evidence for age-related changes in exposure to violence, and the occurrence of exposure to violence within the school context, signify the importance of understanding how exposure to violence changes during the school-age years.[1]

CORRELATES AND ANTECEDENTS OF EXPOSURE TO VIOLENCE

Exposure to violence has detrimental effects on children. Children who experience violence at school are, for example, at risk for developing traumatic symptoms and are more likely to engage in violent behavior. Empirical findings on the risk and **protective factors** underlying exposure to violence also provide evidence for some notable developmental differences. For example, in some of our own work we demonstrated that levels of mental health problems among children experiencing elevated levels of exposure to violence are particularly high during the middle childhood years but decline throughout adolescence. Exposure to violence also affects school achievement as well as **risk factors** for school achievement. Children who experience violence have more difficulty concentrating in the classroom and are also more likely to be truant and to drop out of school. These latter effects are particularly profound for children exposed to bullying at school. Exposure to violence also impacts school functioning indirectly through the adverse effects on social and emotional adjustment, and is associated with depression, suicide, and early alcohol use. Adolescents who have academic difficulties are more likely to experience and engage in high-risk behaviors, and it is likely that exposure to violence has indirect effects on academic achievement by affecting risky behavior and emotional adjustment.[2]

Boys are generally more likely to be victims of violence than girls, although this may somewhat differ by subtype, with females more likely to be victims of sexual violence and males more likely to be victims of physical violence. Other work on juvenile offenders' **comorbidity** for mental health and substance abuse problems found some additional evidence for gender specificity of subtypes of violence victimization: Females report much higher rates of sexual victimization than do males. Among perpetrators of violence, violent females are more likely to have experienced victimization compared with violent males. Females with a history of exposure to violence are also more likely to become violence perpetrators compared with males with a history of violence victimization.[3] Together, these findings illustrate that despite the fact that males may experience, on average, higher levels of exposure to violence than females, exposure to violence has more serious implications for mental health and violence perpetration among females as compared with males. An important future research direction is to investigate the nature of these gender differences.

The aforementioned body of research has provided some great insight into the developmental course of violence exposures, as well as their antecedents and correlates. This body of research has, however, been primarily based on variable-centered techniques that focus on understanding individual differences. These variable-centered techniques do not allow for the possibility that developmental differences regarding violence exposure are not solely based on individual differences, but may also be described by group differences in development. Group differences have been described with regard to violence perpetration, primarily informed by Moffitt's dual taxonomy.[4] According to this dual taxonomy, there are two groups of violence perpetrators: childhood-onset/life-course-persistent offenders and adolescent-limited offenders. The childhood-onset/life-course-persistent group starts engaging in violence perpetration during childhood and continues to be involved in violence perpetration throughout adulthood. The adolescent-limited group starts engaging in violence perpetration during adolescence and experiences declining levels of violence perpetration into adulthood. Empirical studies confirmed the existence of this dual taxonomy, although many studies also identify a number of additional groups that display a unique developmental trajectory of violence perpetration with age. This body of research has been largely limited to the study of males.[5]

Research on developmental trajectories of violence has, however, been primarily limited to the study of violence perpetration. Yet, previous empirical studies do indicate that violence perpetration and violence exposure are associated at high levels. Experiences of violence victimization and witnessing violence are much higher among violent youth compared with nonviolent youth. During the elementary school years, violence exposure accounts for a substantial amount of the variability in violence perpetration among boys (26 percent) and girls (22 percent).[6] Together, these findings demonstrate the importance of considering knowledge on the development of violence perpetration when studying violence victimization.

It is important to consider whether changes in violence exposure can be described using a developmental typology for two reasons. First, investigating whether there is a developmental typology can provide insight into whether all children follow the same trajectory regarding violence exposure. In other words, can changes in violence exposure be "best" described by one growth function or by multiple growth functions? Identifying whether there are multiple trajectories underlying violence exposure has direct implications for theory and conceptualization underlying these phenomena. Second, identifying whether there are multiple trajectories underlying violence exposure informs developmentally tailored interventions. If children follow different trajectories of violence exposure, it would be important to recognize that intervention and prevention efforts should be tailored uniquely to each group.

GROUP-BASED MODELING OF MIDDLE CHILDHOOD EXPOSURE TO VIOLENCE: PREVIOUS FINDINGS AND EMPIRICAL ILLUSTRATION

One notable exception to the lack of research on developmental trajectories of exposure to violence is a Canadian school-based study of fifth through seventh graders that focused on violence victimization.[7] Using latent class growth curve analysis, Goldbaum and colleagues identified four developmental trajectories based on two items pertaining to school **bullying**: nonvictims (low levels of victimization across time, $n = 1089$), desisters (decreasing levels of victimization across time, $n = 76$), late-onset victims (starting with low levels of victimization in grade 5 but increasing across time, $n = 56$), and stable victims (starting with high levels of victimization at grade 5 but increasing across time, $n = 20$). Although there were no statistically significant differences by gender, males were somewhat overrepresented (about 60 percent) in the desister, late-onset, and stable trajectories. Furthermore, it is important to note that the trajectories representing some developmental change were rather small in terms of sample size. More recently, Barker and colleagues applied growth mixture models to the identification of victimization trajectories in a Scottish sample of adolescents aged 12 to 16. They identified three trajectories: low/stable (85 percent), high/decreasing (10 percent), and high/increasing (5 percent).[8]

We wanted to extend the previous findings to investigate whether they would translate to a U.S. sample of middle childhood individuals by using a victimization measure that is more broadly constructed. The items in the aforementioned studies focused solely on school bullying. For the current analyses we relied on items from a violence exposure measure that considers victimization and witnessing in various contexts including school, home, and neighborhood. The Recent Exposure to Violence Scale is a 22-item measure of violence exposure. Eleven items focus on violence victimization and another 11 items focus on witnessing violence. Previous factor analytic studies suggest that the structure of the Recent Exposure to Violence Scale is consistent across different age groups and genders, and that—depending on the goals of the study—it may be important to separate victimization versus witnessing by developmental contexts.[9]

It is important to conduct analyses separately for boys and girls. As previously mentioned, prevalence rates on exposure to violence differ between boys and girls. Furthermore, the literature on violence perpetration is not clear regarding the existence of gender differences related to the developmental typology of violence perpetration. The large majority of studies on violence perpetration focus on boys. There is some theoretical evidence that developmental trajectories of violence perpetration may differ across boys and girls,[10] but studies that only focus on girls tend to find a smaller number of developmental trajectories of violence perpetration than studies that include both males and females. In summary, because the empirical literature on gender differences related to the developmental typology of violence perpetration is inconsistent and these empirical studies with regard to exposure to violence are lacking, it is important to explore whether this typology differs for boys and girls.

Sample and Statistical Analyses

In the current study we used data from a cohort of Arizona elementary school children and adolescents, commonly known as the PeaceBuilders study.[11] For the current analyses we limited ourselves to the grades where a large majority of the students had violence exposure data available, namely grades 5, 6, and 7 ($n = 538$). This sample was predominantly Hispanic (76 percent) with White being the next largest ethnic group (20 percent).

The identification of developmental trajectories was guided by statistical modeling, more specifically latent class growth analysis. It is important to use statistical modeling over assigning individuals to different developmental trajectories for several reasons. Most notably, assigning techniques assume that homogeneous subpopulations of offenders exist (without verifying their existence), and pose the problem of potentially incorrectly specifying the number of subpopulations.[12] Assignment may also involve somewhat random or arbitrary choices. In addition to a range of fit statistics, practical usefulness was used to decide which developmental trajectory solution best represents the data. Practical usefulness consists of the following aspects: trajectory shapes, number of individuals in each class, substantive theory, and predictive validity.

Male Victimization Trajectories

Unconditional latent class growth curve analyses were conducted to determine the number of male violence exposure trajectories that best fit the data. The three-class model represented the following violence-exposure trajectories for males: decreaser (moving from high to low levels of violence exposure, 10.9 percent, $n = 31$), stable (consistently low levels of violence exposure, 77.2 percent, $n = 217$), and increaser (accelerating from low to high levels of violence exposure, 11.9 percent, $n = 33$).

Figure 12-1 depicts the observed means at each time point for the three-class male violence-exposure trajectories. The decreaser trajectory included individuals who had the highest levels of violence exposure at fifth grade, drastically decreased from fifth grade to sixth grade, and then decreased to the second lowest level of violence exposure at seventh grade. The stable trajectory had the lowest violence exposure levels consistently across all three grades. The increaser trajectory had the second lowest intercept, slightly increased to the highest violence exposure levels at sixth grade, and then drastically increased to the highest violence exposure levels at seventh grade. There were no ethnic differences between the trajectory groups.

Female Victimization Trajectories

The same method—unconditional latent class growth curve analyses—was conducted to determine the number of female victimization trajectories that best fit the data. The three-class model was considered the best fit of the data. The three-class model of female violence exposure trajectories included the following trajectories: stable (consistently low levels of violence exposure, 70.8 percent, $n = 182$), decreaser (decreasing from high to average levels of violence exposure 5.4 percent, $n = 14$), and increasers (slight increases in levels of violence exposure across the three grades, 23.8 percent, $n = 61$).

Figure 12-2 depicts the observed means at each time point for the three-class female violence-exposure model. The low stable trajectory had the lowest level of violence exposure across the three grades and did not increase or decrease. The decreaser trajectory had the highest initial level of

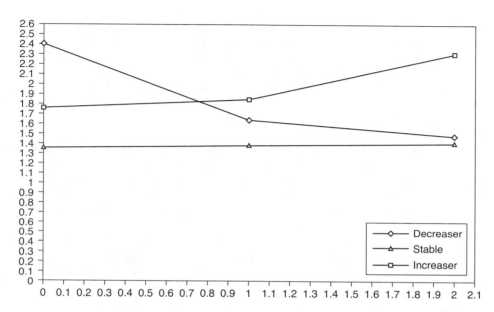

FIGURE 12-1.   Male Victimization Trajectories

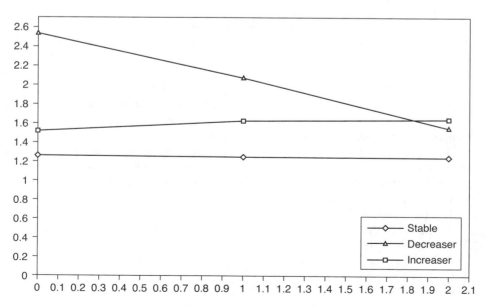

FIGURE 12-2.   Female Victimization Trajectories

violence exposure and then consistently decreased across the three grades. The decreaser trajectory had the second lowest levels of violence exposure at seventh grade. The slight increasers had the second lowest initial violence exposure levels, slightly increased between fifth and sixth grades, and then remained stable between sixth and seventh grades. The slight increasers had the highest levels of violence exposure at seventh grade. There were no ethnic differences between the trajectory groups.

Teachers' Knowledge of Children

We also explored whether teachers knew who these children were. The information on violence exposure was based on children's self-report, but teachers reported for a subset of children on several indicators of school functioning. Cross-sectional findings with the elementary school sample from the current study indicated, for example, that for a large group of children (33 percent), child self-reports and teacher reports of threatening behaviors diverged.[13] Furthermore, findings from the same sample provided evidence that teacher reports of children's behavior—aggression and social competence—affect the child's perceived quality of the teacher–child relationship, with children's aggression having a negative effect on this relationship and social competence a positive effect.

For boys, analysis of variance yielded statistically significant results for grade 7 teacher-reported preferred behavior where the child compromises with peers when a situation calls for it, and teacher-reported school adjustment where the child listens carefully to teacher instructions and directions. At grade 7, teachers reported higher levels of preferred behavior for decreasers compared with increasers of violence exposure. Teachers also reported higher levels of grade 7 preferred behavior for individuals with stable low levels of violence exposure compared with individuals with increasing levels of violence exposure. Teachers reported higher levels of school adjustment at grade 7 for individuals with stable low levels of violence exposure compared with individuals with increasing levels of violence exposure.

For girls, grade 7 teacher reports of school adjustment and preferred behavior were associated with trajectory membership. Group differences were, however, not in the predicted direction. Teachers reported higher levels of school adjustment for girls with slightly increasing levels of violence exposure compared with girls with decreasing levels of violence exposure. Similarly, teachers reported higher levels of preferred behavior for girls with slightly increasing levels of violence exposure compared with girls with decreasing levels of violence exposure. Teachers also reported higher levels of preferred behavior at grade 7 for girls with stable low levels of violence exposure compared with girls with decreasing levels of violence exposure.

We also investigated whether teacher reports of threatening behavior would be linked to violence exposure trajectories. Previous analyses with the current sample indicated that self-reported and teacher-reported threats of interpersonal violence diverge for about one-third of all children, indicating that teachers and children provide unique perspectives on threats of violence. For teacher reports of threatening behavior during fifth grade, there were statistically significant differences comparing girls in the decreaser trajectory and the slight increaser trajectory to the girls in the stable trajectory. Girls in the decreaser trajectory were nearly 12 times more likely to threaten compared with girls in the stable trajectory. Girls in the slight increaser trajectory were nearly four times more likely to threaten compared with girls in the stable trajectory. There were no statistically significant differences in threatening behavior during fifth grade between girls in the decreaser trajectory and girls in the slight increaser trajectory.

For teacher reports of threatening behavior during seventh grade, there were statistically significant differences comparing girls in the decreaser trajectory and the slight increaser trajectory to the girls in the stable trajectory. Girls in the decreaser trajectory were nearly 10 times more likely to threaten compared with girls in the stable trajectory. Girls in the slight increaser trajectory were three times more likely to threaten compared with girls in the stable trajectory. There were no statistically significant differences on threatening behavior during seventh grade between girls in the decreaser trajectory and girls in the slight increaser trajectory.

DISCUSSION

The findings from the current study indicate that children follow various trajectories of exposure to violence through early adolescence. Previous findings suggest that levels of exposure to violence decrease from middle childhood into adolescence. The findings of the current study add to this body of literature by highlighting that some children experience increases in violence exposure with age, whereas other children experience consistently low levels of violence exposure or decreases in violence exposure. These findings highlight the importance of developing timely preventive interventions.

The findings of the current study partially support results of group-based modeling with a Canadian sample in that we also identified a stable–low and desister trajectory. Contrary to the findings with the Canadian sample, however, the current analyses did not identify two, but rather one group of children characterized by increasing levels of violence exposure from fifth through seventh grade. One of the two increaser groups in the Canadian sample consisted, however, of only 1.6 percent of the total sample. It is possible that we were not able to replicate this latter finding because of the small size of that group.

It was somewhat surprising that the percentage of individuals in trajectory groups other than the low–stable group was relatively small. Among boys, 23 percent of individuals experienced either increases or decreases in violence exposure. This percentage was somewhat larger among girls—30 percent. Cross-sectional survey studies show a much higher percentage of youth experiencing violence victimization, with almost half of all adolescents experiencing violence at school. Previous findings with the violence-exposure measure used in this study indicate that about 40 percent of 7- to 15-year olds experience violence victimization at home, and about one-third of these children experience violence victimization at school. It is important to note, however, that further exploration of the data indicated that more than 90 percent of individuals in the low–stable group exceeded a score of one on the exposure to violence measure, indicating that this group on average does experience some level of violence exposure. In other words, even though we found a relatively small group of children who experienced changes in levels of violence exposure across the three grades, a large majority of children in this group did report some level of violence exposure during this time period.

The current study was exploratory in nature, and there are a number of logical next steps. First, it is essential to investigate whether trajectories of exposure of violence differ across victimization and witnessing violence. A much larger number of children witness violence as compared with being a direct victim of violence. In addition to mean-level differences between witnessing and direct victimization, factor analytic work with violence exposure measures also indicates the potential importance of separating items regarding direct victimization from items that measure aspects of witnessing violence. Therefore, a next logical step is to investigate whether the developmental typology differs across these two dimensions of violence exposure.[14]

In addition to investigating whether the developmental typology of violence exposure differs across dimensions of witnessing violence and victimization, it is important to investigate how this typology may differ across developmental contexts, such as home, school, and neighborhood. Work on violence exposure not only highlighted the importance of potentially separating items by type of violence exposure, but also by developmental context. Violence exposure rates differ by developmental context and are somewhat lower in school than in the home/neighborhood context. Despite these mean-level differences—and evidence that specific experiences of violence exposure within a specific context, such as the school, have detrimental effects on mental health and behavioral outcomes—few studies investigate effects of violence exposure within a specific context of individual adjustment and psychopathology. Studying violence exposure by specific developmental context is particularly important because the role, function, and salience of these various contexts changes as children move through middle childhood and adolescence.

A third logical step is to investigate the relationship between these developmental trajectories and individual outcomes. For example, we would expect a strong relationship between trajectory membership on exposure to violence and trajectory membership on violence perpetration (i.e., Moffitt typology), largely based on previous research indicating a strong relationship between exposure to violence and becoming a violent perpetrator. Specific to exposure to violence as related to school-age children, it would be important to explore how these trajectories are related to school functioning and academic performance. Previous empirical findings, as discussed earlier, demonstrated both direct and indirect links between exposure to violence and academic functioning. The data for this project did not allow us to answer this specific question, but we did demonstrate some important associations between (1) teacher perceptions of children's behavior in the classroom as well as teacher reports of threatening behavior, and (2) different violence exposure trajectories based on children's self-reports of violence exposure. These findings demonstrated that among boys, decreasers and children with low levels of exposure to violence were rated more favorably than children experiencing increasing levels of violence exposure. It was somewhat puzzling that the girls who were rated most favorably by their teachers were those who also experienced increasing levels of exposure to violence. These findings do, however, highlight the importance of including both boys and girls in trajectory analyses. It is important for future studies to try to replicate these findings and explore possible reasons as to why these children were rated more favorably by their teachers. With regard to teacher reports of threatening behavior, we generally found that individuals who experienced exposure to violence during the middle school years were also more likely to threaten their peers. These findings provide further support for the previously demonstrated relationship between exposure to violence and violence perpetration.

CONCLUSION

Surprisingly, we did not find gender differences in the current study. The developmental typology of violence exposure during middle childhood was similar for boys and girls, although there were some important prevalence differences between boys and girls. There were much larger percentages of increasers and decreasers for girls as compared with boys. In other words, girls showed more change, either increasing or decreasing, in violence victimization during middle childhood compared with boys. Clearly, these findings require further replication, but are somewhat consistent with the literature on violence perpetration. Empirical findings have demonstrated gender differences with regard to prevalence rates in various developmental trajectories and little empirical evidence for a unique developmental typology underlying violence perpetration among girls.

GLOSSARY

Bullying—the use of intimidation or direct violence to secure some desired outcome

Comorbidity—the co-occurrence of two or more psychiatric disorders

Protective factors—any factor that decreases the likelihood that an antisocial outcome will occur

Risk factors—any factor that increases the likelihood that an antisocial outcome will occur

Violence exposure—refers to either being a victim of violence or witnessing violence

NOTES

1. Flannery, D. J., Wester, K. L., & Singer, M. I. (2004). Impact of exposure to violence in school on child and adolescent mental health and behavior. *Journal of Community Psychology, 32*, 559–573; Hashima, P. Y., & Finkelhor, D. (1999). Violent victimization of youth versus adults in the National Crime Victimization Survey. *Journal of Interpersonal Violence, 14*, 799–820; Moore, M., Petrie, C. V., Braga, A. A., & McLaughlin, B. L. (2003). *Deadly lessons: Understanding lethal school violence.* Washington, DC: National Academies Press; Olweus, D. (1993). *Bullying at school: What we know and what we can do.* Cambridge, MA: Blackwell; Singer, M. I., Anglin, T. M., Song, L., & Lunghofer, L. (1995). Adolescents' exposure to violence and associated symptoms of psychological trauma. *Journal of the American Medical Association, 273*, 477–482.

2. Blum, R. W., Beuhring, T., Shew, M. L., Bearing, L. H., Sieving, R. E., & Resnick, M. D. (2000). The effects of race/ethnicity, income and family structure on adolescent risk behaviors. *American Journal of Public Health, 90*, 1879–1884; Cullingford, C., & Morrison, J. (1995). Bullying as a formative influence: The relationship between the experience of school and criminality. *British Educational Research Journal, 21*, 547–561; Flannery, D. J., Singer, M. I., van Dulmen, M., Kretschmar, K., & Belliston, L. M. (2007). Exposure to violence, mental health and violent behavior. In D. J. Flannery, A. T. Vazsonyi, & I. Waldman (Eds.), *The Cambridge handbook of violent behavior and aggression* (pp. 306–322). Cambridge, UK: Cambridge University Press; Flannery, D., & Wester, K. (2004). Risk factors related to academic achievement in adolescence. In F. Pajeras & T. Urdan (Eds.), *Contemporary practices and challenges in the education of adolescents* (pp. 1–31). New York: Information Age Publishing; Swahn, M. H., Bossarte, R. M., & Sullivent, E. E. (2008). Age of alcohol use initiation, suicidal behavior, and peer and dating violence victimization and perpetration among high-risk, seventh-grade adolescents. *Pediatrics, 121*, 297–305.

3. Finkelhor, D., & Dziuba-Leatherman, J. (1994). Children as victims of violence: A national survey. *Pediatrics, 94*, 413–420; Flannery, D. J., Singer, M. I., & Wester, K. (2001). Violence exposure, psychological trauma, and suicide risk in a community sample of dangerously violent adolescents. *Journal of the American Academy of Child and Adolescent Psychiatry, 40*, 435–442; Flannery, D. J., Singer, M., Williams, L., & Castro, P. (1998). Adolescent violence exposure and victimization at home: Coping and psychological trauma symptoms. *International Review of Victimology, 6*, 29–48; Hussey, D. L., Drinkard, A. M., & Flannery, D. J. (2007). Comorbid substance use and mental disorders among offending youth. *Journal of Social Work Practice in the Addictions, 7*, 117–138.

4. Moffitt, T. E. (1993). "Life-course-persistent" and "adolescent-limited" antisocial behavior: A developmental taxonomy. *Psychological Review, 100*, 674–701.

5. van Dulmen, M. H. M., Goncy, E., Vest, A., & Flannery, D. J. (2009). Group based trajectory modeling of externalizing behavior problems from childhood through adulthood: Exploring discrepancies in the empirical findings. In J. Savage (Ed.), *The development of persistent criminality.* New York: Oxford University Press.

6. Song, L., Singer, M. I., & Anglin, T. M. (1998). Violence exposure and emotional trauma as contributors to adolescents' violent behaviors. *Archives of Pediatrics and Adolescent Medicine, 152*, 531–536.

7. Goldbaum, S., Craig, W. M., Pepler, D., & Connolly, J. (2003). Developmental trajectories of victimization: Identifying risk and protective factors. *Journal of Applied School Psychology, 19*, 139–156.

8. Barker, E. D., Arseneault, L., Brendgen, M., Fontaine, N., & Maughan, B. (2008). Joint development of bullying and victimization in adolescence: Relations to delinquency and self-harm. *Journal of the American Academy of Child and Adolescent Psychiatry, 47*, 1030–1038.

9. van Dulmen, M. H. M., Belliston, L. M., Flannery, D. J., & Singer, M. (2008). Confirmatory factor analysis of the recent exposure to violence scale across three samples from middle childhood through adolescence. *Children & Schools, 30*, 93–102.

10. Silverthorn, P., & Frick, P. J. (1999). Developmental pathways to antisocial behavior: The delayed-onset pathway in girls. *Development and Psychopathology, 11*, 101–126.

11. Embry, D., Flannery, D. J., Vazsonyi, A., Powell, K., & Atha, H. (1996). PeaceBuilders: A theoretically driven, school-based model for early violence prevention. *American Journal of Preventive Medicine, 12*, 91–100; Flannery, D. J., Singer, M. I., & Wester, K. (2003). Violence, coping, and mental health in a community sample of adolescents. *Violence and Victims, 18*, 403–418.

12. Nagin, D. S. (2005). *Group-based modeling of development*. Cambridge, MA: Harvard University Press.

13. Liau, A. K., Flannery, D. J., & Quinn-Leering, K. (2004). A comparison of teacher-rated and self-reported threats of interpersonal violence. *Journal of Early Adolescence, 24*, 231–249.

14. Singer, M. I., Flannery, D. J., Guo, S., Miller, D., & Leibbrandt, S. (2004). Exposure to violence, parental monitoring, and television viewing as contributors to children's psychological trauma. *Journal of Community Psychology, 32*, 489–504; Singer, M. I., Miller, D. B., Guo, S., Flannery, D. J., Frierson, T., & Slovak, K. (1999). Contributors to violent behavior among elementary and middle school children. *Pediatrics, 104*, 878–884.

CHAPTER THIRTEEN

A Partial Test of Social Structure Social Learning: Neighborhood Disadvantage, Differential Association with Delinquent Peers, and Delinquency

Chris L. Gibson, Traci B. Poles, and Ronald L. Akers

KEY TERMS

Disadvantaged neighborhoods
Ethnic diversity
Informal social control
Neighborhood structural elements
Residential instability
Social disorganization theory
Social learning

INTRODUCTION

Historically, research has shown that crime occurs at higher rates in neighborhoods with elevated levels of disadvantage compared with those that are less disadvantaged. **Disadvantaged neighborhoods** are characterized by high poverty rates, high percentages of unemployment, racial and ethnic heterogeneity, frequent residential mobility, and segregation. In their classic study, Shaw and McKay established that areas in Chicago defined by such structural characteristics had higher rates of crime and juvenile delinquency, among other social ills, that they attributed to a breakdown in a neighborhood's ability to maintain effective social control. Although much research since has supported Shaw and McKay, only recently has research emerged on how the delinquent behaviors of youths are impacted by the neighborhood conditions where they grow up.[1]

Studies analyzing neighborhood influences on children's behavioral development suggest that where children reside matters. In fact, several studies have found some support for the notion that children residing in poorer, segregated, and less cohesive neighborhoods are more likely to engage in antisocial behavior, but the mechanisms as to why they do are not empirically well established. Researchers and theorists have recently begun to explore reasons why children growing up under such disadvantaged conditions are more inclined to engage in delinquency and why some engage in such behaviors more frequently than others. Several of these explanations or mechanisms have been directed toward the social conditions of neighborhoods, such as collective efficacy, neighborhood organization, and institutional resources.[2]

In the social structure social learning (SSSL) model, Akers offers an additional explanation for why children engage in antisocial behaviors more often in disadvantaged, socially disorganized neighborhoods, which has been largely neglected in the literature on neighborhood influences on children. Akers argues that **social learning** substantially mediates the link between structural conditions of a neighborhood and delinquency. Specifically, he theorizes that children and adolescents growing up in socially disorganized neighborhoods are more likely to engage in delinquent behaviors because of increased associations with delinquent peers, differential reinforcement for engaging in delinquent behaviors, exposure to more favorable definitions for delinquent behavior, and more delinquent models to imitate.

Social disorganization is a theoretically defined condition included in the SSSL model and hypothesized to affect variation in the social learning variables that, in turn, impact delinquent behavior. One prediction of the SSSL model is that adolescents residing in socially disorganized neighborhoods are more likely to engage in acts of delinquent behavior because they differentially associate with delinquent peers. Delinquent peer networks provide youth with more opportunities to imitate behaviors that are most favorable to crime and delinquency. Differential association with delinquent peers also increases the likelihood that antisocial behaviors will be positively reinforced, and definitions that are more favorable to antisocial behaviors will become more prevalent. To date, only a handful of studies have empirically tested propositions from the SSSL model, none of which has focused on a mediating effect of differential association on the relationship between neighborhood social disorganization and youth involvement in delinquency.[3]

Because theory can best be understood with an application of data, this chapter analyzes data from the Project on Human Development in Chicago Neighborhoods (PHDCN) to test one core proposition from the SSSL model—that is, whether a measure of differential association with delinquent peers mediates the effect of neighborhood disadvantage on children and adolescent's involvement in one form of antisocial behavior (i.e., delinquency). The PHDCN is well suited for exploring such mediating effects given that it is a large-scale, interdisciplinary study on how neighborhood conditions influence developmental outcomes. The PHDCN consists of a longitudinal study of approximately 6000 children, adolescents, and young adults ranging in age from infancy to 18 years old, and a large-scale neighborhood data collection effort from approximately 80 neighborhood areas in which these youth and young adults reside.

Before discussing our data and measures, several aspects of neighborhood influences on behavior are reviewed. We review social disorganization theory and its link to both crime and delinquency, and models that have been offered as mechanisms for why neighborhood structural conditions are related to delinquency. We provide an overview of the SSSL model and studies that have tested various aspects of the model, as well as studies that show support for the model even though they were not specific tests.

SOCIAL DISORGANIZATION, CRIME, AND DELINQUENCY _____

Social disorganization theory is at the core of studies that link neighborhood conditions to child development; therefore, it is important to provide an overview of its main tenets. The relationship between geographical areas of cities (and other geographical units), social control, and crime is the main focus of social disorganization theory. At a time of high immigration, Shaw and McKay studied arrest rates of juveniles from 1900–1906, 1917–1923, and 1927–1933 throughout Chicago. Shaw and McKay studied these three different time periods to identify whether the environment in which immigrants and minorities lived influenced delinquency. If a particular immigrant group had high delinquency rates during his or her migration through Chicago's different ecological environments, then delinquency could be associated with certain cultural or constitutional features. However, if delinquency rates decreased through the different ecological environments, then delinquency had to be connected with the environment and not a constitution of the immigrants and/ or minority groups.

Shaw and McKay used the concentric zone model created by Park, Burgess, and McKenzie in 1925 to study arrest rates in different geographical areas of Chicago. The concentric zone model distinguished five "natural urban areas" of Chicago starting from the center of Chicago to the outer cities: central business district, transitional zone (recent immigrants, deteriorated housing, factories, abandoned housing), working class zone (single family tenements), residential zone (single family homes, yards, and garages), and commuter zone (suburbs). Rates of delinquency were found to decrease as distance from the center of the city increased. No matter what racial or ethnic group lived there, delinquency rates always remained high in a particular region of the city, the "zone in transition."[4]

Shaw and McKay argued that the relationship between poverty and delinquency in urban areas is produced by the connection of poverty with a combination of **residential instability** and **ethnic diversity**. They argued that the combination of these factors produces social disorganization that in turn weakens systems of social relationships in a community. Social disorganization is the consequence of a community's inability to realize common values and to solve problems of its residents; a breakdown of effective community social control results. From research on social disorganization theory, Shaw and McKay determined that within inner-city areas, the course of becoming delinquent occurred through a network of interpersonal relationships involving family, gangs, and the neighborhood. Shaw and McKay assessed the physical appearance of the neighborhoods, the population's income, ethnicity, the percentage of people who rent versus homeowners, and how rapidly the population of the area changed. Moreover, they found that delinquency was not the result of personal characteristics of the people who lived in them, but rather the result of a strong neighborhood or place influence.[5]

Years later, Sampson and Groves's study was the first to address problems of prior studies testing social disorganization theory. Using data from neighborhoods in Great Britain, they formed a model that analyzed both structural and organizational characteristics of neighborhoods, where residential instability, poverty, family disruption, and racial/ethnic heterogeneity have traditionally been assumed to produce social disorganization, which was hypothesized to be the main neighborhood condition that increased the likelihood of crime and delinquency. Prior research used these presumed causes as indicators of social disorganization. Sampson and Groves argued, however, that these are not direct measures of social disorganization; therefore, previous research had failed to offer a direct test of social disorganization theory. In their research, they offered three direct indicators of social disorganization: weak local friendship networks, low organizational participation, and unsupervised teenage groups within a neighborhood. They found, consistent with social disorganization theory, that in neighborhoods in Great Britain where friendship ties were weak, organizational participation was low, and teen groups were unsupervised, crime was higher. Further, their social disorganization variables largely mediated the effects of other structural characteristics on crime.[6]

NEIGHBORHOODS AND CHILD DEVELOPMENT

Research has focused on how neighborhood structure and organization can influence child development, specifically how neighborhood conditions influence delinquent behavior of individual children as opposed to aggregate crime rates. Although researchers found links between **neighborhood structural elements** (e.g., poverty) and negative child and adolescent outcomes, there is less understanding as to why this link exists.[7]

There are several theoretical frameworks in which neighborhood conditions may influence behavioral development. They differentiate between the following: epidemic (or contagion), collective socialization, institutional, competition, and relative deprivation models. The epidemic or contagion model focuses on problem behaviors and is based on the idea that negative behaviors of peers can quickly spread throughout a neighborhood, thus affecting children and adolescents. The collective socialization model suggests that neighborhoods influence children and adolescents through their level of organization and control. If adult role models are present, they can help supervise, monitor, and configure routines and opportunities for children and adolescents in their neighborhoods. In contrast, a lack of monitoring and structuring in neighborhoods may lead to children becoming involved in delinquent and antisocial activities. It is argued in the institutional resource model that neighborhood institutions are believed to influence both positive and negative child development. Specifically, schools and police protection, along with access to resources such as parks, libraries, and community centers, are institutions within neighborhoods that promote positive behavioral development, and their absence may have negative impacts. The competition model suggests that neighbors compete with one another for scarce community resources, which in turn can lead to negative behaviors of children and adolescents. The relative deprivation model hypothesizes that neighborhood conditions and surroundings affect children and adolescents by means of their evaluation of their situation vis-à-vis others in the neighborhood.

SOCIAL STRUCTURE/SOCIAL LEARNING: AN EXPANDED AND COMPLEMENTARY REASON FOR THE LINK BETWEEN NEIGHBORHOOD STRUCTURE AND DELINQUENT BEHAVIOR

Although several models put forth by Jencks and Mayer[8] may indirectly implicate learning as a mechanism for explaining the link between neighborhood conditions and child and adolescent involvement in delinquent behaviors, none focus specifically on the neighborhood as a context for learning delinquent behaviors. For instance, Jencks and Mayer's epidemic or contagion model refers to peer influences as a meditating variable, but no detail is provided as to how this process operates. Social learning theory can provide a complimentary theoretical framework for understanding the link between neighborhoods and delinquency, which should not be seen as a competing explanation but rather as an additional piece of the puzzle. As such, social learning can help explain why children living in neighborhoods that are poor, with limited institutional resources and low collective socialization, are likely to engage in more delinquency.

The contexts in which social exchanges occur are full of opportunities to learn behaviors. The neighborhood is one such context where its level of organization provides varying opportunities to learn both conforming and delinquent behaviors. A given neighborhood's social structure and normative culture provide approval and disapproval of certain types of behavioral acts. The neighborhood context also consists of behavioral models that may increase chances to mimic or imitate both conforming and delinquent behaviors. Further, the application of informal sanctions within a neighborhood may allow children to learn which behaviors are acceptable and which are not. Like these other dimensions, the reinforcement and punishment contingencies vary depending on the social structure of the neighborhood context.[9] In sum, Akers (and colleagues) argues that antisocial, delinquent, and criminal behaviors are partially a function of the norms, social organization, and social control systems within a structural context that provides learning environments, socialization opportunities, and reinforcement schedules that are conducive to conforming and delinquent behaviors. Thus, neighborhoods with low **informal social control**, high crime, a lack of institutional resources, and poverty provide children and adolescents with more opportunities to associate with delinquent others, model delinquent and criminal behaviors, and have their own delinquent behaviors reinforced or at least not punished as often.

In his SSSL model, Akers provided a detailed theoretical framework for why social structure influences involvement in antisocial, delinquent, and criminal behaviors, arguing that structural variables influence crime rates by impacting how people learn to engage in delinquent and conforming behaviors. He put forth four elements of social structure that offer the contexts in which social learning variables (i.e., differential association, differential reinforcement, definitions favorable to antisocial behavior, and imitations) operate: differential social organization, differential location in the social structure, theory-driven aspects of social structure (e.g., social disorganization and collective efficacy), and differential location in primary, secondary, and reference groups. Differential social organization consists of community-level factors such as population, size, and density, as well as demographic characteristics such as racial and gender proportions in the community or neighborhood. These factors indicate how communities are organized and have been related to crime rates of neighborhoods, cities, and other macro-level contexts. Differential location in the social structure refers to social characteristics such as gender, race, age, and socioeconomic status. Akers argues that although these sociodemographic variables are often taken as descriptive of individual characteristics, they also serve to locate people in the larger social structure regarding their roles or social groupings. Theoretically defined constructs from macro-level theories, such as social disorganization and collective efficacy, predict that crime rates will be higher under greater social disorganization and lower collective efficacy. Differential social location refers to the meso-level social contexts of primary or secondary groups of family, friendship/peer groups, and other social groups to which one belongs or is affiliated that can encourage conforming or deviant behavior.

Akers argues that these four categories of social structure are correlated with individual involvement in antisocial, delinquent, and criminal behaviors through social learning processes. That is, dimensions of social learning (principally differential association, differential reinforcement, attitudes/definitions, and imitation) are expected to mediate the relationship between the social structure element and antisocial behaviors. For the SSSL model to be supported, several empirical relationships must be established. First, social structural variables must be correlated with individual antisocial behaviors. Second, social structural variables must be correlated with social learning variables. Finally, learning variables should have substantial and statistically significant effects on antisocial behaviors that in turn reduce the relationship between social structural variables and antisocial behaviors.

Although social learning theory's main propositions have been supported across time with various measures of constructs and deviant behavioral outcomes, diverse samples, and in different cultures, few studies have tested full or partial SSSL models. With some exceptions, the findings from that research have been generally supportive of the main hypothesis that social learning variables mediate the effects of structural variables on crime and deviance. Page found that learning variables mediated virtually all the modest, but statistically significant, main effects of family structure on delinquent behavior. Lee and colleagues reported multivariate analyses of adolescent substance use that showed that the social learning variables mediated "substantial, and in some instances virtually all, of the effects of gender, socio-economic status, age, family structure, and community size on these forms of adolescent deviance."[10] However, the net effects of gender, although very much reduced, remained statistically significant. Jensen found that the social learning variables of differential peer association and neutralizing definitions produced the strongest mediation of gender effects on officially recorded and self-reported minor and serious delinquency. In a study of the structural effects on levels of binge drinking conducted at several universities, others reported mixed but generally favorable findings on the mediation hypothesis. Other researchers also reported findings that were supportive of the theory.[11]

CURRENT FOCUS

In this chapter, data from the PHDCN longitudinal study are used to provide a partial test of the SSSL theoretical model. The focus is on one social structural element: theoretically defined constructs from the neighborhood level used to represent the structural conditions of neighborhoods in the PHDCN study, such as concentrated disadvantage. An important, but secondary, social structural element explored in this chapter is differential location in the social structure, including race, gender, and socioeconomic status. Further, we center on one social learning element: differential association with delinquent peers, which is arguably one of the most empirically influential elements of social learning theory. In sum, the current study tests several main research hypotheses derived from the SSSL model applied to neighborhoods and children's delinquent behaviors: (1) children and adolescents residing in neighborhoods with more concentrated disadvantage will have a higher proportion of delinquent peers than those living in less concentrated disadvantaged neighborhoods; (2) children and adolescents residing in neighborhoods with more concentrated disadvantage engage in more delinquency than those living in less disadvantaged neighborhoods; and (3) peer delinquency will have a direct influence on delinquency and will substantially mediate the effect of neighborhood disadvantage on children's and adolescents' delinquency.

METHODS

The PHDCN is an interdisciplinary study of how families, schools, and neighborhoods impact child and adolescent antisocial and prosocial behaviors. The data have two primary components. First, data on neighborhood structural and social processes were collected beginning in the early 1990s. This particular component of data collection includes measurements of the social, economic, organizational, political, and cultural processes, and the dynamic changes that take place in these structures and processes over time in Chicago neighborhoods. Second, the project had a component that was a longitudinal study of seven different cohorts of children and adolescents. This involved more than 6000 children, adolescents, and young adults, as well as their primary caregivers, who agreed to have their lives and personal characteristics observed over time to understand what combination of neighborhood, family, and individual characteristics put them at risk for antisocial behavior. Although data were collected on seven cohorts of children and adolescents, the current study uses data from the 9-, 12-, and 15-year-old cohorts, whereas the infant, 3-, 6-, and 18-year-old cohorts were excluded from the current study. For the current study, data from waves 1 and 2 are analyzed, largely to establish time ordering between our independent and dependent variables. Time between waves was approximately 2.5 years.

Two aspects of the sampling design are important to discuss: selection of neighborhoods and households for the longitudinal cohort study. First, regarding neighborhood selection, 847 census tracts were combined to create 343 neighborhood clusters. Neighborhood clusters averaged approximately 8000 people each, which are larger than census tracts; typically two census tracts combined that were similar and adjacent to one another. For the longitudinal study of children and adolescents, a stratified probability sample of 80 was chosen for more intensive study that reflected the racial, ethnic, and socioeconomic diversity of Chicago neighborhoods. Neighborhood data consisted of community surveys to measure neighborhoods' political, cultural, and social control structures. Further, systematic social observation data were collected to measure objective signs of physical and social disorder using video and audio recordings. Finally, census data and data on crime from the Chicago police department were linked to each neighborhood cluster. For the current analysis, we only use data from the U.S. census as indicators of neighborhood disadvantage. Second, block groups in the 80 neighborhood clusters were sampled. From these block groups a list of almost 40,000 dwellings were screened. Infants, young children, adolescents, and young adults were recruited to participate if they were within 6 months of the following age categories: 0, 3, 6, 9, 12, 15, and 18 years. The caregivers of these children were also asked to participate.

The current study uses longitudinal data from the 9-, 12-, and 15-year-old cohorts (descriptive statistics for the analysis sample are reported in **Table 13-1**). At wave 1, the first interview, there were 828 children and adolescents in the 9-year-old cohort, 820 in the 12-year-old cohort, and 696 in the 15-year-old cohort. Combined, the sample size of these three cohorts is 2344 subjects. However, after taking into account missing data and nonresponses on each variable, 1686 subjects at wave 1 remained. At wave 2, approximately 2 years after the initial wave 1 interviews, the sample size was further reduced to 1214 subjects. This attrition could have many sources; for example,

TABLE 13-1 Descriptive Statistics				
Variable	**Mean**	**SD**	**Minimum**	**Maximum**
Delinquency (wave 2)	2.16	1.95	0.00	14.00
Delinquent peers	27.23	5.63	19.00	50.00
Concentrated disadvantage	−.52	3.51	−5.36	11.92
White	.15	.35	0.00	1.00
Black	.38	.49	0.00	1.00
Hispanic	.43	.50	0.00	1.00
Other	.04	.20	0.00	1.00
Gender	.50	.50	0.00	1.00
Age (wave 1)	12.34	2.37	8.43	16.91
SES	−.07	1.41	−2.95	3.52

Note: SES = socioeconomic status, SD = standard deviation.

these 472 children may have relocated and decided not to participate in the study, and some of them could have invalid or missing data at wave 2. We decided to impute delinquency scores for the 472 individuals who had missing data at wave 2 by using information on the individual level data included in our models (demographic and peer delinquency measures). Of the analysis sample, 50.12 percent were male, 14.5 percent White, 38.14 percent Black, 43 percent Hispanic, and 4 percent other (Asian, Pacific Islander, etc.). The average age at wave 1 for the analysis sample was approximately 12 years old with a minimum age of 8.43 and a maximum age of 16.91.

Delinquency was measured using the reduced child behavioral checklist administered to primary caregivers at wave 2 to report about their child's behaviors. Specifically, this study uses the delinquency subscale taken from the externalizing behavior scale of the child behavioral checklist.[12] The delinquent behavior subscale at wave 2 contains 11 items that have response categories of "0," "1," or "2," indicating whether a behavioral characteristic is 0, "not true"; 1, "sometimes true"; or 2, "often true." Responses to items were summed so that higher scores on this scale indicated more delinquent involvement. This subscale was shown to have adequate to good reliability or internal consistency. Examples of items include "lies and cheats," "sets fires," "runs away from home," "disobedient at home," and "disobedient at school."

During wave 1 interviews, a 19-item measure of self-reported or perceived delinquency of peers was used to measure differential association with delinquent peers, one of the major social learning constructs that most likely captures several dimensions of social learning. Items include property offenses and theft, drugs, and violence. Items were measured on a 3-point scale, with 1 indicating "none of my friends," 2 "some of my friends," and 3 "all of my friends." Responses to these 19 questions were summed so that higher scores signified more differential association with delinquent peers and lower scores indicated less association with delinquent peers. Concentrated disadvantage serves as an indicator of neighborhood structure that taps into the degree of the concentration of poverty and isolation of a neighborhood, which has been shown to be related to neighborhood crime, collective efficacy, and disorder. This measure consists of six items taken from the 1990 U.S. census that have been shown to be highly correlated and appear to be measuring one construct in Chicago neighborhoods. Items include percentage of people below the poverty line, percentage of people on welfare, percentage of female-headed households, percentage unemployed, percentage younger than 18 years of age, and percentage Black. This measure was standardized so that increasingly more positive scores indicate more neighborhood disadvantage and increasingly more negative scores indicate less disadvantage.

The theoretical model places importance on differential location in the social structure, indicated by variables such as age, gender, race/ethnicity, and socioeconomic status. Furthermore, these variables have been shown to be correlates of delinquency. As such, these variables are important not only as control variables, but also as independent social structural variables to consider when assessing the effects of differential association with delinquent peers on the relationship between neighborhood-concentrated disadvantage and delinquency. In fact, the SSSL model indicates that social learning processes will also mediate measures of differential location in the social structure.

Race is measured using a series of four dummy variables where White is the reference group to which Blacks, Hispanics, and others are compared. The coding of these dummy variable are such where 1 = Black, 1 = Hispanic, and 1 = other. Gender is coded so that 0 = female and 1 = male. Age is measured as a continuous variable. Socioeconomic status is measured using a 3-item composite measure of household income, maximum education level achieved by primary caregiver and partner, and the socioeconomic index for primary caregivers' and partners' jobs. Socioeconomic status is treated as continuous, with higher scores representing higher family socioeconomic status.

RESULTS

Table 13-2 shows a linear regression model predicting wave 1 differential association with delinquent peers. This model is statistically significant and explains approximately 17 percent of the variation in association with delinquent peers. Several variables are significant predictors of differential association with delinquent peers. Importantly, neighborhood-concentrated disadvantage has a significant and positive effect on delinquent peers, indicating that children and adolescents residing in neighborhoods that are increasingly more disadvantaged are likely to associate with friends who are more delinquent compared with those living in less disadvantaged neighborhoods. Race, gender, and age also had significant and positive effects on delinquent peers. That is, Blacks (compared with Whites) were likely to associate with more delinquent friends. Males were more likely than females to associate with delinquent friends, and older subjects were more likely to associate with more delinquent friends than younger subjects. Socioeconomic status had a negative and significant effect, indicating that children and adolescents from families with lower socioeconomic status had more delinquent peers. Overall, results from this model are in line with what the SSSL theoretical model anticipates.

Table 13-3 shows a series of regression models predicting delinquency scores on the child behavioral checklist from wave 2, collected approximately 2.5 years after wave 1 peer delinquency was measured. These models investigate the mediating role of delinquent peers in the link between neighborhood-concentrated disadvantage and delinquency. Model 1 of Table 13-3 shows results from a bivariate linear regression where wave 2 delinquency is regressed on neighborhood-concentrated disadvantage. Overall, the model is statistically significant but explains approximately 3 percent of the variation in delinquency. Neighborhood-concentrated disadvantage has a significant and positive effect on delinquency. As anticipated by the SSSL model and neighborhood research, children and adolescents residing in neighborhoods with more concentrated disadvantage are more likely to engage in delinquency than those from less disadvantaged neighborhoods.

Model 2 shows the effect of neighborhood-concentrated disadvantage on delinquency when controlling for differential association with delinquent peers. As expected by the SSSL theoretical model, the positive effect of neighborhood-concentrated disadvantage on delinquency has been

TABLE 13-2 Regression Model Predicting Differential Association with Delinquent Peers

	b	SE	t
Concentrated disadvantage	.13	.04	2.97*
Black	1.80	.44	4.11*
Hispanic	.40	.40	1.01
Other	.03	.88	.03
Gender	1.07	.29	3.63*
Age (wave 1)	.84	.06	14.42*
SES	−.26	.10	−2.52*

Note: $F = 46.77$; $R^2 = .17$. SE = standard error, SES = socioeconomic status, b = unstandardized regression coefficient, t = t-value; * = significant at the $p < .05$ level.

TABLE 13-3 Regression Model Assessing the Mediating Effect of Differential Association with Delinquent Peers on the Link Between Concentrated Disadvantage and Wave 2 Child Behavioral Checklist Delinquency Scores

	Model 1			Model 2			Model 3		
	b	SE	t	b	SE	t	b	SE	t
Concentrated disadvantage	.10	.02	5.76*	.07	.02	4.68*	.03	.02	1.28
Delinquent peers	.09	.01	8.86*	.08	.01	7.40*	—		
Black							.66	.18	3.77*
Hispanic							.07	.15	.43
Other							.05	.22	.22
Gender							.27	.10	2.77*
Age (wave 1)							.05	.02	2.90*
SES							−.02	.04	−.68
	$F = 33.18*$			$F = 47.87*$			$F = 22.16*$		
	$R^2 = .03$			$R^2 = .10$			$R^2 = .12$		

Note: SES = socioeconomic status. $*p \leq .05$; Models were calculated using robust standard errors (SE) because children are nested within neighborhood, b = Unstandardized regression coefficient.

noticeably reduced, although it still remains statistically significant. In fact, a comparison indicates that the effect of concentrated disadvantage is reduced by approximately 30 percent. Further, the category of delinquent peers has a positive and statistically significant direct effect on delinquency, indicating that children and adolescents who associate with more delinquent peers will engage in more delinquency. Model 2 explains approximately 10 percent of the variation in delinquency, showing an approximately 7 percent increase in explained variance when the delinquent peers category was added to the regression model.

Model 3 shows the effects of concentrated disadvantage and peer delinquency on delinquent behavior when simultaneously taking into account demographic characteristics. Overall, Model 3 is statistically significant and explains approximately 12 percent of the variance in delinquency. Including age, gender, race, and family, socioeconomic status explained 2 percent more variation in delinquency; however, several noteworthy findings emerged. First, once controlling for demographic characteristics, concentrated disadvantage no longer had a statistically significant effect on delinquency. Second, although reduced, the category of delinquent peers maintained its positive, statistically significant effect on delinquency. Gender and age both have positive and significant effects on delinquency. Males are more likely to engage in delinquency than their female counterparts. Age has a significant and positive effect on delinquency, with older subjects more likely to engage in delinquency than younger subjects. Blacks were still significantly more likely to engage in delinquency than Whites. In fact, upon further investigation, it appears that the Black–White difference in delinquency is what explained or reduced the effect of neighborhood-concentrated disadvantage on delinquency to insignificance.

TABLE 13-4 Regression Model Assessing the Mediating Effect of Differential Association on the Link Between Location in the Social Structure and Delinquency						
	Model 1			**Model 2**		
	b	SE	*t*	*b*	SE	*t*
Concentrated disadvantage	.04	.02	1.71	.03	.02	1.28
Delinquent peers	.80	.18	4.36*	.66	.18	3.77*
Black	.10	.16	.61	.07	.15	.43
Hispanic	.05	.23	.22	.50	.22	.22
Other	.35	.10	3.49*	.27	.10	2.77*
Gender	.12	.02	6.09*	.05	.02	2.90*
Age (wave 1)	−.04	.04	−1.16	−.02	.04	−.68
SES	—	—	—	.08	.01	7.40*
	$F = 23.73$*			$F = 22.16$		
	$R^2 = .08$			$R^2 = .12$		

Note: SES = socioeconomic status. *$p ≤ .05$; Models were calculated using robust standard errors (SE) because children are nested within neighborhoods, *b* = Unstandardized regression coefficient.

The SSSL model also suggests that location in the social structure is mediated by social learning variables. **Table 13-4** shows two models: one model that includes all variables related to location in the social structure and concentrated disadvantage, and another that adds differential association with delinquent peers. These models reveal several important results as they relate to the SSSL model. As shown in Model 1, the effect of concentrated disadvantage no longer exists when variables that indicate location in the social structure were controlled. Again, upon further investigation we find that the insignificant effect of concentrated disadvantage can solely be explained by the significant difference in delinquency for Blacks and Whites. In Model 2, regression results show that differential association with delinquent peers is a mediating variable in the effects of race, gender, and age on delinquency.

DISCUSSION

To date, little evidence has been garnered as to how or by what process neighborhoods in which children live can influence their behavioral outcomes. Although a link has been established between delinquency and neighborhoods, research has been less able to identify potential mechanisms by which that link takes place. This chapter has focused on proposing and testing the links proposed by Akers's general theory.[13] He argues that many neighborhood-level explanations can benefit from exploring how social learning processes may differ for children and adolescents residing in neighborhoods that have varying amounts of disadvantage, informal social control, institutional resources, and crime. Neighborhood contexts have different opportunity structures and learning environments for acquiring, maintaining, or changing conforming and delinquent behaviors. The current measure of the learning variables by which these environments affect behavior is limited to differential peer association from the PHDCN data. The theory would expect that not only this variable, but also others such as definitions favorable to delinquency and reinforcement for engaging in delinquency, vary depending on the structural and social conditions of a neighborhood.

This chapter set out to investigate a core proposition of Akers's SSSL theoretical model that social learning constructs should substantially mediate the effect of social disorganization (measured in this study indirectly by concentrated disadvantage) on child and adolescent involvement in delinquency. Although some studies have tested propositions from the SSSL model and found support for the theory, the current study is the first to do so using data from many neighborhoods in combination with data on children and adolescents.

This study tested three hypotheses originating from Akers's main SSSL proposition. First, children and adolescents residing in more concentrated disadvantaged neighborhoods are likely to associate with more delinquent peers than those residing in less disadvantaged neighborhoods. Second, children and adolescents residing in more concentrated disadvantaged neighborhood are likely to be involved in more delinquency than those from less disadvantaged neighborhoods. Third, associations with delinquent peers will mediate the effect of neighborhood-concentrated disadvantage on delinquency. We also assessed how differential peer association mediates differential location in the social structure.

Results showed that neighborhood disadvantage influences the amount of delinquency children and adolescents engage in, as well as the proportion of delinquent peers they have. Further, results show that the link between neighborhood disadvantage and delinquency can be partially explained by the proportion of delinquent peers that children and adolescents report having. Regardless of age, race, gender, and socioeconomic status, differential association with delinquent peers influenced involvement in delinquency approximately 2 years later. The same cannot be said for neighborhood influences on delinquency; once demographic characteristics were taken into account, the influence of neighborhood disadvantage on child and adolescent involvement in delinquency was reduced to statistical insignificance. Further, results show that the effects of race, gender, and age were also partially mediated by differential association with delinquent peers. In sum, results from the PHDCN show preliminary support for Akers's mediation hypothesis.

Several limitations of the current study should be acknowledged. First, the current study used only one measure of social learning—differential association—to investigate the empirical validity of the SSSL model. Although results did show evidence for mediation, the mediating effect of the measure of differential association with delinquent peers was moderate at best and did not fully explain the effect of neighborhood disadvantage, although the SSSL model never does state that full mediation will occur. Acknowledging that imitation, differential reinforcement, and definitions favorable to delinquency and crime are other key constructs in social learning theory, it would be advisable for future studies to include such measures to understand the mediation effect of social learning variables on the link between neighborhoods and delinquency. The PHDCN data were limited in that measures of these other social learning constructs were unavailable.

Another limitation is the delinquency measure used. For the most part, the child behavioral checklist delinquency subscale captures more minor forms of delinquency; therefore, it is yet to be determined if the same pattern of results found in the current study would hold up for more serious behaviors. Findings from the PHDCN data not reported in the current study suggest that a similar pattern of relationships is present for self-reported offending behaviors. Future research on the SSSL model should assess more serious behaviors such as violence and other serious criminal offenses using both official and self-report data. Third, only two waves of data were analyzed. Although this allows temporal order to be established among delinquent peers and delinquency, such a design does not allow for the assessment of how delinquency is changing over time and whether such change is a function of changes in where children and adolescents reside, changes in the proportion of delinquent peers, or both. Fourth, although this study was designed to test one of the SSSL model's core propositions regarding neighborhoods and delinquency, variables other than social learning that were not taken into consideration may influence this relationship. Future studies should determine which individual-level, theoretically derived measures fair better at mediating the relationship between neighborhood disadvantage and delinquency, because currently little is known about the mechanisms driving the link between neighborhoods and delinquent behavior of children and adolescents.

Fifth, several social processes that occur within neighborhoods (e.g., informal social control and social cohesion) have been implicated as mechanisms that can explain why neighborhood disadvantage is related to child and adolescent delinquency. Akers acknowledges these constructs as well when he discusses the conceptual links between social disorganization and social learning theories. He proposes that other mechanisms of neighborhood effects suggested by such concepts as informal social control are parallel to and may be essentially the same as those found in social learning theory.

CONCLUSION

We believe that the SSSL model has merit and can add to our understanding of how neighborhoods impact child and adolescent behavioral development. Others are encouraged to continue to investigate the social learning processes that occur in neighborhoods, and perhaps even study how such processes interact with neighborhood conditions to increase or reduce delinquent behaviors. It would then also be important to assess how other more social aspects of neighborhoods are related to learning variables. For instance, does the amount of informal social control influence social learning and in turn the development of delinquency? The preliminary findings presented in this chapter raise many questions regarding neighborhoods and the learning process that should be explored in future research.

GLOSSARY

Disadvantaged neighborhoods—characterized by high poverty rates, high percentages of unemployment, racial and ethnic heterogeneity, frequent residential mobility, and segregation

Ethnic diversity—heterogeneity in ethnicity among residents from a particular area

Informal social control—ability of residents to intervene for the common good of a particular area

Neighborhood structural elements—characteristics, such as poverty rate and segregation, that are found in the neighborhood

Residential instability—amount of residential turnover in a particular area

Social disorganization theory—theory advanced by Shaw and McKay that focused on explaining why certain areas of cities had high rates of crime that persisted over time

Social learning—social processes by which a person learns

NOTES

1. Sampson, R. J., & Groves, W. B. (1989). Community structure and crime: Testing social disorganization theory. *American Journal of Sociology, 94*, 774–802; Sampson, R. J., Raudenbush, S. W., & Earls, F. J. (1997). Neighborhoods and violent crime: A multilevel study of collective efficacy. *Science, 277*, 918–924; Shaw, C. R., & McKay, H. D. (1942). *Juvenile delinquency in urban areas.* Chicago: University of Chicago Press; Leventhal, T., & Brooks-Gunn, J. (2000). The neighborhoods they live in: The effects of neighborhood residence on child and adolescent development. *Psychological Bulletin, 126*, 309–337.
2. Brooks-Gunn, J., Duncan, G. J., Klebanov, P. K., & Sealand, N. (1993). Do neighborhoods influence child and adolescent development? *American Journal of Sociology, 99*, 309–337; Brooks-Grunn, J., & Leventhal, T. (2003). Children and youth in neighborhood context. *Current Directions in Psychological Science, 12*, 27–31; Edwards, B. (2005). Does it take a village? An investigation of neighbourhood effects on Australian children's development. *Family Matters, 72*, 36–43; Haynie, D. L., Silver, E., & Teasdale, B. (2006). Neighborhood characteristics, peer networks, and adolescent violence. *Journal of Quantitative Criminology, 22*, 147–169; Kohen, D. E., Leventhal, T., Dahinten, S. L., & McIntosh, C. N. (2008). Neighborhood disadvantage: Pathways of effects for young children. *Child Development, 79*, 156–169; Jencks, C., & Mayer, S. (1990). The social consequences of growing up in a poor neighborhood. In L. Laurence, Jr., & M. McGeary (Eds.), *Inner city poverty in the United States* (pp. 111–186). Washington, DC: National Academies Press.
3. Akers, R. L. (1998). *Social learning and social structure: A general theory of crime and deviance.* Boston: Northeastern University Press.
4. Park, R., Burgess, E. W., & McKenzie, R. D. (1925). *The city.* Chicago: University of Chicago Press; Shaw, & McKay (1942), see Note 1.
5. Shaw, C. R., & McKay, H. D. (1969). *Juvenile delinquency and urban areas.* Chicago: University of Chicago Press.
6. Sampson, & Groves (1989), see Note 1.
7. Booth, A., & Crouter, A. (2001). *Does it take a village? Community effects on children, adolescents, and families.* Mahwah, NJ: Lawrence Erlbaum Associates; Duncan, G. J., & Raudenbush, S. W. (1999). Assessing the effects of context in studies of child and youth development. *Educational Psychologist, 34*, 29–41; Sampson, R. J., Morenoff, J., & Gannon-Rowley, T. (2002). Assessing "neighborhood effects": Processes and new directions. *Annual Review of Sociology, 28*, 443–478; Sampson, R. J., Morenoff, J. D., & Raudenbush, S. (2005). Social anatomy of racial and ethnic disparities in violence. *American Journal of Public Health, 95*, 224–232.
8. Jencks, & Mayer (1990), see Note 2.
9. Lee, G., Akers, R. L., & Borg, M. (2004). Social learning and structural factors in adolescent substance use. *Western Criminology Review, 5*, 17–34.
10. Page, E. R. (1998). *Family structure and juvenile delinquency: The mediating role of social learning variables* [PhD Dissertation]. Gainesville, FL: University of Florida; Lee et al. (2004), p. 29, see Note 9.
11. Bellair, P., Roscigno, V. J., & Velez, M. B. (2003). Occupational structure, social learning, and adolescent violence. In R. L. Akers & G. F. Jensen, (Eds.), *Social learning theory and the explanation of crime: A guide for the new century. Advances in criminological theory* (Vol. 11, pp. 197–226). New Brunswick, NJ: Transaction Publishers; Jensen, G. F. (2003). Gender variation in delinquency: Self-images, beliefs, and peers as mediating mechanisms. In R. L. Akers & G. F. Jensen, (Eds.), *Social learning theory and the explanation of crime: A guide for the new century. Advances in criminological theory* (Vol. 11, pp. 151–177). New Brunswick, NJ: Transaction Publishers; Lanza Kaduce, L., & Capece, M. (2003). A specific test of an integrated general theory. In R. L. Akers & G. F. Jensen, (Eds.), *Social learning theory and the explanation of crime: A guide for the new century. Advances in criminological theory* (Vol. 11, pp. 179–196). New Brunswick, NJ: Transaction Publishers; Verrill, S. W. (2008). *Social structure-social learning and delinquency: Mediation or moderation?* New York: LFB Scholarly Publishing.
12. Achenbach, T. M. (1991). *Manual for the youth self-report and 1991 profile.* Burlington, VT: University of Vermont.
13. Akers (1998), see Note 3.

CHAPTER FOURTEEN

Timing Is Everything: Gangs, Gang Violence, and the Life Course

Scott H. Decker and David Pyrooz

KEY TERMS

Continuity
Desistance
Enhancement model
Facilitation model
Gangs
Onset
Selection model

INTRODUCTION

There has been an explosion of interest in **gangs** since the early 1990s. Some of this interest certainly has stemmed from media attention, whereas other sources of interest include law enforcement and criminal justice attention. Researchers have also played a role in this increased attention to gangs, focusing on the role of gangs in homicide, drug sales, disorder, gun use, violence, and disruption of neighborhood patterns of socialization. Notably absent from criminologists' attention to gangs has been a focus on theory. By and large, the academic attention to gangs has examined patterns of behavior or criminal justice responses with only minimal attempts to place these lines of study within the bounds of traditional criminological theory or to test such theories with gang data. This chapter reviews studies of gang violence and places them in the framework of life-course theory.

This chapter examines a number of aspects of gang violence using a life-course perspective as a framework. The definition of gangs and gang violence, the distinction between gang and nongang violence, and the intersection between the major tenets of developmental and life-course theory and gang violence are examined. Attention is paid to a notable form of gang violence—homicide. This chapter also covers the status of knowledge regarding prison gang violence and discusses individuals who leave their gang, emphasizing the role of life-course theory in explanations of this behavior.

DEFINITION

One of the keys to understanding the relationship between gangs and violent crime is to define the problem appropriately. Like many topics in the study of gangs, the definitional issues are complex and deeply rooted. One of the key issues in the study of gangs is the choice of the appropriate unit of analysis. Some argue that the best focus is on the gang member, whereas others argue that it is the acts committed by the gang member or members that should be the main focus.[1]

A crucial issue in understanding youth gangs and violent behavior is whether an act of violence was the product of a gang. There is little consensus on what the definition of a gang crime is or should be. The Federal Bureau of Investigation (FBI) defines a gang in terms of organizational characteristics and tight structure.[2] The approach by local law enforcement agencies to define gangs and gang crimes is different from this approach. There are two groups of definitions. The first defines gang crime based on the participation of a gang member in the act, either as a victim or an offender. This is the definition used by the city of Los Angeles and many other cities in Southern California. This is a broad and inclusive definition that depends only on the determination of whether a victim or offender is a documented gang member. Other cities, such as Chicago, use a narrower definition that relies on motive. For offenses that may involve a gang member both as victim and offender, an offense may not be classified as gang related unless the motive had something to do with the intentions or desires of the gang or furthering their interests. Typically, this behavior includes fights over gang turf, retaliation against rival gangs or gang members, or crimes committed to generate economic gain for the gang.

The second group of definitions focuses on motives and requires more information and review than those that focus on the memberships of individuals. Klein, Maxson, and colleagues reviewed homicides in Los Angeles using both definitions. They found that the member-based definition produced nearly twice as many gang-related homicides than when the narrower motive approach was used. This observation underscores the dramatic difference that definition makes in the study of gang violence. Equally important, however, was the finding that gang homicides share a number of common characteristics, regardless of definitional approach. The demographic characteristics of the participants in homicide (race, age, gender) and the situational characteristics of the events (guns, location, victim–offender relationship) were the same regardless of the definition used.[3]

GANGS, VIOLENCE, AND THE LIFE COURSE

A key issue in gang research is whether gangs attract individuals predisposed to delinquency, or whether mechanisms within the gang culture cultivate an environment that spawns delinquency. This is a vital question in gang research that links it closely to life-course theory. This question has had even larger theoretical implications since Gottfredson and Hirschi's publication of *A General Theory of Crime* and the ongoing debate between control and learning perspectives in explaining crime and delinquency.[4]

Cross-sectional research completed in the 1980s found that gang members were more involved in delinquency than nongang members, but only longitudinal data could properly address the causal issues. Thornberry and colleagues developed a construct to bridge the knowledge gap between the individual, the gang, and aggregate delinquency. In doing so, they identified three models—selection, facilitation, and enhancement—that include the pre-gang, in-gang, and post-gang years of delinquent involvement.[5] In this chapter, we avoid discussion of substance abuse and property delinquency and focus primarily on acts of violence, the acts for which gangs are best known. We focus on studies measuring violent delinquency before joining, while in the gang, and upon desistance in comparison with those who never joined a gang. This has ramifications for two areas: providing empirical support for the explanation of the gang–delinquency link while contributing to the larger theoretical understanding of gang violence.

The **selection model** is an individual-level, or "type-of-person," model. It holds that joining the gang is a conduit that draws together like-minded youth. As such, the gang does not increase involvement in crime; rather, the gang is an unorganized troop, and as Gottfredson and Hirschi put it, it is "inevitable that from time to time they [gang members] will congregate on the streets of U.S. cities."[6] Thus, the gang is more like a magnet for attracting or recruiting youth with a predisposition to delinquency or violence than it is a catalyst for violent crime. This implies that before the onset of and subsequent to desistance from gang membership, participation in delinquency by those individuals should exceed that of nongang adolescents, expressing the individual criminality of the gangs' members.

The **facilitation model** holds that before joining a gang, future gang members' rate of delinquent involvement should be no higher than those who do not go on to join gangs. This is suggestive of a "type-of-group" model. Thus, upon joining, the normative processes occurring within the gang—learning definitions, reinforcement, reciprocation, goals, participation, and motives—"facilitate," whether individually or collectively, delinquent or violent behavior. Delinquent involvement stimulated by the gang should subside to pre-gang levels upon desisting, as the group processes are no longer present.

The **enhancement model** can be viewed as an interaction of the above models. Before joining a gang, the delinquent involvement of future gang members exceeds that of nongang adolescents. Upon joining, the gang "enhances" (rather than facilitates) the frequency and degree of engaging in delinquent actions and desistance should equate to a reduction in delinquent involvement.

Gang violence has been well documented in the discipline, displayed by way of media, and possesses myths and truths of epic proportion. Before discussing the results of longitudinal research that provides empirical support for the models, a discussion of gang violence is warranted. The word *gang* almost always connotes violence. Gang violence is perhaps the most salient and distinctive characteristic of gangs, setting them apart from other groups. This is what draws the attention of policy wonks, city leaders, and academics. The involvement in violence is underscored by the disproportionately high representation of gang members as perpetrators and victims.

The U.S. Department of Justice funds the National Youth Gang Center to conduct national surveys of law enforcement agencies regarding gang problems in their particular jurisdictions. Between 1996 and 1998, of the 968 cities that submitted information to the study, 438 (45 percent) reported at least one gang homicide in their city. Furthermore, of the 16,582 homicides reported in the FBI's *Uniform Crime Report* in 2003, nearly 9 percent were reported by law enforcement agencies as being gang homicides.[7]

Tita and Abrahamse examined homicides more narrowly in California. Of the nearly 62,000 homicides committed in California between 1981 and 2001, they reported that more than 10,000, roughly 16 percent, were gang homicides. Additionally, looking specifically at the beginning and end of the two decades, gang homicides increased in their percentage of the total number (i.e., they found other types of homicides were decreasing while gang homicides were increasing) of homicides in Los Angeles County and the rest of California by over fourfold and threefold, respectively.[8]

Los Angeles and Chicago include more gang homicides than all other jurisdictions combined. In an examination of 132 cities with populations greater than 100,000, Egley, Howell, and Major report that approximately 1300 gang homicides occurred in 2001. Of those, Los Angeles and Chicago accounted for nearly 700 of them, totaling nearly 54 percent. Examining the gang homicide trends between 1996 and 1998, Maxson, Curry, and Howell report that dips in gang homicides in Los Angeles (41 percent) and Chicago (19 percent) resulted in 165 fewer gang homicides and, accordingly, accounted for 71 percent of the decrease in the national gang homicide total within those years.[9]

Gang violence poses additional risks for nonfatal victimizations. In an analysis of National Crime Victimization Survey data between 1993 and 2003, Harrell reports that gang members were disproportionately perceived to be perpetrators in nonfatal violent crimes such as robbery and aggravated assault. She reports that total victimizations by gang members peaked between 1993 and 1996, amounting to roughly 8 to 10 percent of all reported victimizations, while holding steady around 6 percent between 1998 and 2003. These numbers are especially troublesome when considering the total population of gang members throughout the United States. The most recent National Youth Gang Center report estimates there are approximately 760,000 gang members. That loosely equates to 2.5 of every 1000 people in the United States being a gang member, and, troublingly, roughly 60 of every 1000 violent victimizations is committed by a gang member. These ratios underscore dramatic overrepresentation of gang members in violence.[10]

What are the causes of violence in gangs? Is it that these violent victimizations are the result of the gangs' attraction or selection of individuals predisposed to violence, or a result of the socialization processes within the gang that facilitate violence? We turn to the work of longitudinal projects that capture reports of delinquency pre-gang, in-gang, and post-gang. Such a design is necessary to understand the issue of gang violence and allows researchers to establish causality.

Longitudinal studies of gangs undertaken in the cities of Bergen (Norway), Denver, Montreal, Pittsburgh, Rochester, and Seattle provide a wealth of information regarding gang violence over the life course. With the exception of Bergen, these studies began in the mid to late 1980s, capturing the period when city homicide and violence rates were at their peak. These projects generally collected their samples via the public school system and concentrated on youth at high risk for delinquency and gang involvement.

None of the studies provided unequivocal support for either the selection or facilitation model. However, there is more empirical support for the facilitation model than the selection model when examining violent acts reported by gang members. The research in Bergen, Denver, Montreal, Pittsburgh, and Rochester all found strong facilitation effects. Although no support for a strong selection effect was found, Thornberry et al. reported overwhelming support for the facilitation model and little, if any, support for selection. But despite these facilitation effects, the selection effects cannot be ignored, lending credibility to the enhancement, or mixed, model to explain the relationship between gangs and violent delinquency.[11] Krohn and Thornberry suggest, "Perhaps the safest conclusion to draw is that there is a minor selection effect, [and] a major facilitation effect."[12] Accordingly, these longitudinal studies indicate that before entry, future gang members exhibit higher involvement in violent delinquency when compared with youth who do not join gangs. Upon entry, the gang facilitates taking part in even more violent activities. Leaving the gang reduces that participation, but still not to the levels of nongang adolescents.

The preponderance of the research shows that gang membership strongly increases involvement in delinquency, especially violent delinquency. As Thornberry et al. state about their Rochester sample, "clearly, being in the gang is generative of violent behavior among these boys."[13] So, exactly what is it about a gang that facilitates violence? Access to weapons? Reputation or status? Protection? Turf disputes? Drug or money disputes? A qualitative perspective can provide evidence to understand the mechanics of a gang that engender violence.

One attempt to grasp a conceptual appreciation of gang violence was introduced by Decker and Van Winkle from interviews of gang members in St. Louis, Missouri. A sample of 99 core (i.e., regular members knowledgeable with respect to the inner workings of the gang) gang members produced a persuasive depiction of the underpinnings of gang violence. These works highlight the collective behaviors of the group by focusing on the role of threat in the construction of gang violence. The following is central to the concept of threat:

> *Threat is the potential for transgressions against or physical harm to the gang, represented by acts or presence of a rival group. Threats of violence are important because they have consequences for future violence. Threat plays a role in the origin and growth of gangs, their daily activities, and their belief systems.[14]*

Whether threat is actual or perceived, it remains present within the belief system of the gang member. A response to threat—whether on impulse or by calculation—originates from a belief system that determines the severity or escalation.

Paradoxically, Decker recounts that the gang members from the St. Louis sample maintained that they do not initiate the violence. Rather, their violence is in response to the provocations or antagonisms of rival gangs. This statement echoes the work of Vigil, who references an "endless game of revenge," marked with precariousness. Decker goes on to document a perpetual threat cycle of gang violence that escalates and deescalates in a roller coaster fashion (think igniting long-standing rivalries, women, relatively close inhabitation). These escalations and conflicts are central to developing solidarity in the gang. As documented by Thrasher, Klein, and Short, the cohesiveness and solidarity that are developed through conflict are converted into more unabashed exhibitions of violence (i.e., a perpetuating cycle of violence).[15]

Prison gangs are more structured than street gangs and have much stronger leadership. The rank and file membership often reflects several gradations in membership, making prison gangs look more like organized crime groups than gangs on the street. Research on street gangs shows that where profits are at stake, violence is often the outcome. Inside prisons, the same pattern appears, as prison gangs are heavily involved in prison violence.[16]

Stretesky and Pogrebin conducted in-depth interviews of 22 gang inmates in Colorado prisons. Much of the gang violence reported to them revolved around masculinity, most notably in the form of ensuring respect and maintaining a reputation. Membership in a gang leads one to act in ways that preserve his own and his gang's reputation, which is often gauged by a willingness to use violent retaliation against rivals. Attacks on one's masculinity must be responded to; if not, "the notion of disrespect [unto a gang member] is analogous to an attack on the self."[17]

Stretesky and Pogrebin's interviews underscore the reality that is experienced upon confinement in juvenile hall, county jail, or a state or federal prison system. Prior gang membership produces a perilous choice: Remain in the gang and be prepared for attacks from rivals, drop one's affiliation and be prepared for retaliation by the gang, or affiliate with a new group. No matter the case, gang members are disproportionately imprisoned due to their increased involvement in violence. The advent of gang enhancement laws (California Penal Code 186.22), civil gang injunctions, and zero-tolerance policies have not only increased the likelihood of an eventual incarceration by broadening the spectrum of punishable offenses, but has also increased the length of punishment for those violations. These events have increased the salience of understanding gang membership in prison.

Gangs in prison are notorious for their use of violence. As prison markets for profit are highly competitive and circumscribed, competition among gangs often leads to violence for items such as drugs and prostitution.[18] Also, the dichotomous relationship of sustaining or snubbing that relationship with the gang eventually increases the cohesion within the gang. Gang solidarity, especially after exhibitions of power, reinforces the self-perceptions of the gang's hegemony. Indeed, there is a perception that gang members have more control over events in prisons than nongang members do. One study suggested that gang members comprised only 3 percent of the population, yet were responsible for more than 50 percent of the violence, a testament to their institutional involvement in violence.[19] All told, the imbalance of violence involvement on the street carries over into confinement facilities, a statement unsurprising to many.

DEVELOPMENTAL AND LIFE-COURSE THEORY

Developmental and life-course (hereafter, DLC) criminology is a relatively new theoretical concept within the field, especially in its application to the behavior of gang members. Developing since the 1990s, the DLC approach is consistent with research on criminal careers. As structural, process, and control theories use data from observations and surveys, the key source of data for DLC theory is the longitudinal design. Studies funded by the National Institute of Justice and Office of Juvenile Justice and Delinquency Prevention have used longitudinal methodological approaches that provide useful data sets holding gangs, gang members, their backgrounds, and their recent actions as individual units of analysis, something previously unrivaled in the gang arena.

Most important to this chapter's goal is the interactional theory of delinquency introduced by Thornberry in 1987. Thornberry's work underscores the influence of prior theory as it emphasizes a mutually perpetual relationship between structural, control, and process theories. Thornberry argues that structural position plays a momentous role in future delinquency, as it elevates risk for gang involvement. A marker for delinquency is the weakening of bonds to conventional society, though that is not enough, as learning processes conducive to delinquency must be in place and reinforced. A precise interaction of the three concepts could then lead to gang involvement that in turn could lead to violent delinquency. Thornberry et al. call attention to the importance of the variation of risk: "All of these conditions are more likely to emerge for youth who grow up under conditions of structural disadvantage."[20]

A DLC perspective of gangs analyzes individual change across the life course. Central to gang involvement are (1) risk and protective factors that push or prevent one from joining a gang, and (2) the facets of gang membership—**onset**, **continuity**, and **desistance**—from a within–individual perspective. Critical concepts include trajectories, transitions, turning points, and negative life events that define the life course. These terms have considerable importance for encouraging involvement in and desistance from gangs and violence.

The phrase "timing is everything" may ring true for business ventures and photography, but timing is also paramount to DLC gang theory in the sense of producing, sustaining, and reducing gang membership. As Thornberry et al. note, "off-age transitions, especially precocious or early transitions, can create disorder in the developmental sequence and lead to later problems of adjustment because the person is less likely to be prepared for the transition."[21] These transitions (short-term events, such as the act of joining the gang) are positioned within ever-important trajectories (long-term patterns). The relationship between the two essentially characterizes the life course and, in a sense, the propensity for joining or avoiding a gang in the first place.

From the DLC perspective, gang membership may be viewed as a trajectory. Trajectories are typically age graded, evolve concomitantly (e.g., through several dimensions such as school, family, or work), and can be seen as a line of development over the lifespan. Trajectories can be described as pathways that are susceptible to alteration and change. Modification of a trajectory is most notably attributable to transitions that can be described as changes in state over shorter time spans that emphasize the duration of that actual event.[22] Without a doubt, gang membership alters the talk, the articles of clothing, the attitude and mentality, and for many, the outcome of one's life during a period, arguably considered to be the most pivotal—adolescence.

Gang behavior in adolescence can alter transitions later in the life course. During gang membership, gang members are far more involved in violent acts, increasing the likelihood of future imprisonment and its impact on opportunities and trajectories. The data from Rochester and Denver indicate that while gang members comprise a minority (30 percent and 14 percent, respectively) of the sample, they contribute the vast majority (69 percent and 79 percent) of the violent acts. The duration of gang involvement gains theoretical importance upon examination of the mean frequency of violent acts reported during gang membership. Sustained gang involvement translates into more delinquency and crime. Put simply, the longer one remains in the gang, the more violent acts that person commits. But as a whole, gang membership does not last for a long time. The findings from Denver, Seattle, and Rochester indicate that only a small share (3–5 percent) of their gang sample remained in the gang longer than 4 years, whereas the majority (55–69 percent) retained membership for a period of 1 year or less.[23] This suggests short-term trajectories with limited long-term impacts on opportunity.

It is that small group, 3 to 5 percent, who do not transition out of the gang that are of concern. For those ensnared in the gang for longer periods, joining a gang is seen as a turning point (as opposed to a transition), given the redirection of the life course. In that respect, Thornberry et al.'s analysis of transient and stable gang members finds that increased length of gang involvement "indicates a deeper penetration along the trajectory of gang membership and therefore potentially more extensive consequences of membership."[24] Again, timing is everything. It may be more difficult for stable or long-term gang members to turn to a conventional lifestyle, whereas transience may result in the diminution of, but not arrest of, the developmental process. Deeper penetration needs to be explored far more than the extant research supplies. In other words, we need to know more about what sustains gang involvement.

The DLC vehicle is a promising avenue to pursue, as future research needs to focus on the duration of gang membership and better understand the processes in the gang that facilitate violence. Gang violence may be examined using the concept of threat with negative life events over a longer period to document transitions and turning points.

GANG DESISTANCE: LEAVING THE GANGS

It is imperative to acknowledge that gang membership is a transitory state. In other words, nearly all youth who join gangs also exit gangs. This suggests that both entrance and exit processes should warrant equal research ventures. This has not been the case. Although there has been an abundance of research concentrating on risk factors for entrance, examinations of the desistance processes largely have been devoid of empirical support in the literature. Explanations for this shortage are puzzling. There have, however, been two studies devoted to gang desistance dialogue. We use these and the work of broader, qualitatively based studies to help us piece together a discussion of gang desistance, a key feature of life-course transitions and trajectories.[25]

The process of desisting from a gang can be characterized by an ongoing strain of "pulls" and "pushes" nested in the contexts of community and gang. The process of leaving the gang may be as simple as saying, "I am not a gang member," but there are further affinities to the gang that must be addressed. The complex definitional issues of gangs include defining "former gang members," an equally complex process that calls attention to the difficulties of measuring gang desistance. The implications of desisting from gangs are discussed with respect to violence and the special value of in-gang intervention.

Part of the difficulty in leaving the gang is the affinity to the gang (an affinity marked by companionship, entertainment, monetary opportunities, status, and protection) that, taken individually, is possible but difficult to attenuate. Leaving the gang is tantamount to shunning one's own peer group, a particularly difficult task especially given the ages of gang members. Many factors are involved in the leaving of a gang, making it a nuanced process.

Decker and Lauritsen analyzed the results of 24 semistructured interviews of former gang members in St. Louis. These former gang members came from gangs ranging from loosely to highly organized. Decker and Lauritsen then juxtaposed their comments with the perceptions of 99 active, core gang members to develop an assessment of leaving the gang. Their work identifies two different ways in which gang desistance occurs. The first is where a gang member abruptly quits the gang, often moving away or suddenly avoiding contact. The second is when a gang member simply drifts away, a practice marked by a gradual decrease in the association with the gang. The authors found the process of gang desistance to be characterized by some in their sample as being motivated by a single event (typically an act of violence). For others it was more an accretion of negative perceptions (victimization of friends, age-graded maturation, family obligation) damaging the hegemonic sense of the gang, thus leading to desistance.[26]

Vigil's work on barrio gangs in Los Angeles emphasizes the role of negative perceptions in leaving the gang. He found a combination of reasons or a series of events that have a succession quality to them, leading to a gang members' desistance. His accounts reflect the heterogeneity of gang desistance; a number echoed Decker and Lauritsen's findings that desistance stems from what distinguishes gangs from delinquent peer groups: violence, or offshoots of violence (i.e., imprisonment or peer victimization). As Decker and Van Winkle put it, "the ability of violence to motivate individuals to join the gang and strengthen the bonds of membership has an upper limit."[27]

The literature suggests differences between individual members' centrality to the gang that influence desistance. Klein and Maxson reference Yablonsky's analogy of thinking of the gang as an artichoke; the "leaves" on the outside consist of fringe (or peripheral or associate) gang members that, as they are peeled away, reveal the "heart" or the core gang members.[28] Caldwell and Altschuler mention increased difficulty in determining when an associate gang members ends his or her participation in gang activities compared with when a core member leaves the gang.[29] An associate, or someone at the fringe of the gang, may exhibit fewer overt signs of gang membership and drift in and out of gang involvement. For a core gang member, there may be more individual deficiencies. This is highlighted in Vigil's findings in a predominantly Hispanic housing project in Los Angeles. The study consisted of a sample of 35 gang members living in the projects. He found that those at the periphery of the gang were more likely to return to a conventional lifestyle than those at the center of the gang. In fact, not one of the 11 core gang members (granted, some were unaccounted for) 10 years later enlisted in a conventional lifestyle. Of the 24 peripheral and outside members, 4 made the obvious change to a conventional lifestyle (jobs and family), whereas at least 11 moved out of the projects (although that does not necessarily equate to avoiding gangs and crime).

Adding to the difficulty of leaving the gang is the status and symbolism associated with gang membership that does not simply evaporate upon separation. A gang member just does not stop appearing to be a gang member (think clothing, attitude, actions, reputation), as they are still known by fellow gang members, neighbors, community, police, and rivals.[30] The ensuing reputation and labeling can impede job opportunities that prove crucial to exiting the state of gang membership. Gang desistance is a process. This process is similar to desisting from crime. Just as it takes many fights and conflicts to build a reputation, so to it takes time to mold a new conventional or prosocial status.

We return to the work of Decker and Lauritsen for a stronger understanding of defining gang desistance. They break down desistance processes according to two categories within a former gang members' previous gang network: (1) reducing emotional ties, and (2) reducing engagement in activities. Some of the self-proclaimed "former" gang members maintained emotional ties with the gang, claiming they would help if trouble arose for their previous network of gang peers. Some still engaged in activities with their previous network of gang peers. Others maintained both an active and emotional tie to their former gang. In this study sample, all participants had left their gangs. A key question is whether someone who still has emotional ties to the gang and engages in activities with their former gang network is actually no longer a gang member. If Decker and Lauritsen ignored those aspects of behavior (emotional ties and associations), then they would also have to restructure their definition of ex-gang members. They found that the concept of gang desistance made it difficult to make any conclusive statements as to what constitutes a former gang member. Given the lack of consensus on the gang definition, it is not surprising that we do not have a stronger understanding of gang desistance. Certainly, the amorphous nature of gangs and intermittent involvement of gang members in their gangs contribute to this.

When discussing the implications of desisting from gang membership, it is important to know where former gang members end up. The answer offered in some research is dead or imprisoned. Of the 760,000 gang members reported in the National Youth Gang Center study, most leave the gang and transition to a post-gang trajectory that does not involve prison or death. This is a group that we need to understand more about, in both field studies and school-based longitudinal studies.

We do not know much more about the "post-gang" trajectory or adult life of former gang members now than we did when Klein noted the importance of this in 1971. However, there are two implications that should not be ignored when discussing gangs and their role in violence. Reducing gang involvement results in a reduction of violent behaviors and a reduction in violent victimization. Once removed from the gang, an individual is less likely to respond to rivalries and retaliation. Furthermore, such reductions in violence reduce the opportunity for additional violent offending and violent victimization (e.g., hanging out on street corners or porches).[31] As a whole, new conflicts should subside as a former gang member turns to legitimate or nongang (i.e., less violent) activities.

The research thus far indicates that removing oneself from the gang is accompanied by a reduction in violent acts. Thornberry and colleagues found that, for male gang members in Rochester, after adolescents leave the gang, they generally exhibit substantial reductions in their level of violent offending. They found that the mean number of self-reported violent acts decreased by more than 50 percent for four of the five groupings (the fifth grouping decreased by just over 40 percent) after they left the gang. These results underscore the heavy enhancement effect for gangs and involvement in violence.[32]

Similar to the violent offense rates, victimization rates among gang members are higher than those among nongang members. With respect to the most violent of victimizations—being shot or shot at—Curry, Decker, and Egley found from their sample of middle school youth in St. Louis that gang members were more than four times more likely to be injured by a gunshot and six times more likely to be shot at. These numbers indicate the particularly salient hazards of gang involvement.[33]

Decker and Van Winkle's work produced a more frightening statistic: Of the 99 gang members they originally interviewed, 28 died as a result of violence over the next 10 years.

The mixed cross-sectional and longitudinal approach used by Peterson, Taylor, and Esbensen found that victimization rates were higher for gang members. They found that victimizations (such as robbery and simple and aggravated assault) were at their peak during the period of gang membership. Their approach also examined whether victimization rates resemble the theoretical models (selection, facilitation, or enhancement) explaining delinquency rates. They found support for the enhancement model; gang members were more likely to be victimized than nongang members before, during, and after gang involvement. Active gang involvement corresponded to the peak in victimization, fitting the enhancement model.[34]

In sum, the literature identifies a shortcoming in gang research—the need for more research that examines the process of desisting from gangs. In general, gang desistance has not received the attention warranted by the implications of this information for crime and violence control. Caldwell and Altschuler call for an age-specific, multifaceted approach to gang intervention, one that accounts for the individual developmental status of gang youth. This approach would provide a glimpse of reality outside of the gang and be targeted appropriately to age, economic, and education needs. Decker and Lauritsen identify intervention techniques that capitalize on moments of reality when gang youth are in or visiting the hospital as a result of violence, hoping to exploit a "triggering event" to pull an adolescent away from the grip of the gang.[35]

CONCLUSION

DLC theory is an appropriate theoretical approach to use in addressing gangs. There is clearly a progression in gang membership, as youth move from early adolescence and the fringes of gangs to full-fledged membership, prison, or ultimately leaving their gangs. The time spent in a gang, whether brief or lengthy, is marked by an increased propensity for violence. This topic has received considerable attention in the research literature. The way gangs are defined plays an important role not only in understanding the context of the problem, but also in understanding the extent of the problem. Even the most conservative of definitions still indicate gang members are more involved in violence than other types of groups. The gang-related statistics referenced in this chapter uniformly underscore the disproportionate involvement of gang members not only in homicide, but also in person offenses and their own victimization.

The renewed interest in gangs in the 1990s laid the grounds for a theoretical explanation for the violence associated with gangs. Longitudinal studies embedded with gang measures provide the details to whether it is the "type of person" or "type of group" responsible for the violence. They suggest gangs are not filled with people who have criminal traits. The studies show that, as a whole, gang members are less inclined to resort to violence before and after gang involvement than while in the gang. In addition, during the time before and after gang involvement, gang members' account of violent involvement is not too much higher than nongang individuals. These results imply that while in the gang, the ethos and processes facilitate violence and support the findings of more qualitatively based work.

The key to this chapter is that DLC approaches to studying gangs have considerable theoretical promise in accounting for changes that occur over the life course of a gang member. This approach shuns a one-dimensional explanation for gang violence and embraces the complexity of the life course by accounting for off-age transitions, constant threat, negative life events, and the complexity of trajectories. This theoretical approach suggests important directions for future research. It is crucial to develop a better understanding of factors that create persistence in gangs. As important as it is to understand how individuals get into gangs, it is even more important to understand how they get out of gangs. In future research, we hope to call attention to the desistance processes and to view gang involvement within the DLC context. Gang involvement requires a broader lens of study, as the ramifications of being a gang member are not confined to life in the gang, but instead these consequences evolve over the course of a life.

GLOSSARY

Continuity—processes that underlie the stability of some phenomenon

Desistance—process by which there is cessation from some type of behavior

Enhancement model—a model that argues that gang members are more violent than nongang members before joining the gang, but become even more violent once they are initiated into the gang

Facilitation model—a model that argues that joining a gang causes a person to become violent

Gangs—a band of people who share organizational characteristics and are characterized by a tight structure

Onset—emergence of some particular behavior

Selection model—a model that argues that the gang violence nexus is due to violent persons seeking out gangs

NOTES

1. Short, J. F., Jr. (1985). The level of explanation problem in criminology. In R. F. Meier (Ed.), *Theoretical models in criminology* (pp. 51–72). Beverly Hills, CA: Sage; Short, J. F., Jr. (1989). Exploring integration of theoretical levels of explanation: Notes on gang delinquency. In S. F. Messner, M. D. Krohn, & A. E. Liska (Eds.), *Theoretical integration in the study of deviance and crime: Problems and prospects* (pp. 243–260). Albany, NY: State University of New York Press.
2. Federal Bureau of Investigation. (1999). *FBI gang alert.* Washington, DC: U.S. Department of Justice.
3. Klein, M. W., & Maxson, C. L. (2006). *Street gang patterns and policies.* New York: Oxford University Press; Maxson, C. L., Gordon, M. A., & Klein, M. W. (1986). Differences between gang and nongang homicides. *Criminology, 23,* 209–222; Maxson, C. L., & Klein, M. W. (1990). Street gang violence: Twice as great or half as great? In C. R. Huff (Ed.), *Gangs in America* (pp. 71–100). Thousand Oaks, CA: Sage; Maxson, C. L, & Klein, M. W. (1996). Defining gang homicide: An updated look at member and motive approaches. In C. R. Huff (Ed.), *Gangs in America* (2nd ed., pp. 3–20). Thousand Oaks, CA: Sage.
4. Gottfredson, M. R., & Hirschi, T. (1990). *A general theory of crime.* Stanford, CA: Stanford Press.
5. Fagan, J. E. (1989). The social organization of drug use and drug dealing among urban gangs. *Criminology, 27,* 633–669; Thornberry, T. P., Krohn, M. D., Lizotte, A. J., & Chard-Wierschem, D. (1993). The role of juvenile gangs in facilitating delinquent behavior. *Journal of Research in Crime and Delinquency, 30,* 55–87.
6. Gottfredson, & Hirschi (1990), p. 209, see Note 4.
7. Curry, G. D., Egley, A., & Howell, J. C. (2004). Youth gang homicide trends in the National Youth Gang Survey. Paper presented at the American Society of Criminology meeting. Nashville, TN; Maxson, C. L.,

Curry, G. D., & Howell, J. C. (2002). Youth gang homicides in the United States in the 1990s. In W. L. Reed & S. H. Decker (Eds.), *Responding to gangs: Evaluation and research* (pp. 107–137). Washington, DC: U.S. Department of Justice, National Institute of Justice.

8. Tita, G., & Abrahamse, A. (2004). *Gang homicide in LA, 1981–2001*. Sacramento: California Attorney General's Office.

9. Egley, A., Howell, J. C., & Major, A. K. (2006). *National youth gang survey: 1999–2001*. Washington, DC: U.S. Department of Justice, Office of Juvenile Justice and Delinquency Prevention; Maxson, Curry, & Howell (2002), see Note 7.

10. Egley, A., & Ritz, C. E. (2006). *Highlights of the 2004 National Youth Gang Survey*. Washington, DC: U.S. Department of Justice; Harrell, E. (2005). *Violence by gang members, 1993–2003*. Washington, DC: U.S. Department of Justice, Bureau of Justice Statistics.

11. Bendixen, M., Endresen, I. M., & Olweus, D. (2006). Joining and leaving gangs: Selection and facilitation effects on self-reported antisocial behavior in early adolescence. *European Journal of Criminology, 3*, 85–114; Esbensen, F., & Huizinga, D. (1993). Gangs, drugs, and delinquency in a survey of urban youth. *Criminology, 31*, 565–590; Gatti, U., Tremblay, R. E., Vitaro, F., & McDuff, P. (2005). Youth gangs, delinquency and drug use: A test of the selection, facilitation and enhancement hypotheses. *Journal of Child Psychology and Psychiatry, 46*, 1178–1190; Gordon, R. A., Lahey, B. B., Kawai, E., Loeber, R., Stouthamer-Loeber, M., & Farrington, D. P. (2004). Antisocial behavior and youth gang membership: Selection and facilitation. *Criminology, 42*, 55–87; Lacourse, E., Nagin, D. S., Tremblay, R. E., Vitaro, F., & Claes, M. (2003). Developmental trajectories of boys' delinquent group membership and facilitation of violent behaviors during adolescence. *Development and Psychology, 15*, 183–197.

12. Krohn M. D., & Thornberry, T. P. (2008). Longitudinal perspectives on adolescent street gangs. In A. M. Liberman (Ed.), *The long view of crime: A synthesis of longitudinal research* (pp. 128–160). Washington, DC: National Institute of Justice, p. 147.

13. Thornberry et al. (1993), p. 81, see Note 5.

14. Decker, S. H. (1996). Collective and normative features of gang violence. *Justice Quarterly, 13*, 243–264, p. 244; Decker, S. H., & Van Winkle, B. (1996). *Life in the gang: Family, friends, and violence*. Cambridge, UK: Cambridge University Press.

15. Klein, M. W. (1971). *Street gangs and street workers*. Englewood Cliffs, NJ: Prentice Hall; Short, J. F., Jr. (1974). Youth, gangs and society: Micro- and macrosociological processes. *Sociological Quarterly, 15*, 3–19; Thrasher, F. M. (1927). *The gang: A study of 1,313 gangs in Chicago*. Chicago: University Press; Vigil, J. D. (1988). *Barrio gangs: Street life, identity, in southern California*. Austin: University of Texas Press, p. 132.

16. Ingraham, B. L., & Wellford, C. F. (1987). The totality of conditions test in eighth-amendment litigation. In S. D. Gottfredson & S. McConville (Eds.), *America's correctional crisis: Prison populations and public policy* (pp. 13–36). New York: Greenwood Press.

17. Stretesky, P. B., & Pogrebin, M. B. (2007). Gang-related gun violence: Socialization, identity, and self. *Journal of Contemporary Ethnography, 36*, 85–114, p. 105.

18. Fleisher, M. S. (1989). *Warehousing violence*. Newbury Park, CA: Sage; Fong, R. S., Vogel, R. E., & Buentello, S. (1992). Prison gang dynamics: A look inside the Texas Department of Corrections. In P. J. Benekos & A. V. Merlo (Eds.), *Corrections: Dilemmas and directions* (pp. 57–77). Cincinnati: Anderson Publishing.

19. Camp, G. M., & Camp, C. G. (1985). *Prison gangs: Their extent, nature, and impact on prisons*. Washington, DC: U.S. Government Printing Office.

20. Thornberry, T. P. (1987). Toward an interactional theory of delinquency. *Criminology, 25*, 863–891; Thornberry et al. (1993), p. 85, see Note 5.

21. Thornberry et al. (1993), p. 85, see Note 5.

22. Elder, G. H. (1985). Time, human change, and agency: Perspectives on the life course. *Social Psychology Quarterly, 57*, 4–15; Sampson, R. J., & Laub, J. H. (1993). *Crime in the making: Pathways and turning points through life*. Cambridge, MA: Harvard University Press.

23. Esbensen, F., Huizinga, D., & Weiher, A. W. (1993). Gang and non gang youth: Differences in explanatory factors. *Journal of Contemporary Criminal Justice, 9*, 94–116; Hill, K. G., Lui, C., & Hawkins, J. D. (2001).

Early precursors of gang membership: A study of Seattle youth. Washington, DC: Office of Juvenile Justice and Delinquency Prevention; Thornberry, T. P. (1998). Membership in youth gangs and involvement in serious and violent offending. In R. Loeber & D. P. Farrington (Eds.), *Serious and violent offenders: Risk factors and successful interventions* (pp. 147–166). Newbury Park, CA: Sage.

24. Thornberry et al. (1993), p. 44, see Note 5.
25. Hagedorn, J. M. (1994). *People and folks: Gangs, crime, and the underclass in a rustbelt city.* Chicago: Lake View; Sanchez Jankowski, M. (1991). *Islands in the street: Gangs and American urban society.* Berkeley, CA: University of California Press; Vigil, J. D. (2007). *The projects: Gang and non-gang families in East Los Angeles.* Austin: University of Texas Press.
26. Decker, S. H., & Lauritsen, J. L. (1996). Leaving the gang. In C. R. Huff (Ed.), *Gangs in America* (3rd ed., pp. 51–67). Thousand Oaks, CA: Sage.
27. Decker, & Van Winkle (1996), p. 272, see Note 26.
28. Yablonsky, L. (1963). *The violent gang.* New York: Macmillan.
29. Caldwell, L., & Altschuler, D. M. (2001). Adolescents leaving gangs: An analysis of risk and protective factors, resiliency and desistance in a developmental context. *Journal of Gang Research, 8,* 21–34.
30. Lemert, E. M. (1951). *Social pathology: A systematic approach to the theory of sociopathic behavior.* New York: McGraw-Hill.
31. Rosenfeld, R., Bray, T. M., & Egley, A. (1999). Facilitating violence: A comparison of gang-motivated, gang affiliated, and nongang youth homicides. *Journal of Quantitative Criminology, 15,* 495–516.
32. Thornberry et al. (1993), p. 44, see Note 5.
33. Curry, G. D., Decker, S. H., & Egley, A. (2002). Gang involvement and delinquency in a middle school population. *Justice Quarterly, 19,* 275–292.
34. Peterson, D., Taylor, T. J., & Esbensen, F. A. (2004). Gang membership and violent victimization. *Justice Quarterly, 21,* 793–815.
35. Caldwell, & Altschuler (2001), see Note 29; Decker, & Lauritsen (1996), see Note 26.

CHAPTER FIFTEEN

Gangs and Antisocial Behavior: A Critique and Reformulation

Matt DeLisi

KEY TERMS

Antisocial homophily
Enhancement
Facilitation
Selection

INTRODUCTION

A central tension in gang research has centered on competing visions about the antisociality of gang members. From one perspective, gang involvement is analogous to being in a youth group where delinquent activities are a peripheral part of gang member activity compared to socializing and other conventional activities. In this sense, gang members are nominally antisocial and are motivated to commit delinquency because of peer relationships, situational inducements, and normal group dynamics. From another perspective, gang involvement is a pernicious risk factor where delinquent activities are central to gang life. In this sense, gang members have high criminal propensity, are already engaging in antisocial behaviors, and likely become even more antisocial when exposed to similarly deviant peers. Thrasher's seminal research set the stage for these competing views:

> The quest for new experience seems to be particularly insistent in the adolescent, who finds in the gang the desired escape from, or compensation for, monotony. The gang actively promotes such highly agreeable activities as rough house, movement and change, games and gambling, predatory activities, seeing thrillers in the movies, sports, imaginative play, roaming and roving, exploration, and camping and hiking.[1]

Although written 85 years ago, Thrasher's quotation about gang activities captures these competing visions. Within the quotation, most of the behaviors are innocent behaviors of youth, yet also included is something far more sinister, namely predatory activities. Whether gangs are mere youth groups or collectives of severely antisocial individuals continues to the present.[2] The heart of this debate is the heterogeneity of the gang population and their antisocial behaviors and whether person-specific (**selection**) effects, group dynamic (**facilitation**) effects, or elements of both (**enhancement**) drive gang conduct.[3] Despite tremendous advances in the knowledge base about gangs, it remains a research area marked by a lack of consensus about who gang members are, what their gang involvement means, and most importantly for the current study, how gang members are related to the larger universe of criminal offenders.

The current exposition seeks to challenge current thinking about gangs. Here, I suggest that gang scholarship has largely overemphasized the organizational and conventional behavioral aspects of gangs. While gang members occasionally engage in mundane, prosocial activities and display markers of organization, gangs are mostly fluid, and are unquestionably pejorative, which counters the notion that gangs are a form of refuge for adolescents seeking a sense of belonging, security, and friendship. Instead, the current study advances that gangs are the culmination of a biosocial developmental process involving high-risk youth with temperamental and neurocognitive deficits that impair their relationships with parents, teachers, and peers. In this regard, gang members are essentially an adjunct of the severe offender prototype from criminal careers literature, with gangs representing the geographic proximity of similarly antisocial youths.[4] Consistent with the broad literature in developmental psychopathology and building on Howell and Egley's developmental risk factor approach, the biosocial deficits of fledgling gang members contribute to peer rejection and **antisocial homophily**.[5] In this sense, gangs are merely the coming together of the most antisocial individuals whose various deficits and protean antisocial behavior launch them into a criminal career. In turn, gangs exert little organizational pressure on their members because of the same deficits that drive their institutional failure in other life domains. In many respects, gangs and gang members are largely redundant to constructs from the criminal careers literature in that they display an asymmetry in offending where few members account for the bulk of total crime and few members account for the preponderance of the serious violence, such as murder, gun assaults, and armed robbery.[6] Moreover, the control of gang problems in large cities is accomplished by suppressing and incapacitating a relatively small number of gang members, which is precisely the correctional response to the control of serious, violent, and chronic juvenile offenders and/or career criminals.[7]

This chapter employs and integrates constructs from developmental psychopathology, criminal careers, focused deterrence, and, of course, gang literature. The first part of this study critiques organizational aspects of gangs to show that there is often little coherence within gangs to suggest organizational control over the behavior of its members. This is understandable and even predictable when the various psychological, social, and behavioral deficits of gang members are considered. The second part of this study offers a reformulation of the gang concept by focusing particularly on the temperamental and neurocognitive profiles of severely antisocial youth. These deficits contribute to strained interpersonal relations, peer rejection, school failure, and selection processes where antisocial youth are increasingly relegated to associating with similarly situated peers. This discussion is consistent with social information–processing theory in developmental psychopathology and Howell and Egley's developmental risk-factor approach in gang research.[8] The new contribution to this foundation is the targeted focus on temperamental and neurocognitive features in early life. Finally, after showing the convergent validity between gang members and career criminals, I draw on focused deterrence research that demonstrates that the policy response to gang members mirrors that of the policy responses to career criminals.

CRITIQUES OF THE GANG CONCEPT

Low Organization

There is a tremendous looseness to gangs that largely defies the notion that gangs are organizations on any level. Most gangs lack formal rules, lack a formal hierarchy, are unaware of the current status of their members, could not differentiate current from former or active from inactive members, and at times do not even have a name. Gang territory or turf, which has historically been portrayed in popular culture as essential to gang life, usually represents the localized area where gang members live and congregate. Frequently, gang members come and go quickly and there is little continuity in gang membership except for the most embedded members. Recently, Pyrooz and colleagues noted that gang embeddedness had nothing to do with the organizational properties of the gang, but was instead reflective of individual gang member identity.[9] The feckless nature of gang organization is not limited to an American context. Based on data from Canadian gang members, for instance, Bouchard and Spindler found that less than half of gangs have a gang name, specified leadership, a hierarchy, a meeting location, or any rules.[10]

There are sometimes distinctions made between street gangs, which reflect the poor organization described herein, and drug gangs, which are presumed to be more structured organizations. Yet, even drug gangs appear largely disorganized. Based on interview data from 34 federal prisoners convicted of smuggling large quantities (mean weight was nearly 2500 pounds) of cocaine, Benson and Decker examined six organizational features of drug-smuggling gangs. These were hierarchy, statement of rules, communication, adaptability, specialization or coordination, and recruitment and promotion procedures. They found little evidence for any of these organizational features, even in gangs that are supposed to be highly organized.[11] In sum, in an overview of the literature, Decker and Curry assessed gang organization:

Taken together, these studies suggested that gangs were not well organized and had weak control over their members, and that rivalries could lead to violence within and between gangs. In addition, these studies pointed to the transitory nature of gang membership, reinforcing the notion that gangs might not be organizations capable of controlling the behavior of their members.[12]

Although the ceremonial aspects of gangs are examined later, there is also evidence that the desistance process from gangs often lacks the drama of a beating out ceremony, suggesting that gang members do not necessarily care if members leave.[13] Indeed, Bolden's qualitative study of former gang members not only confirmed the largely nonexistent organizational structure of gangs, but also found that gang members switched involvement in various gangs with relative frequency. In a clever example, Bolden proposed, "the ease of switching gangs suggests that gang member status is transferable, somewhat like a professor's tenure."[14] However, unlike universities and their multiple layers of bureaucracy and organization, gangs are transitory and mostly incoherent entities. Given the various temperamental and neurocognitive deficits examined later in this chapter, the disorganization of gangs is to be expected.

Low Commitment and Disloyalty

Given the weak organizational structure of most gangs, it is unsurprising that gang members display little commitment or loyalty to their gang. In some respects, gangs reflect "something to do" until a more worthwhile activity is available. For instance, there is evidence that gang involvement declines sharply during the school year as youth become busy with school, sports, and other extracurricular activities.[15] This suggests little youth investment in gangs other than socialization, particularly for youth who remain invested in conventional activities and life domains. This disloyalty is not limited to street gangs. In her examination of the fluidity between street gangs and prison gangs, Maxson concluded there is little evidence of gang loyalty, membership fidelity, organizational structure, or even racial exclusivity in prison gangs.[16]

One of the clearest indicators of the low commitment and disloyalty among gang members relates to the intragang violence that typifies them. Several studies drawing on gang data from various cities indicated that gang members often victimize members from their own gang rather than a rival gang. In their study of gang homicides in St. Louis, Missouri, Decker and Curry found that most gang members were murdered by nongang–affiliated offenders. Moreover, when gang members were the perpetrators of homicide, the offenders and victims were often from the same gang.[17] Based on data from homicides in Newark, New Jersey, Pizarro and McGloin found that approximately 60 percent of gang homicides involved an offender–victim relationship characterized as acquaintance, friend, or significant other.[18]

In this way, the low commitment and minimal investment that gang members offer to their gang is fully consistent with the low commitment and minimal investment that they offer at school, at work, at home, in structured activities, and to a large extent, to conventional social institutions.

Nonceremonial Onset and Desistance

There is evidence that gang onset involves a violent beating-in ceremony, but mostly entrée into gang life is more gradual and reflects the assortative processes described in the next section. Gang onset corresponds to a larger disengagement from conventional life and a worsening involvement in an antisocial one. In their study of gang onset, Melde and Esbensen found that reduction in prosocial peers, school commitment, and guilt (a temperamental feature of psychopathy), and increases in delinquent neutralizations/rationalizations, negative peer commitment, unstructured socializing, delinquent peer groups, and an anger-based identity were associated with gang onset.[19]

Popular culture and conventional wisdom suggests that once an individual joins a gang, there are largely two ways out: death or ritual violence. In practice, gang members leave gangs, and even desist from gang involvement altogether, often without incident. Approximately 80 percent of youth who desist from gang life are able to walk away without any form of hostility or violence from their former gang colleagues.[20]

In short, while there is some evidence of organizational features to gangs, the bulk of studies suggest little formal organization to gang life. That there is little structure and influence to gangs is understandable and even expected when the deficits of gang members are considered. Gang members lack the wherewithal and overall functioning to maintain formal gang involvement, in many respects because of their deficits that similarly impair their home, school, and work functioning. The current thesis is consistent with a propensity-based approach to antisocial behavior. For example, Gottfredson and Hirschi suggested:

> *People who lack self-control tend to dislike settings that require discipline, supervision, or other constraints on their behavior; such settings include school, work, and, for that matter, home. These people therefore tend to gravitate to "the street" or, at least in adolescence, to the same-sex peer group.*[21]

Given this profile, gangs are not sustainable due to the deficits of their members.

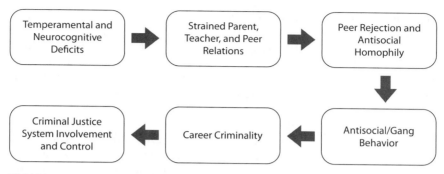

FIGURE 15-1. Conceptual Model

Since its inception, gang theorists have mostly advanced a normative conceptualization of gang formation where gangs represented a deliberately formed, quasi-structured organization that was comparable to a youth group or sports team. Although they were loosely organized, gangs were still organized. But the various psychosocial deficits of gang members make the existence of a structured gang unlikely—and the evidence supports that. Moreover, the decision to create gangs during early and middle adolescence seems at odds with a life-course understanding of gang members themselves. Compared to youths who never join gangs, gang youths display high, and at times extreme, psychopathology that retards their social development while enhancing their antisocial development. This process begins in the first years of life and reflects a systemic network of biosocial processes that are broadly shown in **Figure 15-1**. These developmental steps, with targeted focus on temperamental and neurocognitive deficits, are examined next.

REFORMULATION OF THE GANG CONCEPT

Temperamental and Neurocognitive Deficits

Although classic criminologists often did not explicitly articulate concepts such as temperament, personality, and neurocognitive deficits by name, there is no question that they were thinking of various person-specific deficits when considering the disposition of gang members. Many of the classic descriptions of gang youth from the middle decades of the 20th century focused on the poor self-regulation (e.g., low effortful control, low inhibitory control, low attention control, and overall poor self-control), high negative emotionality (e.g., high anger, high hostility, poor temper, easily frustrated), and high neurocognitive deficits (e.g., low verbal intelligence, speech and language pathology, impulsivity, low planning, poor decision making) of gang youths. These individual-level deficits contributed to their defiance toward and disengagement from conventional society that is the backbone of cultural deviance theory.

There are many examples of this approach. For instance, in describing gang youths, Cohen suggested, "For the child who breaks clean with middle-class morality, on the other hand, there are no moral inhibitions on the free expression of aggression against the sources of his frustration."[22] There was also the suggestion that the most antisocial youth are prone to gang involvement and even gang leadership. Short and Strodtbeck found that "Among conflict gangs the leaders are known to have the capacity to function aggressively against other members when necessary to maintain their dominance."[23]

Miller's focal concerns of trouble, toughness, smartness, excitement, fate, and autonomy clearly relate to modern temperamental and neurocognitive constructs such as fearlessness, aggression, antagonism, sensation seeking, impulsivity, and external locus of control.[24] Short and Strodtbeck offered one of the most comprehensive assessments of the temperament, neurocognitive, and social deficits of gang youths:

> *The failure of individuals to make satisfactory adjustments in any institutional sphere inevitably handicaps their ability to achieve future goals. Our gang boys fail often in school, on the job, in conventional youth-serving agencies, and in the eyes of law enforcement officials (and therefore in the public eye). They fail more often in each of these respects than do the non-gang boys we have studied, both middle and lower class.[25]*

Although classic criminological studies of gang-involved youth almost exclusively focused on adolescents, it is important to note that the seeds of youth violence, externalizing symptoms, and delinquency are planted much earlier in life. Even before entering kindergarten, there is extraordinary variance in temperament, personality functioning, and self-regulation among young children. Some children exhibit pronounced behavioral problems and perpetrate serious delinquent offenses, including assault, strong-arm robbery, and sexual aggression. To illustrate, Keenan and Wakschlag reported that preschoolers as young as 30 months present symptoms that are consistent with serious behavioral disorders, such as attention deficit hyperactivity disorder (ADHD), oppositional defiant disorder (ODD), and conduct disorder (CD). In their study of 79 clinic-referred preschool children 2.5 to 5.5 years of age, nearly half met criteria for CD and 75 percent met criteria for ODD. Nearly 30 percent met diagnostic criteria for all three disorders. Some of these children were so aggressive and out of control that they were expelled from preschool.[26] Other studies have shown strong stability in aggression, low self-regulation, and other social-cognitive problems from the first years of life through late childhood and adolescence[27] and even into adulthood.[28]

Based on data from the largest nationally representative survey of kindergarteners, Vaughn, DeLisi, Beaver, and Wright reported evidence of a severe subgroup of children with self-regulation deficits, cognitive deficits, externalizing symptoms, and interpersonal deficits. This group had difficulty getting along with peers in age-appropriate ways and difficulties handling the demands of appropriate classroom behavior. Moreover, these children were reared by parents characterized by high levels of physical punishment, high levels of parental stress, and low levels of parental involvement. The interaction of their severe deficits and the difficulty with which adults respond to them leads to a cycle of continued negative displays and coercive responses.[29]

Although gang outcomes are rarely the developmental variable of interest in developmental psychopathology, there is evidence of temperamental and neurocognitive problems as precursors of gang involvement. In a longitudinal study of data selected from the Pittsburgh Youth Study, Lahey, Gordon, Loeber, Stouthamer-Loeber, and Farrington reported that onset of gang involvement was in many respects simply a culmination of a set of traits and behaviors relating to behavioral disorders. Specifically, boys with symptoms of CD and/or increased levels of behaviors relating to CD were those most likely to join an antisocial gang by age 19. Although gangs are commonly viewed as a social construct, the diagnostic criteria for CD read like a laundry list of activities that serious delinquents and/or gang youths perpetrate, such as intimidation, physical fights, use of weapons, extortion, armed robbery, destruction of property, burglary, and broadband violations of societal rules.[30]

Fearlessness is the cardinal temperamental feature of psychopathy, and there is evidence that deficits in fear induction and attendant negative emotional states increase the likelihood of gang involvement. For instance, Esbensen, Winfree, He, and Taylor found that guiltlessness was significantly predictive of five types of gang membership.[31] In their study of Mexican American gang members, Valdez, Kaplan, and Codina found that nearly 50 percent of the gang members in their survey scored in the moderate to highly psychopathic range on the Psychopathy Checklist Revised: Screening Version (PCL:SV). Compared to nongang youth, gang members had higher scores on total, affective, and behavioral facets of psychopathy and were significantly less empathic.[32]

Drawing on data from a longitudinal study of boys reared in socioeconomically impoverished neighborhoods in Montreal, Craig, Vitaro, Gagnon, and Tremblay compared stable gang members, unstable gang members, and nongang members on a variety of psychosocial characteristics spanning ages 10 to 14. They found clear differences in their neurocognitive functioning. Stable gang members were the most aggressive, had the least anxiety (suggesting psychopathy), were the least agreeable, were the most hyperactive, had the highest levels of inattention, and were the most oppositional. Conversely, nongang children were the least aggressive, most anxious, most prosocial and agreeable, most self-regulated, had better attentional control, and were the least oppositional.[33] In a related study, boys characterized as fearless, hyperactive, and infrequently prosocial at kindergarten were *seven times* more likely than their peers to join a gang in early adolescence.[34]

It is critical that juvenile justice practitioners and policymakers focus on early developing temperamental and neurocognitive factors to best identify the most behaviorally at-risk individuals. Indeed, cross-national research has shown that these children can be identified as early as in kindergarten.[35] Waiting to identify these deficits among adolescents is problematic. For example, Melde, Gavazzi, McGarrell, and Bynum recently evaluated a prevention and intervention program for youth at risk for gang activity and found that the targeted/intervention groups displayed less risk than the nontargeted group, suggesting that the program was not being delivered to those most in need. The program targeted three broad features of mental health functioning: internalizing behaviors, externalizing behaviors, and ADHD.[36] While these are important constructs, they are less important than more pernicious features such as CD and especially psychopathy. Moreover, serious behavioral problems during adolescence are more normative than the same behavioral problems occurring among children, and it is the latter group that most likely will become chronic and/or gang offenders.

Strained Relations, Peer Rejection, and Antisocial Homophily

A variety of parenting factors, including low warmth, low monitoring, erratic supervision, inconsistent punishment, abuse, neglect, and lax behavioral control, increases the likelihood that children not only will engage in externalizing behaviors, but also ultimately interact with deviant peers who share their family background characteristics. Compared to their nongang peers, gang members are likely to be reared by antisocial parents where there is exposure to home delinquency and high levels of aggressive behaviors. Decades ago, Cloward and Ohlin suggested, "this sequence in the development of delinquent norms and justifying beliefs and values makes it easier to understand the intractable and apparently conscienceless behavior of the fully indoctrinated members of delinquent subcultures."[37] By the end of the first decade of life, children begin to sort themselves according to common interests, values, skills, and behaviors. It is then when the dynamics of peer rejection and self-selection occur to form nascent gangs, and this process is examined next.

The highly antisocial youths who are susceptible to gang life are worn down by their temperamental and neurocognitive deficits and the strains that those deficits put on their family and social relationships.[38] A series of coercive exchanges with parents, teachers, and their own peers contributes to a personality style characterized by high levels of negative emotionality, especially anger and hostility, hostile attribution bias, sensitivity to provocation and judgment, and estrangement from the typical developmental processes of childhood. There is often considerable overlap among children who are aggressive, rejected, and/or delinquent, and their friendship networks overwhelmingly are comprised of similarly troubled peers.[39]

Social information–processing theory establishes a set of interrelated processes that contribute to antisocial development. In this model, children and adolescents' cognitive style is dominated by aggression that is rooted partially in their greater likelihood of child abuse victimization and overall exposure to violence. These experiences create cognitive "scripts" whereby children encode and interpret social cues, social interactions, and social dynamics in generally confrontational, negative terms that warrant a commensurately negative, aggressive response. In a landmark study on this process, Dodge, Bates, and Pettit compared children who had been physically abused to a control group who had not. There were significant differences all disadvantaging the child abuse group in the encoding of relevant social information cues, prevalence of hostile attribution bias, number of responses generated to each social problem, proportion of aggressive responses, proportion of competent responses, positive evaluations of aggression, and positive evaluations of acting competently.[40]

Moreover, Crick and Dodge reported that schoolchildren with social information–processing deficits are prone to be reactively aggressive, proactively aggressive, or display both forms of aggression compared to control groups of peers. More disturbingly, they found that proactively aggressive children reported more positive outcome expectations and greater self-efficacy from aggressive behavior compared to children who were aggressive, but not proactively aggressive.[41] Given this profile, normative children strongly dislike children who are aggressive, and this disliking is even more pronounced when aggressors target nonassertive victims.[42]

The various coercive relationships that acutely antisocial youth experience almost inevitably increase the likelihood of failure at conventional social institutions. What does the long-term developmental course of gang involvement look like? In a study of more than 900 juvenile arrestees who were part of the Arizona Arrestee Drug Abuse Monitoring (ADAM) program, Katz, Webb, Fox, and Shaffer reported high levels of educational failure. Between 33 and 48 percent of gang associates, former gang members, or current gang members had been expelled, suspended, or dropped out of school.[43] Relatedly, Krohn, Ward, Thornberry, Lizotte, and Chu assessed data from the Rochester Youth Development Study and found that gang involvement in early waves culminated in school dropout; teen parent status; economic hardship characterized by unemployment and low income; family strife characterized by arguing, fighting, and poor emotional control; adult crime; and criminal justice system involvement.[44] This is precisely the modal lifestyle and developmental course of a career criminal.

Studies using propensity score matching make clear that compared to nongang youth, gang youth are significantly more antisocial in the first place and thus their attraction to gangs as a selection process is intuitive.[45] Consistent with the sequelae of peer rejection, gang youth have significantly lower self-control and more drug-using peers than nongang youth. In other words, they are prone to misbehavior in school and in social settings that leads to negative labeling and distancing from conventional peers. Even when matched, gang youths commit more nonviolent and violent delinquency, with violence showing longitudinal effects, and are also more likely to be victimized.

In school, nascent gang members constitute a source of fear and intimidation to nongang youth who attend school with them. For instance, Randa and Wilcox found that gang members in school are a major cause of student fear and can contribute to avoidant behaviors by students where pro-social youth simply avoid the possibility of contacting their peers who are antisocial/involved in gangs.[46] Others have likewise found that children with neurocognitive deficits, self-regulation deficits, and related temperamental problems are notably likely to be the most disruptive students at school.[47] Similarly, Huff compared gang members to at-risk nongang members and found many significant behavioral differences between them. Gang members were nearly 10 times more likely to carry knives in school and 4 times more likely to carry guns at school. These are delinquent behaviors that considerably increase the likelihood of expulsion.[48]

Another important factor in the peer rejection and assortative processes of gang formation is personality. Using data from prisoners in the United Kingdom, Egan and Beadman found that antisocial personality features predicted which offenders were more embedded in gangs. Moreover, offenders with high levels of antagonism (or low agreeableness) sought out similar peers in terms of their personality and antisocial views after being rejected by conventional peers. In other words, the antisociality of at-risk youth facilitates the gravitation toward similar peers, and the result is a group in which deviance and violence is likely to ensue.[49] As Gottfredson and Hirschi noted:

Adventuresome and reckless children who have difficulty making and keeping friends tend to end up in the company of one another, creating groups made up of individuals who tend to lack self-control. The individuals in such groups will therefore tend to be delinquent, as will the group itself.[50]

To summarize, the temperamental and neurocognitive features of seriously antisocial children engender a series of coercive interactions that are compatible with social information–processing theories and Howell and Egley's developmental risk factor approach. Importantly, the seeds of this antisocial pathway are established much earlier than adolescence, such that the precursors of gang involvement are essentially the same precursors of serious, violent, and chronic delinquency.

Antisocial/Gang Behavior and Career Criminality

By the time that individuals have endured the rejection from conventional peers and mostly repudiated their involvement in conventional institutions, the versatile antisocial behaviors that gang members display are in many respects indistinguishable from the behaviors of career criminals. Based on data from the National Longitudinal Study of Adolescent Health, Kort-Butler described this behavioral underworld as characterized by high levels of general delinquency, violent delinquency, experienced victimization, witnessed victimization, negative emotions, group fights with peers, and low socioeconomic status. About 14 percent of youth in her sample indicated that they expected to be murdered by age 21.[51] In a study of 504 incarcerated felons between ages 17 and 25, Varano, Huebner, and Bynum found that those involved in unorganized and organized gangs had significantly more delinquent friends, harbored more antisocial attitudes, and were more likely to carry firearms and buy and sell drugs.[52]

In the Pittsburgh Youth Study, approximately 30 percent of boys who perpetrated homicide were also involved in gangs.[53] Their behavioral profile exemplifies the constructs examined earlier.

About 85 percent exhibited the highest behavioral risk score, nearly 80 were suspended from school and 61 percent were chronically truant. Many of the boys had disruptive behavioral disorders (e.g., ODD, CD, ADHD, or all three) and exhibited psychopathic traits, such as callousness and cruelty to others. These youths held favorable attitudes toward delinquency, had delinquent peers, held favorable attitudes toward substance use, engaged in versatile forms of crime, and displayed overall failure in major life domains.

Over the years, gang researchers have consistently documented that although small in number, gang members commit most of the delinquency and serious violence in their samples. In addition to this disproportionate criminal involvement, there is also evidence of a small subgroup of gang members who do not transition out of gang life, remaining deeply embedded.[54] This raises an important empirical question about whether entrenched gang members and career criminals are in fact the same subgroup. For example, in the National Longitudinal Study of Adolescent Health, the prevalence of gang membership is 5 percent, which is the same prevalence as career criminality.[55] Like career criminals, gangs present an asymmetry in offending where very few account for most of the violence and antisocial behavior. According to the most recent National Gang Threat Assessment, gangs account for between 48 and 90 percent of the serious violence in communities across the United States.[56]

There is evidence that the most antisocial individuals are not only most prone to gang activity, but also most prone to extreme violence within gangs. For instance, based on a longitudinal study of nearly 900 adolescents in Sweden, Kerr and colleagues found that psychopathic traits moderated the effect of peers on delinquent conduct. Specifically, psychopathic youths, particularly those with high levels of callous and unemotional traits, were most likely to entice their peers to engage in delinquency. In addition, the most psychopathic youths were the least affected by their peers, suggesting that their sheer antisociality was the driving force in the overall delinquency of the youth group.[57]

In short, versatile criminal offending accompanies failure and/or rejection of conventional activities for gang youths and career criminals alike. The reason that gang activity appears to decline into adulthood is consistent with the general decline in peer associations that typifies emerging adulthood. Hanging out with peers is normative during adolescence but less so among adults. This is the reason gangs have disproportionately been an adolescent phenomenon. Regardless of their gang history, serious offenders continue to engage in crime and amass significant records of arrests and convictions. Ironically, it is their visibility as a serious offender that makes gang-involved offenders prime targets for punitive and effective criminal justice sanctions.

Criminal Justice System Involvement

In a variety of contexts, gang members constitute security problems for the criminal justice system based on their sustained involvement in deviance even while under correctional supervision. In addition to increased risk for institutional misconduct, gang members are also disproportionately involved in the most violent forms of prison violence.[58] For example, a recent study of inmate murders found that 43 percent were gang involved.[59] Prison gang members are also significantly more likely to perpetrate an aggravated assault against a corrections worker or other staff member within prison.[60]

The noncompliance of gang members is not limited to a correctional context. There is evidence that after release from confinement facilities, gang members are more likely than their nongang peers to recidivate.[61] In addition, Caudill found that formerly confined gang members recidivated faster than their nongang peers. However, after 6 months the differences reduced and other predictors, such as race and prior delinquent history, were stronger.[62] For these reasons, gangs are a major target for juvenile justice and criminal justice system policies.

It is from a policy perspective where the empirical overlap between gang members and career criminals is most clear. Several commentators have noted that the antisociality of gang members makes them attractive targets to police. Moreover, suppressing gang member behavior is presumed to net important reductions in crime because of their disproportionate criminal involvement. For example, Spergel noted that a key to gang-reduction programs centered on the control of individual gang members whose "hothead" nature made them most likely to commit violent crime.[63] In other words, gang interventions and gang members themselves recognized that the most acutely antisocial individuals within the gang displayed psychosocial traits that made violent behavior likely. Kennedy suggested, "the very characteristics that define our most worrisome criminal populations—that they offend often, in manifold ways, often in groups, and have frequent contact with criminal justice agencies—also represent vulnerabilities for them, and opportunities for crime control."[64] Braga and Weisburd observed, "The chronic involvement of gang members in a wide variety of offenses made them, and the gangs they formed, vulnerable to a coordinated criminal justice system."[65]

Braga and Weisburd's meta-analytic review of focused deterrence strategies provides the strongest evidence that active gang members and career criminals are ostensibly the same in terms of their contribution to homicide and other serious crime. By targeting small numbers of gang members, focused deterrence strategies have yielded large reductions in homicide. For instance, Operation Ceasefire in Boston produced a 63 percent reduction in youth homicides, the Indianapolis Violence Reduction Partnership produced a 34 percent reduction in total homicides, Operation Peacekeeper in Stockton, California yielded a 42 percent reduction in gun homicides, Project Safe Neighborhoods in Lowell, Massachusetts contributed to a 44 percent reduction in gun assaults, and the Cincinnati Initiative to Reduce Violence produced a 35 percent reduction in homicides.[66] Focused deterrence strategies are a major area of research in the gang literature, but few gang researchers beyond Braga and Kennedy have seen the parallels between gang members and career offenders in terms of their offending asymmetry and the effectiveness of criminal justice initiatives that seek to remove the small subgroup of offenders from society.

CONCLUSION

Like all areas of science, the study of youth violence and juvenile justice is differentiated into mostly discrete research areas where knowledge is produced by a relatively discrete sect of researchers. This research specialization is not intrinsically problematic, but does run the risk of groupthink and a lack of innovation. There is concern that gang research has reached this point, and has become mostly stultified. The current conceptual study seeks to invigorate gang theorizing by suggesting that although real, gangs are also ostensibly mythical given their inability to sustain themselves. Contrary to early gang scholars and the sentiment in the quotation of Thrasher that began this study, gang members are not innocent youths looking for something to do. Gangs are an assemblage of troubled children and adolescents who evince high psychopathology that not only facilitates their antisocial development, but also largely dooms their prosocial development. Parents, teachers, and their own peers recognize this, and sooner or later reject the antisocial tendencies of gang youth, and, in doing so, reject the youths themselves. The result is the assortative processes of antisocial homophily.

As their delinquent careers unfold, it becomes more difficult to discern gang members from career criminals. I fully concur with Braga, who recently advised:

> The research also suggests that the terminology used to describe the types of groups involved in urban violence matters less than their behavior. Gangs, their nature, and their behavior remain central questions for communities, police, and scholars. At the same time, where violence prevention and public safety are concerned, the gang question is not the central one. The more important observation is that urban violence problems are, in large measure, concentrated among groups of chronic offenders; thus the dynamics between and within these groups have major implications for crime rates.[67]

Perhaps it is time to suggest they are the same, because, indeed, behaviorally they are the same.

GLOSSARY

Antisocial homophily—the idea that antisocial people tend to associate with similarly antisocial people

Enhancement—model that suggests that antisocial youths are attracted to gangs, and once they have joined, their deviance will increase

Facilitation—model that asserts that gangs increase delinquency because of the antisocial properties of the group

Selection—model that asserts that the most antisocial youths gravitate toward gangs because they match their behaviors and traits

NOTES

1. Thrasher, F. (1927/1963). *The gang: A study of 1,313 gangs in Chicago.* Chicago: University of Chicago Press, p. 68.
2. Howell, J. C. (2007). Menacing or mimicking? Realities of youth gangs. *Juvenile and Family Court Journal, 58,* 9–20; Howell, J. C. (2012). *Gangs in America's communities.* Thousand Oaks, CA: Sage.
3. Gatti, U., Tremblay, R. E., Vitaro, F., & McDuff, P. (2005). Youth gangs, delinquency and drug use: A test of the selection, facilitation, and enhancement hypotheses. *Journal of Child Psychology and Psychiatry, 46,* 1178–1190; Lacourse, E., Nagin, D. S., Tremblay, R. E., Vitaro, F., & Claes, M. (2003). Developmental trajectories of boys' delinquent group membership and facilitation of violent behaviors during adolescence. *Development and Psychopathology, 15,* 183–197; McGloin, J. M. (2012). Gang involvement and predatory crime. In M. DeLisi & P. J. Conis (Eds.), *Violent offenders: Theory, research, policy, and practice,* (2nd ed., pp. 221–234). Burlington, MA: Jones & Bartlett Learning; Thornberry, T. P., Krohn, M. D., Lizotte, A. J., & Chard-Wierschem, D. (1993). The role of juvenile gangs in facilitating delinquent behavior. *Journal of Research in Crime and Delinquency, 30,* 55–87.
4. DeLisi, M. (2005). *Career criminals in society.* Thousand Oaks, CA: Sage; Moffitt, T. E. (1993). Adolescence-limited and life-course-persistent antisocial behavior: A developmental taxonomy. *Psychological Review, 100,* 674–701; Vaughn, M. G., DeLisi, M., Gunter, T., Fu, Q., Beaver, K. M., Perron, B. E., & Howard, M. O. (2011). The severe 5%: A latent class analysis of the externalizing behavior spectrum in the United States. *Journal of Criminal Justice, 39,* 75–80.
5. Howell, J. C., & Egley, Jr., A. (2005). Moving risk factors into developmental theories of gang membership. *Youth Violence and Juvenile Justice, 3,* 334–354.

6. Pizarro, J. M., Zgoba, K. M., & Jennings, W. G. (2011). Assessing the interaction between offender and victim criminal lifestyles and homicide type. *Journal of Criminal Justice, 39,* 367–377.

7. Braga, A. A. (2012). Focused deterrence strategies and the reduction of gang and group-involved violence. In M. DeLisi & P. J. Conis (Eds.), *Violent offenders: Theory, research, policy, and practice* (2nd ed., pp. 259–279). Burlington, MA: Jones & Bartlett Learning; Kennedy, D. M. (1997). Pulling levers: Chronic offenders, high-crime settings, and a theory of prevention. *Valparaiso University Law Review, 31,* 449–484.

8. Howell, & Egley (2005), see Note 5.

9. Pyrooz, D. C., Sweeten, G., & Piquero, A. R. (2012). Continuity and change in gang membership and gang embeddedness. *Journal of Research in Crime and Delinquency,* doi: 10.1177/00224211434830.

10. Bouchard, M., & Spindler, A. (2010). Groups, gangs, and delinquency: Does organization matter? *Journal of Criminal Justice, 38,* 921–933; Spindler, A., & Bouchard, M. (2011). Structure or behavior? Revisiting gang typologies. *International Criminal Justice Review, 21,* 263–282.

11. Benson, J. S., & Decker, S. H. (2010). The organizational structure of international drug smuggling. *Journal of Criminal Justice, 38,* 130–138.

12. Decker, S. H., & Curry, G. D. (2002). Gangs, gang homicides, and gang loyalty: Organized crimes or disorganized criminals? *Journal of Criminal Justice, 30,* 343–352, p. 346.

13. Pyrooz, D. C., Decker, S. H., & Webb, V. J. (2012). The ties that bind: Desistance from gangs. *Crime & Delinquency,* doi: 10.1177/0011128710372191.

14. Bolden, C. L. (2012). Liquid soldiers: Fluidity and gang membership. *Deviant Behavior, 33,* 207–222, p. 220.

15. Walker-Barnes, C. J., & Mason, C. A. (2001). Ethnic differences in the effect of parenting on gang involvement and gang delinquency: A longitudinal hierarchical linear modeling perspective. *Child Development, 72,* 1814–1831.

16. Maxson, C. L. (2012). Betwixt and between street and prison gangs: Defining gangs and structures in youth correctional facilities. In F. A. Esbensen & C. L. Maxson (Eds.), *Youth gangs in international perspective: Results from the Eurogang program of research* (pp. 107–124). New York: Springer.

17. Decker, & Curry (2002), see Note 12.

18. Pizarro, J. M., & McGloin, J. M. (2006). Explaining gang homicides in Newark, New Jersey: Collective behavior or social disorganization? *Journal of Criminal Justice, 34,* 195–207.

19. Melde, C., & Esbensen, F-A. (2011). Gang membership as a turning point in the life course. *Criminology, 49,* 513–552.

20. Pyrooz, D. C., & Decker, S. H. (2011). Motives and methods for leaving the gang: Understanding the process of gang desistance. *Journal of Criminal Justice, 39,* 417–425.

21. Gottfredson, M. R., & Hirschi, T. (1990). *A general theory of crime.* Stanford, CA: Stanford University Press, p. 157.

22. Cohen, A. K. (1955). *Delinquent boys: The culture of the gang.* New York: The Free Press, p. 132.

23. Short, Jr., J. F., & Strodtbeck, F. L. (1963). The response of gang leaders to status threats: An observation on group process and delinquent behavior. *American Journal of Sociology, 68,* 571–579, p. 578.

24. Miller, W. B. (1958). Lower class culture as a generating milieu of gang delinquency. *Journal of Social Issues, 14,* 5–19.

25. Short, Jr., J. F., & Strodtbeck, F. L. (1965). *Gang process and gang delinquency.* Chicago: University of Chicago Press, p. 230.

26. Keenan, K., & Wakschlag, L. S. (2000). More than the terrible twos: The nature and severity of behavior problems in clinic-referred preschool children. *Journal of Abnormal Child Psychology, 28,* 33–46.

27. Blair, C., & Diamond, A. (2008). Biological processes in prevention and intervention: The promoting of self-regulation as a means of preventing school failure. *Development and Psychopathology, 20,* 899–911; Campbell, S. B. (1994). Hard-to-manage preschool boys: Externalizing behavior, social competence, and family context at two-year follow-up. *Journal of Abnormal Child Psychology, 22,* 147–166; Campbell, S. B. (2002). *Behavior problems in preschool children: Clinical and developmental issues* (2nd ed.). New York: The Guilford Press; Campbell, S. B., & Ewing, L. J. (1990). Follow-up of hard-to-manage preschoolers: Adjustment at age 9 and predictors of continuing symptoms. *Journal of Child Psychology and Psychiatry, 31,* 871–889; DeLisi, M., & Vaughn, M. G. (2011). The importance of neuropsychological deficits relating to self-control and temperament to the prevention of serious antisocial behavior. *International Journal of Child, Youth and Family Studies, 1 & 2,* 12–35.

28. Caspi, A. (2000). The child is father of the man: Personality continuities from childhood to adulthood. *Journal of Personality and Social Psychology, 78,* 158–172; Caspi, A., Elder, G. H., & Bem, D. J. (1987). Moving against the world: Life-course patterns of explosive children. *Developmental Psychology, 23,* 308–313.

29. Vaughn, M. G., DeLisi, M., Beaver, K. M., & Wright, J. P. (2009). Identifying latent classes of behavioral risk based on early childhood manifestations of self-control. *Youth Violence and Juvenile Justice, 7,* 16–31.

30. Lahey, B. B., Gordon, R. A., Loeber, R., Stouthamer-Loeber, M., & Farrington, D. P. (1999). Boys who join gangs: A prospective study of predictors of first gang entry. *Journal of Abnormal Child Psychology, 27,* 261–276.

31. Esbensen, F-A., Winfree, Jr., L. T., He, N., & Taylor, T. J. (2001). Youth gangs and definitional issues: When is a gang a gang, and why does it matter. *Crime & Delinquency, 47,* 105–130.

32. Valdez, A., Kaplan, C. D., & Codina, E. (2000). Psychopathy among Mexican American gang members: A comparative study. *International Journal of Offender Therapy and Comparative Criminology, 44,* 46–58.

33. Craig, W. M., Vitaro, F., Gagnon, C., & Tremblay, R. E. (2002). The road to gang membership: Characteristics of male gang and nongang members from ages 10 to 14. *Social Development, 11,* 53–68.

34. Lacourse, E., Nagin, D. S., Vitaro, F., Côté, S., Arseneault, L., & Tremblay, R. E. (2006). Prediction of early-onset deviant peer group affiliation: A 12-year longitudinal study. *Archives of General Psychiatry, 63,* 562–568.

35. Lacourse et al. (2006), see Note 34; Vaughn et al. (2009), see Note 29.

36. Melde, C., Gavazzi, S., McGarrell, E., & Bynum, T. (2011). On the efficacy of targeted gang interventions: Can we identify those most at risk? *Youth Violence and Juvenile Justice, 9,* 279–294.

37. Cloward, R. A., & Ohlin, L. E. (1960). *Delinquency and opportunity: A theory of delinquent gangs.* New York: The Free Press of Glencoe, p. 132.

38. Cairns, R., & Cairns, B. (1991). Social cognition and social networks: A developmental perspective. In D. Pepler & K. Rubin (Eds.), *The development and treatment of childhood aggression* (pp. 389–410). Hillsdale, NJ: Lawrence Erlbaum Associates.

39. Deptula, D. P., & Cohen, R. (2004). Aggressive, rejected, and delinquent children and adolescents: A comparison of their friendships. *Aggression and Violent Behavior, 9,* 75–104.

40. Dodge, K. A., Bates, J. E., & Pettit, G. (1990). Mechanisms in the cycle of violence. *Science, 250,* 1678–1683.

41. Crick, N. R., & Dodge, K. A. (1994). A review and reformulation of social information processing mechanisms in children's social adjustment. *Psychological Bulletin, 115,* 74–101; Crick, N. R., & Dodge, K. A. (1996). Social information-processing mechanisms on reactive and proactive aggression. *Child Development, 67,* 993–1002.

42. Courtney, M. L., Cohen, R., Deptula, D. P., & Kitzmann, K. M. (2003). An experimental analysis of children's dislike of aggressors and victims. *Social Development, 12,* 46–66.

43. Katz, C. M., Webb, V. J., Fox, K., & Shaffer, J. N. (2011). Understanding the relationship between violent victimization and gang membership. *Journal of Criminal Justice, 39,* 48–59.

44. Krohn, M. D., Ward, J. T., Thornberry, T. P., Lizotte, A. J., & Chu, R. (2011). The cascading effects of adolescent gang involvement across the life course. *Criminology, 49,* 991–1028.

45. Barnes, J. C., Beaver, K. M., & Miller, J. M. (2010). Estimating the effect of gang membership on nonviolent and violent delinquency: A counterfactual analysis. *Aggressive Behavior, 36,* 437–451; DeLisi, M., Barnes, J. C., Beaver, K. M., & Gibson, C. L. (2009). Delinquent gangs and adolescent victimization revisited: A propensity score matching approach. *Criminal Justice and Behavior, 36,* 808–823; Lyon, J-M., Henggeler, S. W., & Hall, J. A. (1992). The family relations, peer relations, and criminal activities of Caucasian and Hispanic American gang members. *Journal of Abnormal Child Psychology, 20,* 439–449.

46. Randa, R., & Wilcox, P. (2010). School disorder, victimization, and general v. place-specific student avoidance. *Journal of Criminal Justice, 38,* 854–861.

47. Way, S. M. (2011). School discipline and disruptive classroom behavior: The moderating effects of student perceptions. *The Sociological Quarterly, 52,* 346–375.

48. Huff, C. R. (1998). *Comparing the criminal behavior of youth gangs and at-risk youths.* Washington, DC: U.S. Department of Justice, Office of Justice Programs, National Institute of Justice.

49. Egan, V., & Beadman, M. (2011). Personality and gang embeddedness. *Personality and Individual Differences, 51,* 748–753.

50. Gottfredson, & Hirschi (1990), p. 158, see Note 21.

51. Kort-Butler, L. A. (2010). Experienced and vicarious victimization: Do social support and self-esteem prevent delinquent responses? *Journal of Criminal Justice, 38*, 496–505.
52. Varano, S. P., Huebner, B. M., & Bynum, T. S. (2011). Correlates and consequences of pre-incarceration gang involvement among incarcerated youthful felons. *Journal of Criminal Justice, 39*, 30–38.
53. Farrington, D. P., Loeber, R., Stallings, R., & Homish, D. L. (2012). Early risk factor for young homicide offenders and victims. In M. DeLisi & P. J. Conis (Eds.), *Violent offenders: Theory, research, policy, and practice*, (2nd ed., pp. 143–159). Burlington, MA: Jones & Bartlett Learning.
54. Decker, S. H., & Pyrooz, D. (2011). Timing is everything: Gangs, gang violence, and the life course. In M. DeLisi & K. M. Beaver (Eds.), *Criminological theory: A life-course approach* (pp. 149–164). Sudbury, MA: Jones & Bartlett Learning.
55. DeLisi, M., & Piquero, A. R. (2011). New frontiers in criminal careers research, 2000–2011: A state-of-the-art review. *Journal of Criminal Justice, 39*, 289–301.
56. National Gang Intelligence Center. (2012). *2011 national gang threat assessment: Emerging trends*. Washington, DC: National Gang Intelligence Center.
57. Kerr, M., Van Zalk, M., & Stattin, H. (2012). Psychopathic traits moderate peer influence on adolescent delinquency. *Journal of Child Psychology and Psychiatry, 53*(8), 826–835.
58. Blackburn, A. G., & Trulson, C. R. (2010). Sugar and spice and everything nice? Exploring institutional misconduct among serious and violent female delinquents. *Journal of Criminal Justice, 38*, 1132–1140; DeLisi, M., Berg, M. T., & Hochstetler, A. (2004). Gang members, career criminals, and prison violence: Further specification of the importation model of inmate behavior. *Criminal Justice Studies, 17*, 369–383; Drury, A. J., & DeLisi, M. (2011). Gangkill: An exploratory empirical assessment of gang membership, homicide offending, and prison misconduct. *Crime & Delinquency, 57*, 130–146; Sorensen, J., & Davis, J. (2011). Violent criminals locked up: Examining the effect of incarceration on behavioral continuity. *Journal of Criminal Justice, 39*, 151–158; Trulson, C. R., DeLisi, M., Caudill, J. W., Belshaw, S., & Marquart, J. W. (2010). Delinquent careers behind bars. *Criminal Justice Review, 35*, 200–219.
59. Cunningham, M. D., Sorensen, J. R., Vigen, M. P., & Woods, S. O. (2010). Inmate homicides: Killers, victims, motives, and circumstances. *Journal of Criminal Justice, 38*, 348–358.
60. Sorensen, J. R., Cunningham, M. D., Vigen, M. P., & Woods, S. O. (2011). Serious assaults on prison staff: A descriptive analysis. *Journal of Criminal Justice, 39*, 143–150.
61. Trulson, C. R., DeLisi, M., & Marquart, J. W. (2011). Institutional misconduct, delinquent background, and rearrest frequency among serious and violent delinquent offenders. *Crime & Delinquency, 57*, 709–731; Trulson, C. R., Haerle, D. R., DeLisi, M., & Marquart, J. W. (2011). Blended sentencing, early release, and recidivism of violent institutionalized delinquents. *The Prison Journal, 91*, 255–278; Trulson, C. R., Marquart, J. W., Mullings, J. L., & Caeti, T. J. (2005). In between adolescence and adulthood: Recidivism outcomes of a cohort of state delinquents. *Youth Violence and Juvenile Justice, 3*, 355–387.
62. Caudill, J. W. (2010). Back on the swagger: Institutional release and recidivism timing among gang affiliates. *Youth Violence and Juvenile Justice, 8*, 58–70.
63. Spergel, I. A. (1986). The violent gang problem in Chicago: A local community approach. *Social Service Review, 60*, 94–131.
64. Kennedy (1997), p. 46, see Note 7.
65. Braga, A. A., & Weisburd, D. L. (2012). The effects of focused deterrence strategies on crime: A systematic review and meta-analysis of the empirical evidence. *Journal of Research in Crime and Delinquency, 49*, 323–358.
66. Braga, & Weisburd (2012), see Note 65.
67. Braga (2012), p. 275, see Note 7.

PART THREE

Crime (Adulthood)

CHAPTER SIXTEEN

Developmental and Life-Course Criminology: Theories and Policy Implications

David P. Farrington

KEY TERMS

Criminal career
Co-offending
Life events
Maturity gap
Offending
Relative stability

INTRODUCTION

Developmental and life-course (DLC) criminology is concerned mainly with three topics: the development of offending and antisocial behavior from the womb to the tomb, the influence of risk and protective factors at different ages, and the effects of life events on the course of development. This chapter reviews the current state of knowledge on these topics, including three major DLC theories by Moffitt, Sampson and Laub, and Thornberry and Krohn and my own theory, the integrated cognitive antisocial potential (ICAP) theory. Finally, major policy implications of DLC research and theories are outlined.

DLC theories aim to explain offending by individuals (as opposed to crime rates of areas, for example). **Offending** refers to the most common crimes of theft, burglary, robbery, violence, vandalism, minor fraud, and drug use, and to behavior that in principle might lead to a conviction in Western industrialized societies such as the United States and the United Kingdom. These theories should explain results on offending obtained from both official records and self-reports. Generally, DLC findings and theories particularly apply to offending by lower-class urban males in Western industrialized societies in the last 80 years or so. To what extent they apply to other types of persons (e.g., middle-class rural females) or offenses (e.g., white collar crimes or sex offenses against children) are important empirical questions I do not attempt to address here.

There are 10 widely accepted conclusions about the development of offending that any DLC theory must be able to explain. First, the prevalence of offending peaks in the late teenage years, between ages 15 and 19. Second, the peak age at onset of offending is between 8 and 14, and the peak age of desistance from offending is between 20 and 29. Third, an early age at onset predicts a relatively long **criminal career** duration and the commission of relatively many offenses. Fourth, there is marked continuity in offending and antisocial behavior from childhood to the teenage years and into adulthood. What this means is that there is **relative stability** of the ordering of people on some measure of antisocial behavior over time and that people who commit relatively many offenses during one age range have a high probability of also committing relatively many offenses during another age range. However, neither of these statements is incompatible with the assertions that the prevalence of offending varies with age or that many antisocial children become conforming adults. *Between-individual* stability in antisocial ordering is perfectly compatible with *within-individual* change in behavior over time. For example, people may graduate from cruelty to animals at age 6 to shoplifting at age 10, burglary at age 15, robbery at age 20, and eventually spouse assault and child abuse later in life. Generally, continuity in offending reflects persistent heterogeneity (the persistence of between-individual differences) more than state dependence (a facilitating effect of earlier offending on later offending), although both processes can occur. There is also continuity in offending from one generation to the next.[1]

Fifth, a small fraction of the population (the "chronic" offenders) commits a large fraction of all crimes. In general, these chronic offenders have an early onset, a high individual offending frequency, and a long criminal career. Sixth, offending is versatile rather than specialized. For example, violent offenders are indistinguishable from frequent offenders in terms of their childhood, adolescent, and adult risk factors. Seventh, the types of acts defined as offenses are elements of a larger syndrome of antisocial behaviors, including heavy drinking, reckless driving, sexual promiscuity, bullying, and truancy. Offenders tend to be versatile not only in committing several types of crimes, but also in committing several types of antisocial behavior.[2]

Eighth, most offenses up to the late teenage years are committed with others, whereas most offenses from age 20 onward are committed alone. This aggregate change is not caused by dropping out processes or by group offenders desisting earlier than lone offenders. Instead, there is change within individuals; people change from group offending to lone offending as they get older. Ninth, the reasons given for offending up to the late teenage years are quite variable, including utilitarian ones (e.g., to obtain material goods or for revenge), for excitement or enjoyment (or to relieve boredom), or because people get angry (in the case of violent crimes). In contrast, from age 20 onward utilitarian motives become increasingly dominant. Tenth, different types of offenses tend to be first committed at distinctly different ages. For example, shoplifting is typically committed before burglary, which in turn is typically committed before robbery. In general, there is increasing diversification of offending up to age 20; as each new type of crime is added, previously committed crimes continue to be committed. Conversely, after age 20 diversification decreases and specialization increases.[3]

The main risk factors for the early onset of offending before age 20 are well known: individual factors (low intelligence, low school achievement, hyperactivity-impulsiveness and risk taking, antisocial child behavior including aggression and bullying), family factors (poor parental supervision, harsh discipline and child physical abuse, inconsistent discipline, a cold parental attitude and child neglect, low involvement of parents with children, parental conflict, broken families, criminal parents, delinquent siblings), socioeconomic factors (low family income, large family size), peer factors (delinquent peers, peer rejection, and low popularity), school factors (a high-delinquency-rate school), and neighborhood factors (a high-crime neighborhood).

The main **life events** that encourage desistance after age 20 are getting married, getting a satisfying job, moving to a better area, and joining the military.[4] The distinction between risk factors and life events is not clear cut, because some life events may be continuing experiences whose duration is important (e.g., a marriage or a job), whereas some risk factors may occur at a particular time (e.g., loss of a parent). Other life events (e.g., converting to religion) may be important but have been studied less. Although the focus in DLC criminology is on the development of offenders, it is important not to lose sight of factors that influence the commission of offenses. It is plausible to assume that offenses arise out of an interaction between the person (with a certain degree of criminal potential) and the environment (including opportunities and victims). Existing evidence suggests that people faced with criminal opportunities take account of the subjectively perceived benefits and costs of offending (compared with other possible activities) in deciding whether to offend. DLC theories should explain the commission of offenses as well as the development of offenders.

TWO IMPORTANT ISSUES

This section explores between-individual differences and within-individual changes as these concepts relate to the development of delinquency over the life course.

Between-Individual Differences Versus Within-Individual Change

Much research in criminology focuses on between-individual differences in biological, individual, family, peer, school, community, and socioeconomic factors influencing offending. Implications are then drawn from this research for the prevention or reduction of offending. For example, it is often found that offenders are more impulsive than nonoffenders, and so it is deduced that cognitive-behavioral skills training programs that reduce impulsivity should lead to a reduction in offending. As another example, it is often found that offenders experienced more inconsistent parental child-rearing methods than nonoffenders, and so it is deduced that parent training techniques that make child rearing more consistent should lead to a reduction in offending.

The problem with these seemingly logical deductions is that they depend on within-individual change: If a person becomes less impulsive over time, that person will as a consequence decrease offending over time. And yet knowledge about risk factors for offending is overwhelmingly based on between-individual research, showing for example that offenders are more impulsive than non-offenders. Can we really draw valid conclusions about within-individual change from research on between-individual differences? This is a key question that needs to be addressed.

More research is needed on the relationship between within-individual changes in risk factors and within-individual changes in offending. For example, in the Cambridge Study in Delinquent Development, which is a prospective longitudinal survey of more than 400 South London males from ages 8 to 48, the relationship between unemployment and crime was investigated by seeing whether each male committed more offenses during his periods of unemployment than during his periods of employment.[5] The results showed that the males did indeed commit more offenses while unemployed than while employed. Furthermore, the difference was only found for offenses involving material gain, such as theft, burglary, robbery, and fraud, which the males committed at a higher rate during periods of unemployment. They did not commit more offenses of other types (violence, vandalism, or drug use) during periods of unemployment than during periods of employment.

These results suggested that the key link in the causal chain between unemployment and crime was a shortage of money: Unemployment caused a shortage of money, which in turn caused offending to get money. It seemed very unlikely that unemployment caused boredom or frustration, which in turn caused offending to reduce boredom or frustration, such as violence, vandalism, or drug use.[6]

This type of within-individual research is more causally compelling than the corresponding between-individual research. For example, in the Cambridge Study we also showed that currently unemployed males committed more offenses than currently employed males; in other words, between-individual differences in unemployment predicted and correlated with between-individual differences in offending. This demonstrates that unemployment is a risk factor for offending but not, of course, that unemployment is a cause of offending. The problem is that unemployed males differ from employed males in many respects, and it is very difficult to disentangle the influence of unemployment from the influence of other explanatory variables.

For example, more antisocial males are more likely to be unsatisfactory employees and more likely to be fired (and hence unemployed) than less antisocial males. Lower-class males by definition have unskilled manual jobs. These jobs tend to be short term, so these males tend to have periods of unemployment between jobs and therefore are more likely to be unemployed than upper- or middle-class males. It is difficult to know whether unemployment causes offending, whether being antisocial causes unemployment, or whether lower-class males are more likely to be unemployed and more likely to offend without there being any causal influence of unemployment on crime.

In within-individual research, long-term persisting individual characteristics such as antisociality or social class are held constant (and hence controlled), because the comparison is between an individual while unemployed and the same individual while employed. By carrying out quasi-experimental analyses addressing threats to internal validity, convincing conclusions about causes can be drawn from such research. It follows that more within-individual research is needed in criminology. This of course requires more prospective longitudinal studies.

Research is also needed that systematically compares results obtained in within-individual analyses and between-individual analyses. In the Pittsburgh Youth Study, which is a prospective longitudinal survey of more than 1500 Pittsburgh boys, Farrington et al. found that poor parental supervision predicted a boy's delinquency both between and within individuals, but peer delinquency predicted a boy's delinquency between individuals but not within individuals.[7] In other words, changes in peer delinquency within individuals (from one assessment to the next) did not predict changes in a boy's delinquency over time. This suggested that peer delinquency might not be a cause of a boy's delinquency but might instead be measuring the same underlying construct (perhaps reflecting co-offending). The message is that risk factors that predict offending between individuals might not predict offending within individuals, so that implications drawn from between-individual comparisons about causes and interventions might not be valid.

Protective and Promotive Factors

Most research seeks to identify risk factors: variables associated with an increased probability of offending. It is also important to identify protective factors that reduce offending, which may have more implications than risk factors for prevention and treatment. However, there are three different ways in which the term *protective factor* has been used.

The first suggests that a protective factor is merely the opposite end of the scale (or the other side of the coin) to a risk factor. For example, if low intelligence is a risk factor, high intelligence may be a protective factor. The value of this statement depends, however, on whether there is a linear relationship between the variable and offending. To the extent that the relationship is linear, little is gained by identifying the protective factor of high intelligence as well as the risk factor of low intelligence. This is essentially creating two variables out of one.

The second definition specifies protective factors that are free standing, with no corresponding, symmetrically opposite risk factor. This especially occurs when variables are nonlinearly related to offending. For example, if high nervousness was associated with a low risk of offending while medium and low nervousness were associated with a fairly constant average risk, nervousness could be a protective factor but not a risk factor (because the probability of offending was not high at low levels of nervousness). In the Pittsburgh Youth Study, Farrington and Loeber discovered a number of variables that were nonlinearly related to delinquency, of which the most important was the age of the mother at her first birth. Boys with teenage mothers had a very high risk of delinquency, but the risk was fairly constant when the mother was older than 20.[8]

The third definition of a protective factor identifies variables that interact with risk factors to minimize or buffer their effects. These protective factors may or may not be associated with offending themselves. To facilitate the exposition here, a risk variable (e.g., family income) can be distinguished from a risk factor (e.g., low family income). Interaction effects can be studied in two ways, by focusing either on the effect of a risk variable in the presence of a protective factor or on the effect of a protective variable in the presence of a risk factor. For example, the effect of family income on offending could be studied in the presence of good parental supervision, or the effect of parental supervision on offending could be studied in the presence of low family income.

Most studies focusing on the interaction of risk and protective factors identify a subsample at risk (with some combination of risk factors) and then search for protective variables that predict successful members of this subsample. In a classic project, Werner and Smith in Hawaii studied children who possessed four or more risk factors for delinquency before 2 years of age but who nevertheless did not develop behavioral difficulties during childhood or adolescence. They found that the major protective factors included being first born, active, and affectionate infants; small family size; and receiving a high amount of attention from caretakers.[9]

To avoid the ambiguity in the meaning of the term *protective factor*, Loeber et al. proposed that factors that predict a low rate of offending should be termed *promotive factors*; the term protective factor should be used only for variables that interact with a risk factor to buffer its effects. In the Pittsburgh Youth Study they found that several variables conventionally viewed as risk factors actually operated mainly as promotive factors. These included low hyperactivity, low parental physical punishment, good parental supervision, and living in a good neighborhood. Promotive effects were most common at younger ages (7–12 years old).[10]

THREE DLC THEORIES

This section explores three of the major theoretical perspectives that conceptualize delinquent development over the life course.

Moffitt

Moffitt proposed two qualitatively different categories of antisocial people (differing in kind rather than in degree), namely *life-course-persistent* (LCP) and *adolescence-limited* (AL) offenders. As indicated by these terms, the LCPs start offending at an early age and persist beyond their 20s, whereas the ALs have a short criminal career largely limited to their teenage years. The LCPs commit a wide range of offenses including violence, whereas the ALs commit predominantly "rebellious" nonviolent offenses such as vandalism. This theory aims to explain findings in the Dunedin longitudinal study.[11]

The main factors that encourage offending by the LCPs are cognitive deficits, undercontrolled temperament, hyperactivity, poor parenting, disrupted families, teenage parents, poverty, and low socioeconomic status. Genetic and biological factors, such as low heart rate, are also important. There is not much discussion of neighborhood factors, but it is proposed that the neuropsychological risk of the LCPs interacts multiplicatively with a disadvantaged environment. The theory does not propose that neuropsychological deficits and a disadvantaged environment influence an underlying construct such as antisocial propensity; rather, it suggests that neuropsychological and environmental factors are the key constructs underlying antisocial behavior. The main factors that encourage offending by the ALs are the **maturity gap** (their inability to achieve adult rewards such as material goods during their teenage years) and peer influence (especially from the LCPs). Consequently, the ALs stop offending when they enter legitimate adult roles and can achieve their desires legally. The ALs can easily stop because they have few neuropsychological deficits.

The theory assumes labeling effects of "snares" such as a criminal record, incarceration, drug or alcohol addiction, and (for females) unwanted pregnancy, especially for the ALs. However, the observed continuity in offending over time is largely driven by the LCPs. The theory focuses mainly on the development of offenders and does not attempt to explain why offenses are committed. However, it suggests that the presence of delinquent peers is an important situational influence on ALs and that LCPs seek out opportunities and victims.

Decision making in criminal opportunities is supposed to be rational for the ALs (who weigh likely costs against likely benefits) but not for the LCPs (who largely follow well learned, "automatic" behavioral repertoires without thinking). However, the LCPs are mainly influenced by utilitarian motives, whereas the ALs are influenced by teenage boredom. Adult life events such as getting a job or getting married are hypothesized to be of little importance, because the LCPs are too committed to an antisocial lifestyle and the ALs desist naturally as they age into adult roles.

The theory suggests that offending peaks in the teenage years because the ALs join in offending during that age range. The ALs tend to commit crimes in groups, whereas the LCPs often offend alone. Moffitt extended her original theory to propose other types of offenders: abstainers and low-level offenders with mental health problems.[12] However, she suggests that there are no true late-onset offenders; she predicts that persons first convicted in the adult years have demonstrated their antisociality earlier in life.

Sampson and Laub

The key construct in Sampson and Laub's theory is age-graded informal social control, which means the strength of bonding to family, peers, schools, and later adult social institutions such as marriages and jobs. Sampson and Laub primarily aim to explain why people do not commit offenses, on the assumption that why people want to offend is unproblematic (presumably caused by hedonistic desires) and that offending is inhibited by the strength of bonding to society. This theory is influenced by their analyses of the Glueck follow-up study of male delinquents and nondelinquents.[13]

The strength of bonding depends on attachments to parents, schools, delinquent friends, and delinquent siblings, and also on parental socialization processes such as discipline and supervision. Structural background variables (e.g., social class, ethnicity, large family size, criminal parents, disrupted families) and individual difference factors (e.g., low intelligence, difficult temperament, early conduct disorder) have indirect effects on offending through their effects on informal social control (attachment and socialization processes).

Sampson and Laub are concerned with the whole life course. They emphasize change over time rather than consistency and the poor ability of early childhood risk factors to predict later life outcomes. They focus on the importance of later life events (adult turning points) such as joining the military, getting a stable job, and getting married in fostering desistance and "knifing off" the past from the present. They also suggest that neighborhood changes can cause changes in offending. Because of their emphasis on change and unpredictability, they deny the importance of types of offenders such as "life-course persisters." They suggest that offending decreases with age for all types of offenders.

Sampson and Laub do not explicitly include immediate situational influences on criminal events in their theory and believe that opportunities are not important because they are ubiquitous. However, they do suggest that having few structured routine activities is conducive to offending. They focus on why people do not offend rather than on why people offend and emphasize the importance of individual free will and purposeful choice in the decision to desist. They do not include strain theory ideas, but they propose that official labeling influences offending through its effects on job instability and unemployment. They argue that early delinquency can cause weak adult social bonds, which in turn fail to inhibit adult offending.

Thornberry and Krohn

The interactional theory of Thornberry and Krohn particularly focuses on factors encouraging antisocial behavior at different ages. The theory is influenced by findings of the Rochester Youth Development Study.[14] They do not propose types of offenders but suggest that the causes of antisocial behavior vary for children who start at different ages. At the earliest ages (birth to 6 years) the three most important factors are neuropsychological deficit and difficult temperament (e.g., impulsiveness, negative emotionality, fearlessness, poor emotion regulation), parenting deficits (e.g., poor monitoring, low affective ties, inconsistent discipline, physical punishment), and structural adversity (e.g., poverty, unemployment, welfare dependency, a disorganized neighborhood). They also suggest that structural adversity might cause poor parenting.

Neuropsychological deficits are less important for children who start antisocial behavior at older ages. At ages 6 to 12, neighborhood and family factors are particularly salient, whereas at ages 12 to 18, school and peer factors dominate. Thornberry and Krohn also suggest that deviant opportunities, gangs, and deviant social networks are important for onset at ages 12 to 18. They propose that late starters (ages 18 to 25) have cognitive deficits such as low IQ and poor school performance but a supportive family and school environment protected them from antisocial behavior at earlier ages. At ages 18 to 25 they find it hard to make a successful transition to adult roles such as employment and marriage.

The most distinctive feature of this interactional theory is its emphasis on reciprocal causation. For example, it is proposed that the child's antisocial behavior elicits coercive responses from parents and rejection by peers, making antisocial behavior more likely in the future. The theory does not postulate a single key construct underlying offending but suggests that children who start early tend to continue because of the persistence of neuropsychological and parenting deficits and structural adversity. Interestingly, Thornberry and Krohn predict that late starters (ages 18–25) show more continuity over time than earlier starters (ages 12–18) because the late starters have more cognitive deficits. In an earlier exposition of the theory, they proposed that desistance was caused by changing social influences (e.g., stronger family bonding), protective factors (e.g., high IQ and school success), and intervention programs. Hence, they do believe that criminal justice processing has an effect on future offending.

ICAP THEORY

This theory was primarily designed to explain offending by lower-class males, and it was influenced by results obtained in the Cambridge Study. The integrated cognitive antisocial potential (ICAP theory)[15] integrates ideas from many other theories, including strain, control, learning, labeling, and rational choice approaches; its key construct is antisocial potential (or, simply AP); and it assumes that the translation from AP to antisocial behavior depends on cognitive (thinking and decision-making) processes that take account of opportunities and victims. **Figure 16-1** is deliberately

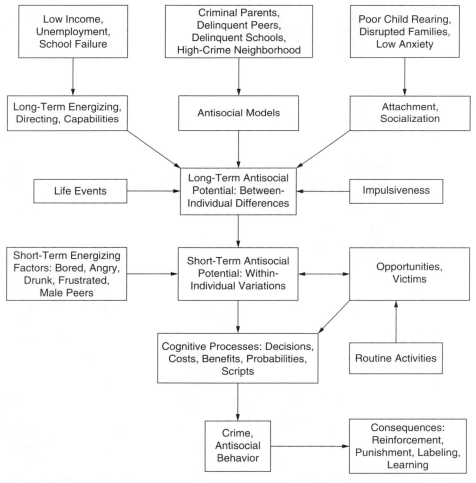

FIGURE 16-1. Integrated Cognitive Antisocial Potential (ICAP) Theory

simplified to show the key elements of the ICAP theory on one page; for example, it does not show how the processes operate differently for onset compared with desistance or at different ages.

The key construct underlying offending is AP, which refers to the potential to commit antisocial acts. Long-term persisting, between-individual differences in AP are distinguished from short-term, within-individual variations in AP. Long-term AP depends on impulsiveness; on strain, modeling, and socialization processes; and on life events, whereas short-term variations in AP depend on motivational and situational factors.

Regarding long-term AP, people can be ordered on a continuum from low to high. The distribution of AP in the population at any age is highly skewed; relatively few people have relatively high levels of AP. People with high AP are more likely to commit many different types of antisocial acts, including different types of offenses. Therefore, offending and antisocial behavior are versatile, not specialized. The relative ordering of people on AP (long-term, between-individual variation) tends to be consistent over time, but absolute levels of AP vary with age, peaking in the teenage years, because of changes within individuals in the factors that influence long-term AP (e.g., from childhood to adolescence, the increasing importance of peers and decreasing importance of parents).

A key issue is whether the model should be the same for all types of crimes or whether different models are needed for different types of crimes. Because of their focus on the development of offenders, DLC researchers concluded that because offenders are versatile rather than specialized, it is not necessary to have different models for different types of crimes. For example, it is believed that the risk factors for violence are essentially the same as for property crime or substance abuse. However, researchers who have focused on situational influences have argued that different models are needed for different types of crimes. It is suggested that situational influences on burglary may be very different from situational influences on violence.

One possible way to resolve these differing viewpoints is to assume that long-term potential is very general (e.g., a long-term potential for antisocial behavior), whereas short-term potential is more specific (e.g., a short-term potential for violence). The top half of the model in Figure 16-1 could be the same for all types of crimes, whereas the bottom half could be different (with different situational influences) for different types of crimes.

In the interests of simplification, Figure 16-1 makes the DLC theory appear static rather than dynamic. For example, it does not explain changes in offending at different ages. Because it might be expected that different factors would be important at different ages or life stages, it seems likely that different models would be needed at different ages. Perhaps parents are more important in influencing children, peers are more important in influencing adolescents, and spouses and partners are more important in influencing adults.

Long-Term Risk Factors

A great deal is known about risk factors that predict long-term persisting, between-individual differences in AP. For example, in the Cambridge Study the most important childhood risk factors for later offending were hyperactivity-impulsivity-attention deficit, low intelligence or low school attainment, family criminality, family poverty, large family size, poor childrearing, and disrupted families.[16] More efforts should be made in the ICAP theory to specify promotive and protective factors as well as risk factors.

Measures of antisocial behavior (e.g., aggressiveness or dishonesty) as risk factors are not included because of concern with explanation, prevention, and treatment. These measures do not cause offending; they predict offending because of the underlying continuity over time in AP.

Measures of antisocial behavior are useful in identifying risk groups but less useful in identifying causal factors to be targeted by interventions. Similarly, not included are variables that cannot be changed, such as gender or ethnicity.

In the risk factor prevention paradigm, the basic idea is very simple: Identify the key risk factors for offending and implement prevention methods designed to counteract them.[17] Although the emphasis is on prevention, knowledge about risk factors can also be used to guide interventions after offending. Of course, a key question is whether the risk factors for offending (or for the onset of offending) are the same as or different from the risk factors for reoffending (or for the persistence of offending).

A major problem is to decide which risk factors are causes and which are merely markers or correlated with causes.[18] Ideally, interventions should be targeted at risk factors that are causes. Interventions targeted at risk factors that are merely markers will not necessarily lead to any decrease in offending. Unfortunately, when risk factors are highly correlated (as is usual), it is very difficult to establish which are causes in between-individual research. For example, the particular factors that appear to be independently important as predictors in any analysis may be greatly affected by measurement error and by essentially random variations between samples.

It is also important to establish how risk factors or causes have sequential or interactive effects on offending. Figure 16-1 shows how risk factors are hypothesized to influence long-term AP. Following strain theory, the main energizing factors that potentially lead to high long-term AP are desires for material goods, status among intimates, excitement, and sexual satisfaction. However, these motivations only lead to high AP if antisocial methods of satisfying them are habitually chosen. Antisocial methods tend to be chosen by people who find it difficult to satisfy their needs legitimately, such as people with low income, unemployed people, and those who fail at school. However, the methods chosen also depend on physical capabilities and behavioral skills; for example, a 5 year old would have difficulty in stealing a car. For simplicity, energizing and directing processes and capabilities are shown in one box in Figure 16-1.

Long-term AP also depends on attachment and socialization processes. AP is low if parents consistently and contingently reward good behavior and punish bad behavior. (Withdrawal of love may be a more effective method of socialization than hitting children.) Children with low anxiety are less well socialized, because they care less about parental punishment. AP is high if children are not attached to (prosocial) parents—for example, if parents are cold and rejecting. Disrupted families (broken homes) may impair both attachment and socialization processes.

Long-term AP is also high if people are exposed to and influenced by antisocial models, such as criminal parents, delinquent siblings, and delinquent peers, for example, in high-crime schools and neighborhoods. Long-term AP is also high for impulsive people, because they tend to act without thinking about the consequences. Also, life events affect AP; it decreases (at least for males) after people get married or move out of high-crime areas, and it increases after separation from a partner. There may also be interaction effects between the influences on long-term AP. For example, people who experience strain or poor socialization may be disproportionally antisocial if they are also exposed to antisocial models. In the interests of simplicity, Figure 16-1 does not attempt to show such interactions.

Many researchers have measured only one risk factor (e.g., impulsivity) and have shown that it predicts or correlates with offending after controlling for a few other "confounding factors," often including social class. The message is: Don't forget the big picture. The particular causal linkages shown may not be correct, but it is important to measure and analyze all important risk (and promotive and protective) factors in trying to draw conclusions about the causes of offending or the development of offenders.

Explaining the Commission of Crimes

According to the ICAP theory, the commission of offenses and other types of antisocial acts depends on the interaction between the individual (with his or her immediate level of AP) and the social environment (especially criminal opportunities and victims). Short-term AP varies within individuals according to short-term energizing factors such as being bored, angry, drunk, or frustrated, or being encouraged by male peers. Criminal opportunities and the availability of victims depend on routine activities. Encountering a tempting opportunity or victim may cause a short-term increase in AP, just as a short-term increase in AP may motivate a person to seek out criminal opportunities and victims.

Whether a person with a certain level of AP commits a crime in a given situation depends on cognitive processes, including considering the subjective benefits, costs, and probabilities of the different outcomes and stored behavioral repertoires or scripts. The subjective benefits and costs include immediate situational factors such as the material goods that can be stolen and the likelihood and consequences of being caught by the police. They also include social factors such as likely disapproval by parents or female partners and encouragement or reinforcement from peers. In general, people tend to make decisions that seem rational to them, but those with low levels of AP will not commit offenses even when (on the basis of subjective expected utilities) it appears rational to do so. Equally, high short-term levels of AP (e.g., caused by anger or drunkenness) may induce people to commit offenses when it is not rational for them to do so.

The consequences of offending may, as a result of a learning process, lead to changes in long-term AP and in future cognitive decision-making processes. This is especially likely if the consequences are reinforcing (e.g., gaining material goods or peer approval) or punishing (e.g., receiving legal sanctions or parental disapproval). Also, if the consequences involve labeling or stigmatizing the offender, this may make it more difficult for that person to achieve his or her aims legally and hence may lead to an increase in AP.

A further issue that needs to be addressed is to what extent types of offenders might be distinguished. Perhaps some people commit crimes primarily because of their high long-term AP and others primarily because of situational influences and high short-term AP. Perhaps some people commit offenses primarily because of situational influences (e.g., getting drunk frequently), whereas others offend primarily because of the way they think and make decisions when faced with criminal opportunities. From the viewpoint of both explanation and prevention, research is needed to classify types of people according to their most influential risk factors and most important reasons for committing crimes.

It is desirable to link up studies of the development of offenders with studies of situational influences on offending by asking about influences on criminal events in prospective longitudinal studies. In the Cambridge Study, investigators asked about why people committed crimes. The most common reasons given for property offenses were utilitarian, rational, or economic ones: Offenses were committed for material gain.[19] The next most common reasons might be termed hedonistic: Offenses were committed for excitement, enjoyment, or to relieve boredom. In general, utilitarian motives predominated for most types of property offenses such as burglary and theft, except that vandalism and motor vehicle theft were committed predominantly for hedonistic reasons, and shoplifting was partly utilitarian and partly hedonistic. Offenses at younger ages (under 17 years old) were relatively more likely to be committed for hedonistic reasons, whereas offenses at older ages (17 or older) were relatively more likely to be committed for utilitarian reasons.

Reasons for aggressive acts (physical fights) were also investigated.[20] The key dimension was whether males fought alone or in groups. In individual fights the male was usually provoked, became angry, and hit to hurt his opponent and to discharge his own internal feelings of tension. In group fights the male often said that he became involved to help a friend or because he was attacked, and rarely said that he was angry. The group fights were more serious, occurring in bars or streets, and they were more likely to involve weapons, produce injuries, and lead to police intervention. Fights often occurred when minor incidents escalated, because both sides wanted to demonstrate their toughness and masculinity and were unwilling to react in a conciliatory way.

POLICY IMPLICATIONS

DLC findings and theories have many policy implications for the reduction of crime. First, it is clear that children at risk can be identified with reasonable accuracy at an early age. The worst offenders tend to start early and have long criminal careers. Often, offending is preceded by earlier types of antisocial behavior in a developmental sequence, including cruelty to animals, bullying, truancy, and disruptive school behavior. It is desirable to intervene early to reduce the later escalation into chronic or LCP offending. For example, programs to prevent bullying in schools are effective.[21] It is desirable to develop risk–needs assessment devices to identify children at risk of becoming chronic offenders—usually children with specific needs. These devices could be developed soon after school entry, at 6 to 8 years of age.

Many effective programs have been developed to tackle important childhood risk factors. For example, home visiting programs and parent training programs are effective in tackling family risk factors such as poor parental supervision and erratic parental discipline. Preschool intellectual enrichment programs are effective in tackling low school attainment. Child skills training is effective in tackling individual risk factors such as high impulsiveness and low empathy.[22] In general, children with more risk factors tend to have worse outcomes. Therefore, it is desirable to implement multicomponent programs that tackle several risk factors at once. For example, Hawkins and colleagues implemented a program combining parent training, teacher training, and child skills training. Generally, multiple-component interventions are more effective than single-component ones.[23]

Moffitt's theory suggests that different programs are needed for AL and LCP offenders. For AL offenders it is especially important to limit contact with delinquent peers. This is one of the main aims of Treatment Foster Care programs, although they are targeted at more serious juvenile offenders.[24] Research on **co-offending** suggests that it is important to identify and target "recruiters," offenders who repeatedly commit crimes with younger, less experienced offenders, and who seem to be dragging more and more young people into the net of offending. Programs that put antisocial peers together may have harmful effects.[25] Moffitt also suggests that to target the maturity gap of AL offenders, it is important to provide opportunities for them to achieve status and material goods by legitimate means.

The main implication of Sampson and Laub's theory is that bonding to the family, the school, and the community should be increased, for example, by providing job training and structured routine activities in adulthood. They also suggest that desistance can be encouraged by fostering bonding to adult institutions such as employment and marriage. Another suggestion is that informal social control in communities could be improved by increasing community cohesiveness or "collective efficacy." They also argue that it is important to minimize labeling or stigmatization of offenders by minimizing the use of incarceration.

The most distinctive feature of Thornberry and Krohn's theory is the suggestion that adult-onset offenders are similar to early-onset offenders in having neuropsychological deficits (e.g., low intelligence), but that the adult offenders were previously insulated by their family and school environment. In the adult years they offend because they have difficulties in making a successful transition to adult roles such as employment and marriage. The clear implication is that these vulnerable young males should be identified and given special help to make the transition from living with their parents to living independently.

The ICAP theory suggests that situational factors have important influences on offending. It follows that attempts should be made to tackle short-term energizing factors such as boredom, drunkenness, and anger. Also, efforts should be made to increase offenders' subjective probability of being caught, to make it harder for them to commit offenses, and generally to reduce the supply of criminal opportunities.

Because of their emphasis on development through life, DLC theories suggest that it is "never too early, never too late" to intervene successfully to reduce offending.[26] In other words, it is highly desirable to focus not only on early intervention to prevent the adolescent onset of offending, but also on later programs to prevent adult onset and to prevent continuation and encourage early desistance.

CONCLUSION

More efforts should be made to compare and contrast the different DLC theories in regard to their predictions and their agreement with empirical results. For example, Farrington et al. studied the development of AL, late-onset, and persistent offenders from 8 to 48 years of age in the Cambridge Study. They found that, contrary to Moffitt's theory, AL offenders had several of the same risk factors as persistent offenders. Contrary to Sampson and Laub's theory, early risk factors were important in predicting which offenders would persist or desist after age 20. Many other results were relevant to DLC theories, but there is not space here to review the empirical adequacy of these theories in detail.

To advance knowledge about DLC theories and test them, new prospective longitudinal studies are needed with repeated self-report and official record measures of offending. Future longitudinal studies should follow people up to later ages and focus on desistance processes. Past studies have generally focused on onset and on ages up to 30. Future studies should compare risk factors for early onset, continuation after onset (compared with early desistance), frequency, seriousness, later onset, and later persistence versus desistance. DLC theories should make explicit predictions about all of these topics. Also, future studies should make more effort to investigate promotive and protective factors and biological, peer, school, and neighborhood risk factors, because most is known about individual and family factors. And future research should compare development, risk factors, and life events for males versus females and for different ethnic and racial groups in different countries.

Because most previous analyses of risk factors for offending involve between-individual comparisons, more within-individual analyses of offending are needed. These should investigate to what extent within-individual changes in risk and promotive factors are followed by within-individual changes in offending and other life outcomes. These analyses should provide compelling evidence about causal mechanisms. More information is needed about developmental sequences and about the predictability of future criminal careers to know when and how it is best to intervene.

It is desirable to derive implications for intervention from DLC theories and to test these in randomized experiments. In principle, conclusions about causes can be drawn more convincingly in experimental research than in nonexperimental longitudinal studies. The results summarized here have clear implications for intervention. The main idea of risk-focused prevention is to identify key risk factors for antisocial behavior and implement prevention methods designed to counteract them. In addition, attempts should be made to enhance key promotive and protective factors.

The fact that offenders tend to be antisocial in many aspects of their lives means that any measure that succeeds in reducing offending will probably have wide-ranging benefits in reducing, for example, accommodation problems, relationship problems, employment problems, alcohol and drug problems, and aggressive behavior. Consequently, it is very likely that the financial benefits of successful programs will greatly outweigh their financial costs. The time is ripe to mount a new program of research to compare, contrast, and test predictions from different DLC theories in the interests of developing more accurate theories and more effective policies.

GLOSSARY

Criminal career—longitudinal pattern of offending

Co-offending—commission of criminal acts with at least one other person

Life events—significant changes in local life circumstances, such as being married

Maturity gap—disjuncture between biological maturity and social maturity

Offending—most common crimes of theft, burglary, robbery, violence, vandalism, minor fraud, and drug use, and behavior that in principle might lead to a conviction

Relative stability—relative rank ordering of people on some particular measure

NOTES

1. Farrington, D. P. (1986). Age and crime. In M. Tonry & N. Morris (Eds.), *Crime and justice* (Vol. 7, pp. 189–250). Chicago: University of Chicago Press; Farrington, D. P. (1989). Self-reported and official offending from adolescence to adulthood. In M. W. Klein (Ed.), *Cross-national research in self-reported crime and delinquency* (pp. 399–423). Dordrecht, Netherlands: Kluwer; Farrington, D. P. (1990). Age, period, cohort, and offending. In D. M. Gottfredson & R. V. Clarke (Eds.), *Policy and theory in criminal justice: Contributions in honour of Leslie T. Wilkins* (pp. 51–75). Aldershot, England: Avebury; Farrington, D. P. (1992). Criminal career research in the United Kingdom. *British Journal of Criminology, 32*, 521–536; Farrington, D. P., Lambert, S., & West, D. J. (1998). Criminal careers of two generations of family members in the Cambridge Study in Delinquent Development. *Studies on Crime and Crime Prevention, 7*, 85–106; Nagin, D. S., & Farrington, D. P. (1992). The stability of criminal potential from childhood to adulthood. *Criminology, 30*, 235–260.
2. Farrington, D. P. (1991a). Antisocial personality from childhood to adulthood. *The Psychologist, 4*, 389–394; Farrington, D. P. (1991b). Childhood aggression and adult violence: Early precursors and later life outcomes. In D. J. Pepler & K. H. Rubin (Eds.), *The development and treatment of childhood aggression* (pp. 5–29). Hillsdale,

NJ: Erlbaum; Farrington, D. P. & West, D. J. (1993). Criminal, penal, and life histories of chronic offenders: Risk and protective factors and early identification. *Criminal Behaviour and Mental Health, 3*, 492–523.

3. Farrington, D. P. (1993). Motivations for conduct disorder and delinquency. *Development and Psychopathology, 5*, 225–241; LeBlanc, M., & Frechette, M. (1989). *Male criminal activity from childhood through youth.* New York: Springer-Verlag; Piquero, A., Paternoster, R., Mazerolle, P., Brame, R., & Dean, C. W. (1999). Onset age and offense specialization. *Journal of Research in Crime and Delinquency, 36*, 275–299; Reiss, A. J., & Farrington, D. P. (1991). Advancing knowledge about co-offending: Results from a prospective longitudinal survey of London males. *Journal of Criminal Law and Criminology, 82*, 360–395.

4. Farrington, D. P., & West, D. J. (1995). Effects of marriage, separation and children on offending by adult males. In J. Hagan (Ed.), *Current perspectives on aging and the life cycle. Vol. 4: Delinquency and disrepute in the life course* (pp. 249–281). Greenwich, CT: JAI Press; Horney, J., Osgood, D. W., & Marshall, I. H. (1995). Criminal careers in the short-term: Intra-individual variability in crime and its relation to local life circumstances. *American Sociological Review, 60*, 655–673; Laub, J. H., & Sampson, R. J. (2001). Understanding desistance from crime. In M. Tonry (Ed.), *Crime and justice* (Vol. 28, pp. 1–69). Chicago: University of Chicago Press.

5. Farrington, D. P., Coid, J. W., Harnett, L., Jolliffe, D., Soteriou, N., Turner, R., & West, D. J. (2006). *Criminal careers up to age 50 and life success up to age 48: New findings from the Cambridge Study in Delinquent Development.* London: Home Office (Research Study No. 299).

6. Farrington, D. P., Gallagher, B., Morley, L., St. Ledger, R. J., & West, D. J. (1986). Unemployment, school leaving and crime. *British Journal of Criminology, 26*, 335–356.

7. Farrington, D. P., Loeber, R., Yin, Y., & Anderson, S. J. (2002). Are within-individual causes of delinquency the same as between-individual causes? *Criminal Behaviour and Mental Health, 12*, 53–68.

8. Farrington, D. P., & Loeber, R. (2000). Some benefits of dichotomization in psychiatric and criminological research. *Criminal Behaviour and Mental Health, 10*, 100–122.

9. Werner, E. E., & Smith, R. S. (1982). *Vulnerable but invincible: A longitudinal study of resilient children and youth.* New York: McGraw-Hill.

10. Loeber, R., Farrington, D. P., Stouthamer-Loeber, M., & White, H. R. (2008). *Violence and serious theft: Development and prediction from childhood to adulthood.* New York: Routledge.

11. Moffitt, T. E. (1993). Adolescence-limited and life-course-persistent antisocial behavior: A developmental taxonomy. *Psychological Review, 100*, 674–701.

12. Piquero, A. R., & Moffitt, T. E. (2005). Explaining the facts of crime: How the developmental taxonomy replies to Farrington's invitation. In D. P. Farrington (Ed.), *Integrated developmental and life-course theories of offending* (pp. 51–72). New Brunswick, NJ: Transaction.

13. Laub, J. H., & Sampson, R. J. (2003). *Shared beginnings, divergent lives: Delinquent boys to age 70.* Cambridge, MA: Harvard University Press; Sampson, R. J., & Laub, J. H. (1993). *Crime in the making: Pathways and turning points through life.* Cambridge, MA: Harvard University Press; Sampson, R. J., & Laub, J. H. (1995). Understanding variability in lives through time: Contributions of life-course criminology. *Studies on Crime and Crime Prevention, 4*, 143–158; Sampson, R. J., & Laub, J. H. (2005). A general age-graded theory of crime: Lessons learned and the future of life-course criminology. In D. P. Farrington (Ed.), *Integrated developmental and life-course theories of offending* (pp. 165–181). New Brunswick, NJ: Transaction.

14. Thornberry, T. P., & Krohn, M. D. (2001). The development of delinquency: An interactional perspective. In S. O. White (Ed.), *Handbook of youth and justice* (pp. 289–305). New York: Plenum; Thornberry, T. P. & Krohn, M. D. (2005). Applying interactional theory to the explanation of continuity and change in antisocial behavior. In D. P. Farrington (Ed.), *Integrated developmental and life-course theories of offending* (pp. 183–209). New Brunswick, NJ: Transaction; Thornberry, T. P., Lizotte, A. J., Krohn, M. D., Smith, C. A., & Porter, P. K. (2003). Causes and consequences of delinquency: Findings from the Rochester Youth Development Study. In T. P. Thornberry & M. D. Krohn (Eds.), *Taking stock of delinquency: An overview of findings from contemporary longitudinal studies* (pp. 11–46). New York: Kluwer/Plenum.

15. Farrington, D. P. (2005). The integrated cognitive antisocial potential (ICAP) theory. In D. P. Farrington (Ed.), *Integrated developmental and life-course theories of offending* (pp. 73–92). New Brunswick, NJ: Transaction.

16. Farrington, D. P. (2003). Key results from the first 40 years of the Cambridge Study in Delinquent Development. In T. P. Thornberry & M. D. Krohn (Eds.), *Taking stock of delinquency: An overview of findings from contemporary longitudinal studies* (pp. 137–183). New York: Kluwer/Plenum.

17. Farrington, D. P. (2000). Explaining and preventing crime: The globalization of knowledge. The American Society of Criminology 1999 Presidential Address. *Criminology, 38*, 1–24.

18. Murray, J., Farrington, D. P., & Eisner, M. P. (2009). Drawing conclusions about causes from systematic reviews of risk factors: The Cambridge Quality Checklists. *Journal of Experimental Criminology, 5*, 1–23.

19. West, D. J., & Farrington, D. P. (1977). *The delinquent way of life*. London: Heinemann.

20. Farrington, D. P., Berkowitz, L., & West, D. J. (1982). Differences between individual and group fights. *British Journal of Social Psychology, 21*, 323–333.

21. Ttofi, M. M., Farrington, D. P., & Baldry, A. C. (2008). *Effectiveness of programmes to reduce school bullying*. Stockholm, Sweden: National Council for Crime Prevention.

22. Olds, D. L., Henderson, C. R., Cole, R., Eckenrode, J., Kitzman, H., Luckey, D., et al. (1998). Long-term effects of nurse home visitation on children's criminal and antisocial behavior: 15-year follow-up of a randomized controlled trial. *Journal of the American Medical Association, 280*, 1238–1244; Sanders, M. R., Markie-Dadds, C., Tully, L. A., & Bor, W. (2000). The Triple-P Positive Parenting Program: A comparison of enhanced, standard and self-directed behavioral family intervention for parents of children with early onset conduct problems. *Journal of Consulting and Clinical Psychology, 68*, 624–640; Schweinhart, L. J., Montie, J., Zongping, X., Barnett, W. S., Belfield, C. R., & Nores, M. (2005). *Lifetime effects: The High/Scope Perry Preschool Study through age 40*. Ypsilanti, MI: High/Scope Press; Tremblay, R. E., Pagani-Kurtz, L., Vitaro, F., Masse, L. C., & Pihl, R. D. (1995). A bimodal preventive intervention for disruptive kindergarten boys: Its impact through mid-adolescence. *Journey of Consulting and Clinical Psychology, 63*, 560–568.

23. Hawkins, J. D., Catalano, R. F., Kosterman, R., Abbott, R., & Hill, K. G. (1999). Preventing adolescent health risk behaviors by strengthening protection during childhood. Archives of Pediatrics and Adolescent Medicine, 153, 226–234; Wasserman, G. A., & Miller, L. S. (1998). The prevention of serious and violent juvenile offending. In R. Loeber & D. P. Farrington (Eds.), *Serious and violent juvenile offenders: Risk factors and successful interventions* (pp. 197–247). Thousand Oaks, CA: Sage.

24. Chamberlain, P., & Reid, J. B. (1998). Comparison of two community alternatives to incarceration for chronic juvenile offenders. *Journal of Consulting and Clinical Psychology, 66*, 624–633.

25. Dishion, T. J., McCord, J., & Poulin, F. (1999). When interventions harm: Peer groups and problem behavior. *American Psychologist, 54*, 755–764.

26. Loeber, R., & Farrington, D. P. (1998). Never too early, never too late: Risk factors and successful interventions for serious and violent juvenile offenders. *Studies on Crime and Crime Prevention, 7*, 7–30.

CHAPTER SEVENTEEN

Self-Control Theory and Antisocial Behavior

George E. Higgins and Margaret Mahoney

KEY TERMS

Empirical literature
Gender differences
Inhibition
Parental management
Self-control

INTRODUCTION

Determining the causes of antisocial behavior has been the chief task of criminologists for several decades. Criminologists have consistently worked to understand why individuals perform antisocial behaviors. This has led to a substantial amount of theorizing and empirical testing. Through the 1980s, Gottfredson and Hirschi worked to develop a singular cause of antisocial behavior that would not drain the scarce resources required for longitudinal studies. That is, they worked to develop a single measure that could be used to understand antisocial behavior that allowed criminologists to use cross-sectional data rather than longitudinal data. Gottfredson and Hirschi's theory, now known as self-control, was published in *A General Theory of Crime*, and was the beginning of the self-control tradition in criminology. Since the presentation of the theory, it has received a substantial amount of research attention, including empirical attention and support, and has led some to consider it as one of the leading crime theories.[1]

This chapter reviews the research on self-control theory in the context of antisocial behavior. The chapter begins with a presentation of self-control theory, reviews the studies that examined this theory, outlines the recent theoretical developments of self-control theory, and concludes with a summary of the key findings and identifies future research directions. The studies reviewed are organized around the specific hypothesis that they are testing.

SELF-CONTROL THEORY

In its simplest form, Gottfredson and Hirschi's self-control theory assumes that individuals are rational decision makers. That is, individuals make decisions using their present knowledge about a behavior. Thus, the individual's present knowledge about the antisocial behavior is likely to be limited. Therefore, when weighing the potential pleasure of the act against the potential pain of the antisocial act, the individual is likely to be using only limited information. In Gottfredson and Hirschi's view, individuals choose to perform antisocial behaviors when they perceive that the pleasure of the act outweighs the perceived pain of the act.

Criminal behavior is one form of antisocial behavior that the individual may be drawn to because it may provide more pleasure than pain. Gottfredson and Hirschi detached themselves from the traditional definition that crime is an act against the legal code. They do so because they believe this definition of a crime does not afford social scientists with an opportunity to learn as much about criminal behavior. Gottfredson and Hirschi go as far as to say that the definition is good for the state's business but does little to help social scientists. Thus, they redefine crime as a potentially pleasurable behavior that is an act of force or fraud that the individual may pursue to satisfy his or her own interests. Crimes have specific forms of characteristics. That is, crimes are generally short-lived, immediately gratifying, simple, easy, and exciting behaviors. These characteristics are typically attractive to individuals with deficits or low self-control.

Individuals with low **self-control** typically have the same characteristics as those who commit crime: risk taking, impulsiveness, lack of empathy, preference for simple and easy tasks, and preference for physical tasks. For individuals with low self-control to commit a crime, they do not need to have specific forms of motivation, just an evaluation that the criminal behavior will provide more pleasure than pain. However, an individual's level of self-control may influence the perception of the pleasure and the pain of a criminal behavior. According to the theory:

> The dimensions [characteristics] of self-control are, in our view, factors affecting calculation of the consequences of one's acts. The impulsive or shortsighted person fails to consider the negative or painful consequences of his acts, the insensitive person has fewer negative consequences to consider; the less intelligent person also has fewer negative consequences to consider (has less to lose).[2]

From this view it is apparent that individuals with low self-control are likely to overestimate the pleasure of a criminal act because it provides an immediate benefit, without necessarily seeing or considering the long-term pain. Thus, self-control is thought to be the singular propensity for criminal behavior. Importantly however, individuals with low self-control are usually the products of poor or ineffective **parental management**.

Based substantially on Patterson's view of parental management, Gottfredson and Hirschi argued that individuals with low self-control are the product of poor socialization. The chief source of socialization comes from parents and how well they perform the tasks of parenting. To be clear, Gottfredson and Hirschi posited that if parents were to instill high levels of self-control, they had to consistently perform four activities with their child on a consistent basis. First, parents needed to have an emotional bond or attachment to their child. This allows parents to invest the time and energy to perform the additional tasks. Second, parents should vigilantly monitor their child's behavior. This allows parents to gather behavioral information about their child. Third, the parent must evaluate the behavioral information to determine if the behavior is criminal or deviant. The evaluation of the deviant behavior leads to punishment of the behavior. Fourth, parents who deem their child's behavior as criminal or deviant are to apply some form of noncorporal punishment. Parents must perform these acts before the age of 8 years; otherwise, parents who do a poor job or inconsistently apply these parenting tasks are likely to instill lower levels of self-control. This increases the likelihood that their child will seize an opportunity for crime or deviant behavior according to Gottfredson and Hirschi, because they see low self-control as a relatively stable propensity for criminal behavior.

Although this is a very simple presentation of Gottfredson and Hirschi's theory, the theorists go on to provide additional hypotheses that attempt to take into account different variations on criminal offending. Keeping to this general theory, they attempt to take into account biological variations in criminal offending, the role of school and peers, differences in male and female offending, and differences between races and ethnicities in offending.

Gender Differences in Crime and Self-Control Theory

Gottfredson and Hirschi's theory has implications for criminological debate on the utility of general theories to explain **gender differences** in criminal behavior. The debate centers on the view that specific criminological theories are necessary to take into account the needs of the different genders. Some have argued that general theories such as self-control theory do not take this view into account.[3] Other criminologists disagree, arguing that general crime theories do address the gap in offending between the genders.[4] Gottfredson and Hirschi contended that the same variables that are related to differences in criminality among boys are the same as those for girls.

Gottfredson and Hirschi's position is that females are less likely to commit crime because they have higher levels of self-control. The distribution of self-control differs for girls from boys because parents tend to be more vigilant in monitoring their female children than their male children. Parents are likely to be more consistent in the application of their parental management tasks for females than they are for males. Gottfredson and Hirschi claim that parents are more likely to closely watch the behaviors of female children because they are fearful of what misbehavior may produce for their female children than for their male children. To clarify, by watching their female child more closely, they are more likely to detect misbehavior. Recognizing the misbehavior in female children provides parents with more opportunities to impose noncorporal discipline to correct it. Thus, more attentive parents are able to recognize and discipline their female children's misbehavior more effectively and consistently. To that end, Gottfredson and Hirschi argued for a distribution difference in the application of the parental management tasks and not a difference in the causal model. This means that for males and females, poor or inconsistent parenting leads to low self-control, and low self-control is likely to lead to criminal behavior.

Racial and Ethnic Differences in Crime and Self-Control Theory

Gottfredson and Hirschi provide a similar view when addressing differences in criminal activity among different races and ethnicities. Similar to their arguments about gender differences in criminal behavior, Gottfredson and Hirschi claimed that racial and ethnic differences of criminal behavior may be traced to differences in parental management practices. Without implicating any individual racial or ethnic group, they argued that minority group members are less likely to monitor their child's behavior, to recognize their child's criminal or deviant behavior, and to noncorporally punish the child's criminal or deviant behavior. This results in differences in self-control levels across the different races. To be clear, Gottfredson and Hirschi's position is as follows:

> *There are differences among racial and ethnic groups . . . in levels of direct supervision by family, and thus there is a "delinquency" component to the racial differences in delinquency rates, but as with gender, differences in self-control probably outweigh differences in supervision accounting for racial or ethnic variations.*[5]

Role of Schools in Self-Control Theory

Thus far the presentation of Gottfredson and Hirschi's theory has focused on the role of the family in the development of self-control. They also maintained that other entities and institutions have an opportunity to develop self-control. Gottfredson and Hirschi contended:

> *We do not restrict the meaning of "familial institution" to the traditional family unit composed of a natural father and mother. The socialization function does not, in our view, require such an institution. It does, however, require responsible adults committed to the training and welfare of the child.*[6]

This view recognizes that other institutions outside the family unit may also have important socializing input in a child's level of self-control. Gottfredson and Hirschi asserted that the school is a socializing institution that imposes restraints that do not allow for the uncontrollable pursuit of self-interests and requires accomplishments. This allows teachers to be the "responsible adults" that may provide socialization for the proper levels of self-control.

Schools are advantageous for the development of self-control for several reasons. First, teachers are in a position to monitor a child's behavior. Second, teachers are able to recognize antisocial behavior (i.e., criminal or deviant behavior). Third, in many schools teachers are given the authority to impose discipline to maintain order in the educational environment. Thus, teachers are able to carry on the parental management tasks in the school environment that parents cannot perform when their child is away at school. To that end, they stated:

> *Most people are sufficiently socialized by familial institutions to avoid involvement in criminal acts. Those not socialized sufficiently by the family may eventually learn self-control through the operation of other sanctioning systems or institutions. The institution given principal responsibility for this task in modern society is the school.*[7]

Peer-Group Participation

The discussion thus far has focused on the socialization parts of self-control theory—parental management and schools—but now let us shift to the consequences of low self-control. One consequence of self-control is peer-group participation. Gottfredson and Hirschi understand that peers are a substantial correlate to criminal and deviant activity; however, they have some important views when it comes to group participation and crime participation in a group that are related to self-control.

For Gottfredson and Hirschi, group participation is determined by self-control. Individuals with low self-control are not usually good friends. These individuals are typically unreliable, untrustworthy, selfish, and thoughtless. However, these individuals may be "fun" to be with because they are risk takers, adventuresome, and reckless; individuals with low self-control may be the "life of the party." Thus, while these individuals may participate in peer groups, their participation is not because they are good friends; rather, their participation is due to their reckless nature, putting them in the company of other reckless individuals. This creates groups that have low levels of self-control. From this reasoning, Gottfredson and Hirschi purport that individuals in these groups are likely to be delinquent, and so is the group.

Crime participation in a group has implications for those with low self-control. Peers in this context are seen as facilitators of crime. Gottfredson and Hirschi theorized that the group serves as a tool to help perform criminal behavior. This is not meant to be understood that the group allows for the learning of self-control, rather that participation in such a group is an indication of low self-control and a lack of concern for long-range consequences.

REVIEW OF SELF-CONTROL THEORY LITERATURE

Most studies of self-control theory have shown empirical support for Gottfredson and Hirschi's views. Pratt and Cullen's meta-analysis of more than 20 studies showed that self-control had a moderate link with crime and deviance, supporting the major hypothesis from Gottfredson and Hirschi's theory. Since the publication of this meta-analysis, researchers have shown that low self-control is linked with several different types of behaviors. One point of contention in this literature has been the measurement of self-control. Hirschi and Gottfredson argued that self-control should be measured using a behavioral set of items independent of criminal activity, because it emphasizes that self-control is an individual propensity rather than a personality trait. However, others have argued that self-control should be measured using a personality instrument following the development of the characteristics of low self-control. Regardless of the type of measurement of self-control, low levels have been shown to be linked with a propensity for crime and deviance.[8]

Researchers have tested Gottfredson and Hirschi's arguments that an individual's level of self-control depends on the consistency of parents' application of the parental management tasks, which would lead to higher levels of self-control and thus lower levels of crime. Gibbs, Giever, and Martin's study used a sample of college students to examine this hypothesis. In their study, they used retrospective measures of parenting practices that asked the college students to remember how their parents treated them in the ninth grade. This was used to correlate with self-control and shown to be associated with delinquency. Other researchers have used a similar format for testing the model and found support for this premise. Some researchers showed that parenting practices have a link with self-control by using current measures of parental management and self-control.[9] Although these researchers showed that self-control and parenting are connected, Wright and Beaver challenged this assumption, arguing that biology matters more than parenting practices in the development of self-control. This view is discussed later in this chapter in the context of biological studies that examine Gottfredson and Hirschi's assumptions.[10]

Although the presentation of self-control theory thus far has focused on the connection between self-control and crime and the role of parenting, the focus of this review now shifts to a smaller hypothesis that is often unclear in the **empirical literature**. Researchers scrutinized Gottfredson and Hirschi's hypothesis that self-control remained relatively stable after age 8. Turner and Piquero used a national probability sample that contained behavioral and personality (i.e., attitudinal) measures of self-control to test this hypothesis. They showed mixed results for the hypothesis that were contingent on the type of measure used. To be clear, behavioral measures were more stable than personality measures of self-control. From these data, Turner and Piquero argued that self-control remained relatively stable. However, other researchers continued to examine this proposition using different measures and have not found support for Gottfredson and Hirschi's claims. For instance, Mitchell and MacKenzie used self-reports to a personality measure of self-control from participants in boot camps to examine this hypothesis. They were not able to show stability in self-control. Other researchers using personality measures were also not able to show that self-control remained stable.[11]

Hay and Forrest used a behavioral measure of self-control from a national probability sample to examine the stability of self-control hypothesis. In contrast to previous studies, they examined the stability of self-control using a class analysis. They found that self-control remained relatively stable for a large portion of the sample (more than 80 percent), but stability did not hold for a smaller portion of the sample (16 percent). Overall, the results on stability indicate that self-control is stable across the life course when using behavioral rather than personality measures.[12] The stability of self-control using behavioral measures and the instability of self-control using personality measures is in accord with the views of Hirschi and Gottfredson.

Gottfredson and Hirschi were not only bold in their claims about stability of self-control, but they were equally as bold in their claims about their theory being able to account for gender differences in offending. They argued that the difference in criminal behavior among males and females was due to differences in levels of self-control that were the proximal result of differences in parenting practices for the genders. Researchers examined these issues in several ways. First, researchers only considered gender differences in self-control as they related to crime.[13] Second, researchers considered the causal modeling differences that take place in self-control theory. For instance, Higgins used responses regarding parental management, self-control, and deviance from a nonrandom sample of college students to show that all these measures have different distributions for males and females. To that end, females scored higher for parental management and self-control but lower for deviance than males. Higgins also showed that the model of "parental management is associated with self-control which in turn is associated with deviance" held for both males and females.[14] This indicates that Gottfredson and Hirschi are correct that gender differences do exist in the context of parenting practices that have important implications for self-control.

Researchers also tested other hypotheses concerning self-control, crime, and gender. For instance, Higgins and Tewksbury used a cross-sectional sample of juveniles to show opportunity, self-control, and delinquency differences for males and females. Further, they showed that opportunity mediated the link between self-control and delinquency for each gender. This indicates that the link between self-control and delinquency may be more complex for males and females.[15] Researchers also examined the hypotheses, as outlined by Gottfredson and Hirschi, that there are racial and ethnic differences in parental management, self-control, and delinquency. For instance, Higgins and Ricketts used a sample of juveniles to show that low self-control had a link for Blacks and Whites with delinquency. However, they showed that the links were not statistically different from each other. This indicates that racial difference in the self-control link with delinquency is not supported, but they used a rather unconventional measure of self-control (i.e., impulsivity and risk taking only). Other researchers using more conventional measures (i.e., personality measures) have shown that low self-control has different links for Blacks and Whites. Vazsonyi and colleagues used data from juveniles from several different countries (Hungry, Japan, the Netherlands, Switzerland, and the United States) to show support for the structure of self-control theory: parental management is associated with self-control which in turn is associated with delinquency. Further, researchers also support self-control theory using international samples that can account for different ethnicities.[16] Although an early test suggests no difference in the racial and ethnic effect of self-control on crime, later studies are able to substantiate Gottfredson and Hirschi's hypotheses.

Thus far, the empirical literature is rather complete with a number of studies that indicate support for the view that parental management has import for self-control development and subsequent criminal activity. However, Gottfredson and Hirschi are clear that schools also have a socializing role in the development of self-control. Turner, Piquero, and Pratt used a national probability sample to examine the influence of the school context in the development of self-control. They showed that school effects on self-control varied based on the parental and neighborhood context. The school only had an effect in neighborhoods that had few problems. However, Turner et al. showed that the school effect on developing self-control remained significant even when parental management was included in their statistical models. This means that the development of self-control is not solely based on parenting practices.[17] This reinforces Gottfredson and Hirschi's view that schools may be able to support parents in developing a child's level of self-control.

Although the literature shows that parental management and school context have implications for self-control and later delinquency, Gottfredson and Hirschi have two positions on peer influence in their theory. First, those with low self-control are not likely to make good friends but are still in groups of peers because they would be among others who have low self-control within the group. Second, those with low self-control may commit crime in a group because others with low self-control may make crime easy to perform. A number of researchers attempted to address these issues. Chapple used national-level probability longitudinal data to show that individuals with low self-control are likely to have delinquent associations. This supports Gottfredson and Hirschi's view. However, Chapple's most notable finding is that those with low self-control experience peer rejection. This supports Gottfredson and Hirschi's view that those with low self-control are likely to make poor friends. Higgins and Makin used a sample of college students to show that peer association conditioned the link between low self-control and pirating (i.e., illegally downloading) software. They interpreted their findings that pirating software is generally a behavior that occurs individually, but belonging to a peer group that pirates software makes the behavior easier to perform.[18] Thus, in the empirical literature, researchers are able to find support for Gottfredson and Hirschi's contentions about the connection between self-control and peer associations.

At this point, the review of the empirical literature has focused on the social factors that either develop self-control in an individual (i.e., parents and schools) or the social factors that are a result of low levels of self-control (i.e., peer associations). Gottfredson and Hirschi were rather forceful about their stance on the role of biology and criminal activity in their theory. For example, they argued that genetics had a near zero effect on crime and that the biological interpretations of crime were missing a key component—a measure of criminality (i.e., self-control). Wright and Beaver used a national data set to determine the role of biology in self-control development beyond that of parents. Their findings revealed that biology meant more in the development of an individuals' self-control level, calling into question the utility of parents in the same process. Beaver, Wright, and DeLisi used a sample of 3000 children to show that neuropsychological deficits are connected to a child's level of self-control. Other empirical studies revealed similar results to those of Beaver and his colleagues:[19] These results directly question the role of biology that Gottfredson and Hirschi purported. It should be noted that the results concerning the biology of self-control may be contributed to the measurement of self-control. These studies used a personality measure of self-control, and personality traits have strong biological tendencies and are heritable. Although casting doubt on the veracity of Gottfredson and Hirschi's claims, these results are exciting and should be considered with other measurements of self-control.

The use of the personality measures flummoxed Hirschi in the development of self-control theory. This has led Hirschi to provide an alternative perspective of self-control. Alternative conceptualizations and measurements of self-control are important to the literature as well. One alternative conceptualization takes the focus away from individual's characteristics and moves researchers away from viewing self-control as a personality trait or a predisposition for crime. In Hirschi's new view, the linkage of personality and self-control is incorrect because it: (1) searches for motives of crime and delinquency that are counter to their original theory, (2) shows little value in the explanation of crime, (3) does not provide an explanation of how self-control operates but intimates that individuals become criminal because they are who they are, and (4) produces a measure that does not view more as better than less. Thus, Hirschi sees self-control not as a personality trait or predisposition for crime, but as the tendency to consider the full range of potential costs (i.e., inhibitions) of a particular act. Under this view, self-control is an **inhibition** that individuals carry with them wherever they go. This removes the focus from long-term costs, and it allows any set of costs to be inhibitors while placing an emphasis on the contemporaneous nature of the inhibitions. Individuals are consistently considering the inhibitions for a behavior in any given situation. Thus, crime and delinquent acts are self-perpetuating, but they are possible due to the absence of an enduring tendency to avoid them (i.e., the inability to see the full range of the inhibitions). Typical inhibitions that an individual considers are consonant with the bonds from social control theory (i.e., commitment, involvement, belief, and attachment) that provide a target for dishonor if a transgression is perpetrated. Because individuals become criminal or delinquent when they feel relatively free from intimate attachments, aspirations, and moral beliefs, noncriminal or nondeviant individuals are exercising self-control by recognizing and adhering to inhibitions so as not to dishonor those they admire. Therefore, self-control is akin to a self-imposed physical restraint on behavior.[20]

To test this view, Hirschi examined data from the Richmond Youth Survey and used nine items to capture a variety of social bonds (i.e., attachment, commitment, and belief). He showed that his conceptualization of self-control has a negative link with delinquency. This is supportive of the reconceptualization of self-control that individuals add up the negative costs of an act and behave in accord. The important issue with this study was Hirschi's new measures. His use of nine items that reflect social bonds is consistent with his view that self-control and social control are one and the same. More recently, Piquero and Bouffard used data from college students to examine the reconceptualization of self-control. They interpreted Hirschi's recent work to be more from the rational choice tradition than from the social bonding tradition. Their approach to operationalizing self-control was to ask students to provide a list of seven "bad things" and the percentage of the likelihood of these bad things occurring. The researchers added together the products of each student's responses, and a higher score on a measure indicated more inhibitions. In comparison, the bad things measure of self-control has a stronger link with drunk driving and sexual aggression than the original scale created by Grasmick et al. Higgins, Wolfe, and Marcum also used data from college students to examine the alternative view of self-control, but in the context of digital piracy (i.e., the illegal downloading of digital media, including music, movies, and software). The results from their study indicated that the alternative conceptualization of self-control does have merit and needs to be tested in a broader perspective.[21]

CONCLUSION

In its most basic form, self-control theory can be thought of as poor or ineffective parental management that leads to lower levels of self-control, which have an influence on the inclination toward criminal activity. Those with deficits in self-control are more likely to be attracted to crime than those with higher levels of self-control. The review of the extant literature has revealed some key findings. Self-control, whether measured using behavioral or personality measures, has a link with criminal behavior. Those who have experienced poor or inconsistent parenting are likely to have lower levels of self-control, but the school context can also assist in the development of self-control for an individual. Researchers have shown that the structure of self-control theory holds across genders, but females have higher levels of parental management, self-control, and lower levels of crime than males. The structure of the theory also holds for different races and ethnicities, with some minorities having lower levels of parental management, lower levels of self-control, and higher levels of criminal behavior. Researchers have shown that Gottfredson and Hirschi are correct in their views that those with low self-control are likely to make poor friends and to be rejected by peers. However, the available literature indicates that the biological assumptions that Gottfredson and Hirschi make may be incorrect.

The review of the extant literature suggests that researchers should continue to develop self-control theory. One avenue is that researchers need to continue to develop self-control theory in light of Hirschi's more recent alternative conceptualization. Given that self-control theory is a control theory, and control theories address the question, "Why don't people commit crime?" the measure of self-control should reflect this conceptualization. It should also be noted that the reconceptualization of self-control theory places an emphasis on self-control being an individual propensity for criminal behavior rather than a personality trait that happens to be correlated with criminal behavior. This suggests that additional work in the areas outlined above should be ripe for development and reassessment. For instance, does the new version of self-control remain stable over time? How different is self-control for males and females? To that end, how different is self-control across races and ethnicities? What are the implications of the new version of self-control for peer association? What are the biological implications for self-control using the alternative measure of self-control? To what behaviors does this new version of self-control relate? These are a few questions that are relevant given the recent developments within self-control theory. The replication of present findings is important as well. This speaks to the fundamental debates that Gottfredson and Hirschi are attempting to address.

GLOSSARY

Empirical literature—studies conducted by using objective standards of science and that often require the use of quantitative analyses

Gender differences—disparities between males and females on some particular measure

Inhibition—central to Hirschi's new definition of self-control that pertains to the ability to consider the full range of consequences associated with a behavior

Parental management—believed to be the main contributing factor to variation in levels of self-control

Self-control—measured through individuals' penchant for risk taking, their impulsiveness, their ability to be empathetic, their preference for simple and easy tasks, and their preference for physical tasks

NOTES

1. Gottfredson, M. R., & Hirschi, T. (1990). *A general theory of crime*. Stanford, CA: Stanford University Press; Pratt, T. C., & Cullen, F. (2000). The empirical status of Gottfredson and Hirschi's general theory of crime: A meta-analysis. *Criminology, 38*, 931–964.

2. Gottfredson, & Hirschi (1990), p. 95, see Note 1; Patterson, G. R. (1986). Performance models for antisocial boys. *American Psychologist, 41*, 432–444.

3. Berger, R. (1989). Female delinquency in the emancipation era: A review of literature. *Sex Roles, 21*, 375–399; Campbell, A. (1990). On the visibility of the female delinquent peer group. *Women & Criminal Justice, 2*, 41–62; Chesney-Lind, M. (1997). *The female offender: Girls, women and crime*. Thousand Oaks, CA: Sage; Daly, K., & Chesney-Lind, M. (1988). Feminism and criminology. *Justice Quarterly, 5*, 497–538; Naffine, N. (1997). *Feminism and criminology*. Philadelphia, PA: Temple University Press.

4. Akers, R. L. (1985). *Deviant behavior: A social learning approach* (3rd ed.). Belmont, CA: Wadsworth; Akers, R. L. (1998). *Social learning and social structure: A General theory of crime and deviance*. Boston: Northeastern University Press; Cernkovich, S., & Giordano, P. (1979). Delinquency, opportunity, and gender. *Journal of Criminal Law and Criminology, 70*, 145–151; Hirschi, T. (1969). *Causes of delinquency*. Berkeley, CA: University of California Press; Merton, R. (1957). *On theoretical sociology*. New York: Free Press; Smith, D., & Paternoster, R. (1987). The gender gap in theories of deviance: Issues and evidence. *Criminology, 24*, 140–172.

5. Gottfredson, & Hirschi (1990), p. 153, see Note 1.

6. Gottfredson, & Hirschi (1990), p. 273, see Note 1.

7. Gottfredson, & Hirschi (1990), p. 105, see Note 1.

8. Hirschi, T., & Gottfredson, M. R. (Eds.). (1994). *The generality of deviance*. New Brunswick, NJ: Transaction; Grasmick, H. G., Tittle, C. R., Bursik, R. J., & Arneklev, B. J. (1993). Testing the core empirical implications of Gottfredson and Hirschi's general theory of crime. *Journal of Research on Crime and Delinquency, 30*, 5–29; Tittle, C., Ward, D. A., & Grasmick, H. (2003). Self-control and crime/deviance: Cognitive vs. behavioral measures. *Journal of Quantitative Criminology, 19*, 333–365.

9. Gibbs, J. J., Giever, D. M., & Higgins, G. E. (2003). A test of Gottfredson and Hirschi's general theory using structural equation modeling. *Criminal Justice and Behavior, 30*, 441–458; Gibbs, J. J., Giever, D. M., & Martin, J. (1998). Parental management and self-control: An empirical test of Gottfredson and Hirschi's general theory. *Journal of Research in Crime and Delinquency, 35*, 42–72; Hay, C. (2001). Parenting, self-control, and delinquency: A test of self-control theory. *Criminology, 39*, 707–736; Higgins, G. E. (2004). Gender and self-control theory: Are there differences in the measures and the theory's causal model? *Criminal Justice Studies, 17*, 33–55; Hope, T., Grasmick, H. G., & Pointon, L. J. (2003). The family in Gottfredson and Hirschi's general theory of crime: Structure, parenting, and self-control. *Sociological Focus, 36*, 291–311; Lynskey, D. P., Winfree, L. T., Esbensen, F. A., & Clason, D. L. (2000). Linking gender, minority group status, and family matters to self-control theory: A multivariate analysis of key self-control concepts in a youth gang context. *Juvenile and Family Court Journal, 51*(3), 1–19; Pratt, T., Turner, M. G., & Piquero, A. (2004). Parental socialization and community context: A longitudinal analysis of the structural sources of low self-control. *Journal of Research in Crime and Delinquency, 41*, 219–243; Unnever, J. D., Cullen, F. T., & Pratt, T. C. (2003). Parental management, ADHD, and delinquent involvement: Reassessing Gottfredson and Hirschi's general theory. *Justice Quarterly, 20*, 471–501.

10. Wright, J. P., & Beaver, K. M. (205). Do parents matter in creating self-control in their children? A genetically informed test of Gottfredson and Hirschi's theory of low self-control. *Criminology, 43*, 1169–1202.

11. Arneklev, B., Cochran, J. K., & Gainey, R. (1998). Testing Gottfredson and Hirschi's "low self-control" stability hypothesis: An exploratory study. *American Journal of Criminal Justice, 23,* 107–127; Beaver, K. M., & Wright, J. P. (2007). The stability of low self-control from kindergarten through first grade. *Journal of Crime and Justice, 30,* 63–86; Mitchell, O., & MacKenzie, D. L. (2006). The stability and resiliency of self-control in a sample of incarcerated offenders. *Crime & Delinquency, 52,* 432–449; Turner, M. G., & Piquero, A. R. (2002). The stability of self-control. *Journal of Criminal Justice, 30,* 457–477; Winfree, L. T., He, N., & Esbensen, F. A. (2006). Self-control and variability over time: Multivariate results using 5-year, multisite panel of youths. *Crime & Delinquency, 52,* 253–286.

12. Hay, C., & Forrest, W. (2006). The development of self-control examining self-control theory's stability thesis. *Criminology, 44,* 739–774.

13. Burton, V. S., Cullen, F. T., Evans, T. D., Alarid, L. F., & Dunaway, R. G. (1998). Gender, self-control, and crime. *Journal of Research in Crime and Delinquency, 35,* 123–147; LaGrange, T. C., & Silverman, R. A. (1999). Low self-control and opportunity: Testing the general theory of crime as an explanation for gender differences in delinquency. *Criminology, 37,* 41–72.

14. Higgins, G. E. (2006). Gender differences in software piracy: The mediating roles of self-control theory and social learning theory. *Journal of Economic Crime Management, 4,* 1–22.

15. Higgins, G. E., & Tewksbury, R. (2006). Sex and self-control theory: The measures and causal model may be different. *Youth & Society, 37,* 479–503.

16. Tittle, C. R., & Botchkovar, E. V. (2005a). The generality and hegemony of self-control theory: A comparison of Russian and U.S. adults. *Social Science Research, 34,* 703–731; Tittle, C. R., & Botchkovar, E. V. (2005b). Self-control, criminal motivation, and deterrence: An investigation using Russian respondents. *Criminology, 43,* 307–353; Vazsonyi, A. T., & Belliston, L. M. (2007). The family → low self-control → deviance: A cross-cultural and cross-national test of self-control theory. *Criminal Justice and Behavior, 343,* 505–530; Vazsonyi, A. T., & Crosswhite, J. M. (2004). A test of Gottfredson and Hirschi's general theory of crime in African-American adolescents. *Journal of Research in Crime and Delinquency, 41,* 407–432; Vazsonyi, A. T., Pickering, L. E., Junger, M., & Hessing, D. (2001). An empirical test of general theory of crime: A four nation comparative study of self-control and the prediction of deviance. *Journal of Research in Crime and Delinquency, 38,* 91–131; Vazsonyi, A. T., Wittekind, J. E., Belliston, L. M., & von Loh, T. D. (2004). Extending the general theory of crime to "the East": Low self-control in Japanese adolescents. *Journal of Quantitative Criminology, 20,* 189–201.

17. Turner, M. G., Piquero, A. R., & Pratt, T. C. (2005). The school context as a source of self-control. *Journal of Criminal Justice, 33,* 327–339.

18. Baron, S. W. (2003). Self-control, social consequences and criminal behavior: Street youth and the general theory of crime. *Journal of Research in Crime and Delinquency, 40,* 403–425; Chapple, C. L. (2005). Self-control, peer relations, and delinquency. *Justice Quarterly, 22,* 89–106; Evans, D. T., Cullen, F. T., Burton, V. S., Dunaway, R. G., & Benson, M. L. (1997). The social consequences of self-control: Testing the general theory of crime. *Criminology, 35,* 475–501; Higgins, G. E., & Makin, D. A. (2004). Does social learning theory condition the effects of low self-control on college students' software piracy? *Journal of Economic Crime Management, 2,* 1–22; Wright, B. R. E., Caspi, A., Moffitt, T. E., & Paternoster, R. (2004). Does the perceived risk of punishment deter criminally prone individuals? Rational choice, self-control, and crime. *Journal of Research in Crime and Delinquency, 41,* 180–213.

19. Beaver, K. M., Wright, J. P., & DeLisi, M. (2007). Self-control as an executive function: Reformulating Gottfredson and Hirschi's parental socialization thesis. *Criminal Justice and Behavior, 34,* 1345–1361; Vaughn, M. G., DeLisi, M., Beaver, K. M., Wright, J. P., & Howard, M. O. (2007). Toward a psychopathology of self-control theory: The importance of narcissistic traits. *Behavioral Sciences and the Law, 25,* 803–821.

20. Hirschi, T. (2004). Self-control and crime. In R. F. Baumeister & K. D. Vohs (Eds.), *Handbook of self-regulation: Research, theory, and applications.* New York: Guilford Press.

21. Higgins, G. E., Wolfe, S. E., & Marcum, C. D. (2008). Digital piracy: An examination of multiple conceptualizations and operationalizations of self-control. *Deviant Behavior, 29,* 440–460; Piquero, A. R., & Bouffard, J. (2007). Something old, something new: A preliminary investigation of Hirschi's redefined self-control. *Justice Quarterly, 24,* 1–27.

CHAPTER EIGHTEEN

Serial Crime: Psychology of Behavioral Consistency and Applications to Linking

C. Gabrielle Salfati

KEY TERMS

Linking
Offender profiling
Signature
Situational factors
Traits
Victim

INTRODUCTION

When looking at the pathway of an offender's criminal career, investigators can do so from a number of different angles and for different purposes. One may look at how a young adolescent becomes a persistent delinquent, and how he consequently develops into the mature and experienced offender. One may also look at offenders' criminal careers, and see whether they are generalists or specialists. Throughout, the key underlying psychological issue is whether the offender remains the same or changes over his or her life course or criminal career. And the goal is to attempt to identify how this change or consistency exhibits itself so that it may be measured, studied, and understood.

In criminal investigations a particular interest is to identify whether a set of crimes is committed by different people or the same person and ultimately how the behavioral pattern exhibited at these crimes may tell something about who the offender is. Consistency can be divided into two different but related fields of study. The first focuses on the consistency across an individual's series of crimes. This first aspect is normally referred to as **linking**. The second focus is the consistency between what an offender does at the crime scene and what an offender does in other aspects of his or her life, including criminal (i.e., previous criminal activities and how this relates to the current crime) and noncriminal (i.e., actions in his or her personal life) aspects. This is normally referred to as **offender profiling**.[1] Both aspects have at their core the understanding of behavioral consistency. This chapter specifically focuses on the issue of linking and the key issues that should be addressed when attempting to theoretically link a series of crimes.

BEHAVIORAL CONSISTENCY AND INDIVIDUAL DIFFERENTIATION _____

To test whether an offender is consistent across his or her offending career (i.e., linking different crime scenes to one another), it is important to look at whether offenders always engage in the same behaviors across their series. But before testing whether an offender is consistent, one needs to decide how to measure this consistency. More specifically, one must define whether it is necessary to look at the individual behavior (e.g., "binding" or "gagging") or whether to look at the psychological type of behavior for a theme (e.g., "controlling").[2] Of equal importance is also to establish not only whether the individual offender remains consistent, but whether he or she can be distinguished in his or her pattern from another individual's series (individual differentiation) and ultimately identifying how many different types of series there are (typologies).

An example that illustrates the issue of consistency with regard to individual differentiation is the example of the electrician and the plumber. The plumber has a toolbox made up of tools specifically made for plumbing. However, the plumber will not know which exact tool is needed until he or she gets to the job. Once at the house where the job needs to be done, the plumber will use the most appropriate tool to get the job done. In terms of criminal behavior, if one uses the example of rape, the investigator can then understand how an offender might use the physical behavior of gagging if the victim is screaming and binding if the victim is trying to escape. Both behaviors illustrate the offender trying to control the victim so he can complete his "job" of raping the victim. In the same way one may compare the plumber to the electrician. The electrician also has his or her own toolbox, but this toolbox is filled with a different set of tools from the plumber, each of which has its own purpose but all relating to electrical work rather than plumbing. Put another way, the offender possesses a variety of offending behaviors with the same function but which are used in a different context. Relying on psychological themes allows for establishing context–victim connections that are typically ignored in offender-focused approaches when analyzing behavior. In this way an understanding of the interpersonal interaction between the offender and **victim** is more reliable and allows for the measurement of this interaction while keeping the focus on the offender. In addition, investigators can identify different offender characteristics (plumber versus electrician) depending on what actions (tools) they use. This results in the creation of typologies of criminal behavior.

An example that pertains specifically to linking is to imagine a situation in which a number of rapes have been committed and have been linked because in each case the offender used the single identifying behavior of gagging his victim. Similarly, another series has been identified by the offender binding the victim. By using a single behavior like this, two series have been identified. However, by focusing on the individual behavior, investigators are excluded from linking these two series to each other and so to the same offender. If instead one looks at the underlying psychological meaning of these behaviors, one can identify a similarity based on the offender's need to control the victim and the situation. By moving from the individual behavior to the thematic, one can thus expand the unit of analysis from the descriptive to the psychological and thereby increase the chances of linking. This specifically becomes important when the individual level of analysis is affected by the situational factors that introduce an element of inconsistency that affects the resulting behavior (e.g., the offender may bind the victim in situations where she tries to run away, yet gag her in situations where she screams). In both situations, although the individual behaviors are different, they both illustrate the offender attempting to control the situation. The importance of taking this situational aspect of the event into account has been discussed extensively in the literature; however, it has not been discussed in relation to linking crimes per se.

The empirically based research on linking as it pertains to criminal investigations is still in its first stages. However, the general literature on the concepts that underpin linking has been ongoing for decades, particularly in the fields of personality and social psychology. This research has been especially important in the quest to establish whether an individual's behavior is based on personality traits or on the situation the person finds oneself in, or, more recently, the dynamics of the interaction between the two.[3]

CONTEXTUAL AND SITUATIONAL INFLUENCES ON BEHAVIOR

When analyzing a situation, it is important to determine whether the actions that occurred were due to only the personality or characteristics of the offender (**traits**) or were equally influenced by the physical situation or even the interaction with the victim during the crime (**situational factors**). This is part of the age-old debate in psychology, and even philosophy, known as the nature–nurture debate. Are people born the way they are (biology, genetics), or are they made into who they are through experiences? The debate is a lengthy and complex one, but must be discussed briefly to understand the offender's actions at a crime and how this may influence our ability to identify a consistent pattern, or series, over time as belonging to one offender.

Social learning theory states that an individual's experiences are determined by his or her social environment. Indeed, its main advocate, B. F. Skinner, contended that behaviors and attitudes are acquired through learning from the world around us. Hans Toch suggested that violent behaviors are rooted in well learned, systematic strategies of violence that have proven to be effective in dealing with interpersonal conflict. Toch postulated that the life histories of violent persons reveal surprising consistency in their approaches to interpersonal relationships. In all likelihood such individuals learned in childhood that violence is an effective way of dealing with conflict; they see violence as a means to obtain rewards and avoid costs.[4]

Huesmann and Eron state that the manner in which an offender demonstrates aggression as well as the intensity of that aggression can be attributed to social learning. They hypothesize that social behavior is largely controlled by responses learned early in life. Learned responses for social behavior in general, and for aggressive behavior in particular, are largely controlled by cognitive scripts. Cognitive scripts are stored in a person's memory and used as guides for behavior and social problem solving, suggesting how one should respond to events and what the likely outcomes of these responses are. They further suggest that these are formed during childhood and persist throughout life and that regular mental retrieval of these violent cognitive scripts results in the accumulation of numerous aggressive scripts stored in memory, making them readily available for use as particular situations arise. Additionally, continuously drawing on these scripts makes them easier to retrieve when faced with a problematic situation. Studies in psychology have also shown that when people are under high states of emotional arousal, as in the case of aggression, they are less able to form clear thinking patterns where they logically evaluate their options and outcomes. Instead, their capacity for calm and logical thought is diminished, and they therefore more automatically resort to strongly established habits such as cognitive scripts to guide their behavior.[5]

Not only does aggression as a characteristic way of solving problems emerge early in life, it is also likely that each individual develops a particular type of aggressiveness that is fairly consistent across social situations and throughout the life course. In a landmark study spanning 22 years, Huesmann, Eron, Lefkowitz, and Walder collected data on the behavior of more than 600 subjects, their parents, and their children. They found that those who were the more aggressive 8 year olds at the beginning of the study were also the more aggressive 30 year olds at the end of the study. They further found that early aggressiveness was predictive of later serious antisocial behavior, including criminality, spouse abuse, self-reported physical aggression, and traffic violations.[6]

The importance of this work is clear: If one can understand the link between an offender's early life experiences and actions during the commission of his or her crime, one can better determine the offender's psychological characteristics and behavioral patterns in general. Studies specifically centered on offender profiling, notably in the field of investigative psychology, have tried to empirically link offenders' criminal actions at the crime scene with other aspects of their lives, including the details of their home life, social patterns, lifestyle, how they relate to other people, and any other activities that might aid police investigating a crime, especially the persistence, severity, range, and specialization of previous crimes committed by suspects. In this way a connection can then be established between crime scene behaviors and the general behaviors in the offender's life—a link that can aid police in narrowing suspects or identifying geographical areas for their investigation and, ultimately, help link crimes committed by the same offender at different times during his or her criminal career.

Interestingly, Furr and Funder found that behaviors that are more automatic and impulsive are generally more consistent than behaviors that are more controlled and cognitively mediated. Studies such as these lay the foundation for work that is important in understanding what needs to be the focus of studies on behavioral and situational consistency, such as work pertaining to linking serial crimes.[7] Funder and Colvin suggested that some behaviors are more consistent than others and that behaviors that occur in response to specific stimuli are more stimulus specific, whereas behaviors that do not require specific, eliciting stimuli are more likely to occur across a broad range of situations and may reflect more stable properties, such as personality/disposition (i.e., they are more intrinsic to the offender).[8]

Behaviors manifest differently at different ages due to the range of behaviors available to the person at that time. For instance, a child might exhibit a behavior such as truancy, being disruptive in class, or being disrespectful to the teacher. An older person may engage in behaviors such as repeatedly losing jobs and getting into frequent fights with bosses, shop owners, and possibly even the police. If one focuses consistency analysis on the actual behavior, then the adolescent would look different from the adult. However, if one looks at the psychological meaning of the behaviors and the context in which they are being elicited, a pattern of someone who has problems with people in authority can be seen. By understanding how behaviors are exhibited in relation to the age of the person as well as in what context, investigators can link that person's behavior and establish consistencies at different times in his or her life. In fact, what may make a situation similar may have more to do with internal processes, such as the psychological meaning of the situation to the individual, in which case the understanding of what constitutes an actual physical context needs to be reevaluated.[9]

In moving from person-based theories to context-based theories, some of the contextual or situational influences that are important to study include the physical location of the crime. Interpreting a behavior needs to be done within its context to understand whether it occurred because it is part of an offender's repertoire that relates to his or her internal needs or whether it is something the offender engaged in due to external influences arising from the situation. Consider again the example discussed earlier of a rapist using binding and gagging: If the offender engaged in this action in a public place where the victim screaming may bring help, the action is externally caused; that is, the offender is reacting to the situation in order to control it, to engage in what is internally important to the offender, notably the rape itself. However, if the offender committed the rape inside or somewhere very remote, where there are no witnesses, and he still gags the victim, this may signify a more internal reason, related to the offender's own agenda, such as personally being distracted by the victim's screaming or even finding the action of gagging the victim an excitement in itself and part of his rape agenda.

But time itself may cause a change in the offender's agenda, which could be mistaken for inconsistency. The offender may experiment as he or she goes through the crime, which may be a consistent pattern in some types of offenders: They may develop and learn from previous crimes, or they may mature, and may develop and refine their agenda over time. Indeed, the issue of time as it relates to the development of the criminal, and where he or she is in life, may also have an effect on consistency. It is suggested in the general literature that younger offenders are less specialized than older and more experienced offenders and may be more opportunistic in the way they commit crimes. This may also translate to implicating behavioral consistency as it pertains to linking. Personality may be essential to establishing and explaining consistency patterns; because behavior will likely be less consistent during periods of personality development, it will have an effect on linking crimes committed by juveniles.[10]

Overall, it is important to establish what behaviors are stable and at what stages of personality development and from there establish what behaviors are the most important ones to evaluate for consistency and ultimately use for linking a series of crimes. Researchers in the study of serial murder and investigative psychology stress that identifying and then correctly labeling and measuring specific behaviors are key in collecting the most accurate and useful information for testing and establishing behavioral patterns.[11]

SIGNATURES VERSUS PSYCHOLOGICAL THEMES

Behavioral scientists are just beginning to study linking. Researchers and practitioners who have written about linking have not explained in detail how they link crimes; thus, the validity of their statements is an open question. Indeed, much of the early literature specifically on linking as it pertains to criminal investigations is not supported by empirical studies but instead is a collection of theoretical and practice-led studies that propose linking patterns.[12] Although this wealth of experience provides a great basis for discussing behavior, a detailed examination and definition of linking and behavioral consistency needs to be developed and empirically tested. More recent research focuses on methodological issues relating to testing and establishing patterns of behavioral consistency. Some of these do not focus on behavioral consistency specifically, but instead evaluate the methodological and statistical processes of linking.[13]

Signatures

Douglas and Munn define a **signature** as the "calling card" left by the offender that is unique to the offender's pattern of behaviors throughout his or her offenses. For example, this signature may be related to the ritualistic acts committed by an offender, perhaps to fulfill a specific fantasy. They explain that the signature aspect of the violent offense is a constant behavior performed by an offender during his or her offenses but that it may evolve and become more developed throughout the series of offenses.[14] The idea behind using individual behaviors is that an investigator should be able to look at a crime—in this scenario, a homicide—and identify individual crime scene behaviors that are unique to a particular offender. From here, if a second homicide occurs, the investigator can look to see if these same individual behaviors exist in the new homicide, and, if so, infer that there is a strong possibility that the two homicides in question are linked. Unfortunately, if researchers do not explain in detail how they arrived at their conclusions, the validity of their statements is left untested, which highlights why much of the literature on signatures has not been supported by empirical studies. Instead, it is largely a collection of theoretical and practice-led studies that proposes these patterns.

Salfati and Bateman provided the first empirical study in serial homicide that tested the concept of signatures by looking at how it can be used to link cases of serial homicide across time. They looked at 90 serial homicide series with 5 offenses in each and identified 6 behavioral categories from the literature on signatures and serial homicide, including body disposal behaviors, forensic awareness behaviors, mutilation behaviors, weapons used, theft behaviors, and sexual behaviors. Each of these categories in turn contained diverse numbers of objectively defined and physically observable behaviors from the crime scene adhering to the description of the category. In determining consistency, they used the most stringent criteria they could, whereby an offender was only deemed consistent if he or she used each individual behavior in four of five crime scenes.

The results show inherent difficulties in using individual behaviors to identify consistency patterns across a series and consequently to link cases. The results from the study further indicated that serial homicide offenders do not seem to show consistency in their use of the same individual crime scene behaviors throughout their series of homicides, and this consistency does not increase when behaviors are looked at as general groups, such as offenders showing the use of any kind of sexual behavior, any kind of body disposal behavior, and so on. This study thus provides the first wave of empirical testing to question current theories in the literature on serial homicide that have suggested that signatures or specific key behaviors are not only obvious, but key in understanding the consistency in offenders' behaviors across their crimes. The study also highlights the importance in identifying which of these key behaviors are more indicative of offenders and their psychology, rather than behaviors that change depending on the situation, such as the environment, the interaction with the victim, and through learning and maturation that occur throughout an offender's series.[15]

Themes

On the other side of consistency research is the theme-based approach. Using a type or theme approach allows investigators and researchers to examine "pools" of behaviors that all encompass the same psychological meaning rather than simply examining the individual behaviors displayed throughout the course of the crime. As explored earlier in the example of sexual assault, it may be more effective because offenders may exploit many different behaviors due to situational factors, but these different behaviors may still be similar (i.e., consistent) because they belong to the same theme of offending style. In this case, using the theme-based approach rather than an individual behavior approach would help to alleviate the problem of finding inconsistency in behaviors due to the situation dependency of behaviors.

Salfati and colleagues tested whether the expressive/instrumental typology found useful in distinguishing single-homicide crime scenes was valuable in distinguishing among different subtypes of serial homicides. Discovering that the expressive/instrumental typology was useful in distinguishing different serial homicide crime scenes, they went on to test the consistency with which offenders offended according to the same theme across their series. Results from this study showed that although this general homicide classification was more productive in analyzing consistency than the single behaviors in their previous study, consistency levels were still not high. Indeed, using a stringent criterion, only 13 percent of cases could be seen to have each crime scene demonstrate a consistency in dominance—that is, where each of the crime scenes showed not only a dominant theme, but also the same dominant theme across the series.

One of the issues the study highlighted was that the individual behaviors that make up each theme still need to be refined to factor out behaviors that are more dependent on the situation than on the offender's psychology. The study therefore highlighted the importance of further investigating the nature of individual behaviors themselves so that the understanding of the consistency and stability of these can be fed back into a more thematic model, which in the current literature has been shown to be more effective and reliable at classifying crime scenes.[16] Although there are some patterns to be seen, scholars are still a long way away from understanding consistency and how it is displayed. Further study is needed to develop an understanding of what consistency is and how it may be displayed and ultimately measured.

LEGAL VERSUS PSYCHOLOGICAL DEFINITIONS OF BEHAVIORS AND CRIMES

Another important factor to consider is how crimes are conceptualized within the criminal justice system and how this has a direct impact on the understanding of crime. Research adhering to legal distinctions accepts a restricted perspective for examining patterns in offenders' behaviors and may consequently overlook important similarities and differences across the various forms. For instance, to fully look at sexual violence it is necessary to extend beyond a research perspective shaped by legal classifications and instead identify the actual psychological variations in crime that differentiate between sexually violent offenses.

This becomes especially important when following the progression of an offender's crime development and attempting to understand the consistency pattern over his or her series as well as looking at the offender's broader development in the context of his or her whole criminal career, such as cases where an offender progresses from sexual assaults to rapes to sexual homicides. Each of these crimes is a separate legal crime type, but psychologically the offender is committing a series of sexual assaults, which shows psychological consistency. In addition, the offender is showing development and change as well as escalation in both physical invasion of the victim and escalation in violence. By going beyond mere legal definitions, different methodology for the analysis of consistency and change can be established. The issue of development becomes additionally important when we select the specific samples upon which we conduct analysis and draw conclusions in relation to crime patterns and consistency in criminal behaviors across time, such as in the case of linking serial crime. Clearly, defining the crime under investigation and putting this in context of the timeline of an offender's criminal activities becomes essential. Normally, in crime research investigators select subjects based on the last crime they committed. In this way a "homicide offender" is one whose last crime was a homicide. This is done because offenders are generally identified in the criminal justice system by their last crime for which they were convicted, and focusing on their last crime makes it easier for researchers to locate offenders within prisons, the probation service, and so on. However, this selection process essentially classifies offenders to a certain legally defined crime type, which influences all consequent analysis and conclusions about patterns in criminal behavior.

CONCLUSION

In criminal investigations, it is of particular interest to investigators to identify whether a set of crimes is committed by different people or the same person and ultimately what the behavioral pattern exhibited at these crimes indicates about the offender.[17] This chapter has explored the key theoretical issues pertaining to linking serial crimes, notably the issue of whether offenders are behaviorally consistent across time and, if they are, how this may manifest. Specifically, we looked at what is known about a person's behavior, how it is based on his or her personality, and how much of it is influenced by situational contexts. Part of this investigation of whether offenders are consistent needs to be done in light of how offenders change and develop over time. To fully understand this, investigators need to situate their series in the context of their criminal career as a whole and look at consistency beyond just legal boundaries, toward understanding behavior in the more psychological realm.

GLOSSARY

Linking—consistency in crimes, including how they were committed and whether they were perpetrated by the same offender

Offender profiling—a field of study that attempts to use evidence at the crime scene and evidence about the crime to create a profile of the offender

Signature—specific piece of evidence that is consistently left behind at the crime scene and that is unique to the offender

Situational factors—factors present at the scene of a crime that may provide information about the nature of the offender/offense

Traits—characteristics, such as shyness and aggressiveness, that contribute to certain behavioral patterns

Victim—the person against whom violence or aggression is perpetrated

NOTES

1. Canter, D. (1994). *Criminal shadows*. London: Harper Collins; Canter, D. V. (2000). Offender profiling and criminal differentiation. *Legal and Criminological Psychology, 5*, 23–46; Salfati, C. G. (2009). Profiling. In B. L. Cutler (Ed.), *Encyclopedia of psychology and law*. Thousand Oaks, CA: Sage.
2. Salfati, C. G. (2008). Linking serial crimes. In J. Fossi & L. Falshaw (Eds.), *Issues in forensic psychology* (Vol. 8). Leicester, United Kingdom: The British Psychological Society.
3. Greene, J. O. (1989). The stability of nonverbal behavior: An action-production approach to problems of cross-situational consistency and discriminativeness. *Journal of Language and Social Psychology, 8*, 193–220; Mischel, W., & Shoda, Y. (1995). A cognitive-affective system theory of personality: Reconceptualizing situations, dispositions, dynamics and invariance in personality structure. *Psychological Review, 102*, 246–268.
4. Toch, H. (1969). *Violent men: An inquiry into the psychology of violence*. Chicago: Aldine.
5. Huesmann, L. R., & Eron, L. D. (1989). Individual differences and the trait of aggression. *European Journal of Personality, 3*, 95–106.

6. Huesmann, L. R., Eron, L. D., Lefkowitz, M. M., & Walder, L. O. (1984). The stability of aggression over time and generations. *Developmental Psychology, 20,* 1120–1134.

7. Furr, R. M., & Funder, D. C. (2003). Situational similarity and behavioral consistency: Subjective, objective, variable-centered, and person-centered approaches. *Journal of Research in Personality, 38,* 421–447.

8. Funder, D. C., & Colvin, C. R. (1991). Explorations in behavioral consistency: Properties of persons, situations and behaviors. *Journal of Personality and Social Psychology, 60,* 773–794.

9. Wright, J. C., & Mischel, W. (1987). A conditional approach to dispositional constructs: The local predictability of social behavior. *Journal of Personality and Social Psychology, 53,* 1159–1177.

10. Woodhams, J., & Toye, K. (2007). An empirical test of the assumptions of case linkage and offender profiling with serial commercial robberies. *Psychology, Public Policy and Law, 13,* 59–85.

11. Alison, L. J., Snook, B., & Stein, K. L. (2001). Unobtrusive measurement: Using police information for forensic research. *Qualitative Research, 1,* 241–254; Canter, D. V., Alison, L. J., Alison, E., & Wentink, N. (2004). The organized/disorganized typology of serial murder: Myth or model? *Psychology, Public Policy, & Law, 10,* 293–320; Salfati, C. G. (2000). Profiling homicide: A multidimensional approach. *Homicide Studies, 4,* 265–293; Salfati, C. G. (2007). The homicide profiling index (HPI): A tool for measurements of crime scene behaviors, victim characteristics, and offender characteristics. In C. G. Salfati (Ed.), *Homicide research: Past, present and future.* Chicago: Homicide Research Working Group.

12. Bateman, A. L., & Salfati, C. G. (2007). An examination of behavioral consistency using individual behaviors or groups of behaviors in serial homicide. *Behavioral Sciences and the Law, 25,* 527–544.

13. Lin, S., & Brown, D. E. (2006). An outlier-based data association method for linking criminal incidents. *Decision Support Systems, 41,* 604–615.

14. Douglas, J. E., & Munn, C. (1992). Violent crime scene analysis: Modus operandi, signature, and staging. *FBI Law Enforcement Bulletin, 61,* 2–5.

15. Salfati, C. G., & Bateman, A. L. (2005). Serial homicide: An investigation of behavioral consistency. *Journal of Investigative Psychology and Offender Profiling, 2,* 121–144.

16. Salfati, C. G., & Canter, D. V. (1999). Differentiating stranger murders: Profiling offender characteristics from behavioral styles. *Behavioral Sciences and the Law, 17,* 391–406; Salfati, C. G., & Taylor, P. (2006). Differentiating sexual violence: A comparison of sexual homicide and rape. *Psychology, Crime and Law, 12,* 107–126.

17. More recent research has improved upon earlier limitations. The following studies are illustrative: Tonkin, M., Woodhams, J., Bull, R., Bond, J. W., & Palmer, E. J. (2011). Linking different types of crime using geographical and temporal proximity. *Criminal Justice and Behavior, 38,* 1069–1088; Trojan, C., & Salfati, C. G. (2011). Linking criminal history to crime scene behavior in single-victim and serial homicide: Implications for offender profiling research. *Homicide Studies, 15,* 3–31.

CHAPTER NINETEEN

Symbolic Interactionism and Crime in the Life Course

Jeffery T. Ulmer

KEY TERMS

Agency
Commitment portfolio
Dialectic
Normative conflict
Stigma

INTRODUCTION

Life-course perspectives on crime emphasize the idea of pathways into and out of crime, biographical elements, life-course trajectories, turning points, and developmental sequences. The perspectives usually may draw on the notion of "career" to chart the onset and persistence of crime, as well as pathways to desistance. Proponents claim that the life-course perspective departs from earlier criminology by: (1) focusing on the relationship between age and crime, (2) focusing on the relationship between prior and future criminal activity, and (3) adopting a process versus static orientation in focusing on the dynamics of offending over age. However, these concerns were also central to the work of more traditional theories in criminology, such as differential association and labeling theory. Furthermore, the emphases of the life-course approach have long characterized symbolic interactionism.

In fact, symbolic interactionists were among the first sociologists to think seriously about the life course in general (e.g., the work of Anselm Strauss and Barney Glaser) and deviant careers in particular (e.g., the work of Erving Goffman, Howard Becker, and John Lofland). Furthermore, many criminological theories that are relevant to crime across the life course are either derived from symbolic interactionism or have borrowed key concepts from it. This is no accident, because symbolic interactionism and early strands of the sociological study of crime (e.g., the work of Edwin Sutherland and Edwin Lemert) have common roots in the distinctive Chicago School approach to sociology.

This chapter provides a general overview of symbolic interactionism's key features, including special linkages between symbolic interactionism and two prominent criminological theories: differential association/social learning and labeling. Major conceptual and policy implications that flow from a symbolic interactionist view of the life course or criminal careers are also examined.

WHAT IS SYMBOLIC INTERACTIONISM?

It is surprisingly difficult to state precisely what symbolic interactionism is and is not in contemporary sociology, for two reasons. First, as some have argued, sociology has increasingly adopted into its mainstream concepts and propositions, such concepts as self, identity, definitions of the situation, and **agency**, whose meanings are at the heart of symbolic interactionism.[1] Second, researchers in the symbolic interactionist tradition have increasingly broadened their empirical focus to a wide range of topics and have engaged with a broad range of theoretical traditions and even other disciplines (cognitive psychology, cultural studies, communication studies, feminist theory, postmodernist perspectives, etc.).

The narrowest answer to the question, "What is symbolic interactionism?" is that it is the social psychology of the self and identity derived from the pioneering theories of George Herbert Mead, Charles Horton Cooley, William James, and John Dewey. Symbolic interactionism emphasizes that humans have the capacity to regard their own minds and behavior as objects through a process known as "reflexivity." This reflexive experience of one's own existence and self-awareness is what interactionists refer to as the self. Mead, following William James, famously conceived of the impulsive, creative, energetic phase of the self as the "I" and the reflexive, socialized phase of the self as the "me." The "me" phase of the self assesses the "I" through an ongoing internal conversation, in light of the views and norms of larger groups that the individual has internalized.

Lonnie Athens later usefully conceptualized this internalized collection of social norms and directives as one's "phantom community."[2] Following the lead of Herbert Blumer, Mead's most authoritative early interpreter for sociology, interactionists have made the reflexive self central to any understandings of human behavior and have emphasized the multifaceted and dynamic nature of the self. Starting in the 1950s interactionists such as Anselm Strauss, Howard Becker, Gregory Stone, and Erving Goffman extended this focus on the self to the notion of identity. In the works of these authors, identities are social locations that are transacted in social interaction. In other words, identities are social identifications that enable individuals to place one another in social space. Identities are also the social interactional face of the self—that is, identities are dimensions of self that we present in interaction and by which other parties to the interaction know us. In an important implication for labeling theory in criminology, deviant stigma is, in the words of Erving Goffman, "spoiled identity." A **stigma** is a devalued identity that is imposed upon a person. Stigma strongly tends to mobilize negative behavioral expectations and attributions of character among others who meet the stigmatized person.

Anselm Strauss's treatise on identity, *Mirrors and Masks* (1959), and Strauss and Barney Glaser's book, *Status Passage* (1971), explicitly put a process-oriented, life-course spin on the concept of identity. In fact, these authors were among the first sociologists to use terms like *turning points* to speak of crucial identity changes in the life course and to speak of identities and selves as having "careers." Their work emphasized how people negotiated stability and change in self and identity across the life course, as related to people's networks of social relationships, and the opportunities and constraints presented by their social worlds.

Many symbolic interactionist scholars would argue for a much broader definition of their perspective, however. This broader definition certainly includes, but goes considerably beyond, the social psychology of self and identity described earlier. From this broader perspective, interactionism is a generic sociological perspective that is based on the philosophy of American pragmatism (Mead, Dewey, James, and Cooley, mentioned earlier, were noted pragmatist philosophers as well as psychologists and sociologists). Indeed, Herbert Blumer, who was seen as the authoritative spokesperson of the perspective until his death in 1987, coined the term *symbolic interactionism* in 1937 to identify a broad approach to sociology that incorporated the insights of Mead and other pragmatists as well as early Chicago School sociologists W. I. Thomas and Robert Park. As such, symbolic interactionism is not so much a formalistic theory, with predictive propositions and hypotheses (though one could perhaps derive such hypotheses) as it is a paradigm or overarching conceptual framework that shapes one's approach to theorizing about and researching any sociological topic. In this view, symbolic interactionism has at least four defining features. These four features—an emphasis on interpretation and meaning, an emphasis on process, an emphasis on human agency, and dialectical thinking and a rejection of dualisms or false dichotomies—are crucial to understanding a symbolic interactionist position on crime and the life course.

An Emphasis on Interpretation and Meaning

Symbolic interactionism rests on three simple propositions distilled from the thoughts of Mead, Dewey, James, Cooley, and other early pragmatists who believed people act on the basis of meanings that objects, situations, or other people have for the actor; meanings emerge, or are created from, interactions between social actors (which, importantly, can include not just individuals, but also groups such as organizations, communities, and the like—Blumer often used the terms *acting units*, *group actors*, and *collective actors*), and meanings are maintained and/or changed through a dynamic process in which people keep or adjust their definitions based on responses and consequences.

According to Blumer, these three propositions are the absolute bedrock of symbolic interactionism. This means that interactionism is at its heart an interpretive perspective—that is, it emphasizes the way people and groups interpret the world around them and act on the basis of those interpretations. Thus, W. I. Thomas and D. S. Thomas's concept of the definition of the situation is central to symbolic interactionism (and, an interactionist would argue, all of sociology). The famous "Thomas Theorem" says that what people define as real becomes real in its consequences. That is, people act in social situations toward objects and other people on the basis of their interpretations and definitions. Note also that social actors' definitions can vary a great deal in accuracy and effectiveness (at least as perceived by others), but the crucial point is that it is those definitions, and not some "objective" reality or stimuli, that drives behavior. As shown later in this chapter, this interactionist principle is at the heart of the social learning theoretical perspective in criminology, as well as labeling/social reaction theory, and is at least implicit in other criminological theories.[3]

An Emphasis on Process

Process, change, evolution, and emergence were central concerns of the early pragmatist philosophers and their early sociological interpreters, such as Park, Thomas, and Blumer. Interactionists focus on how individuals and their social worlds are always active, in process, ongoing, becoming, and changing. In other words, all of society, social organization, and social structure rest on, are made up of, and are transformed by myriad and complex ongoing processes of social interaction, in much the same way that a solid object such as a table or steel bar is made up of atomic and sub-atomic particles in interactive motion.[4] Nothing is seen as static. In fact, even stability and continuity do not "just happen" but require action to maintain. In fact, symbolic interactionist conceptions of social organization and structure rest on these views of structure and organization as ongoing interaction process.

An Emphasis on Human Agency

Humans do not just automatically or passively respond to stimuli (as older behaviorist psychological imagery would have it), nor is human behavior predetermined by instinct. People define the objects, events, and others around them and act with intention. People create, innovate, make choices, formulate goals, and attempt to achieve those goals (with varying degrees of success). People can also flexibly adapt and adjust their lines of action to the changing demands and opportunities of the situations around them. This emphasis on agency therefore explains the centrality of the self and identity in interactionist thought. The self is the agent who acts and interacts and, in doing so, constitutes society.

What this means for crime in the life course is that interactionism does not view the individual as being passively buffeted about throughout his or her life by external forces. Rather, individuals actively adapt and adjust to the situations they find themselves in.[5] People make choices between criminal and conforming actions, albeit within social and physical constraints. Over the course of life, people pursue or aspire to some identities, and reject others, with varying degrees of success. People also attempt to exercise at least some choice as to the others with whom they associate and the groups with whom the affiliate, though they are not always successful. To emphasize agency is not to say, however, that interactionism ignores constraint, including constraint from social structures. Instead, interactionists see agency as being in a dialectical relationship with biological and social constraints of various kinds.

Dialectical Thinking and a Rejection of Dualisms or False Dichotomies

Interactionism emphasizes relationships of mutual interdependence and mutual determinism. Dialectical thinking of this sort was a key feature of American pragmatist thought, and it fundamentally shapes symbolic interactionist sociology. For interactionists, to say that two phenomena form a **dialectic** is to say that they mutually interrelate in a manner that is practically inseparable. In fact, two dialectical phenomena can be said to mutually create one another, in the sense that one cannot exist without the other and one has no meaning without reference to the other. Thus, interactionists generally do not believe in many of the dualisms (i.e., opposed dichotomies) commonly discussed in social science, such as "body versus mind," "nature versus nurture," "individual versus society," "micro versus macro" social phenomena, "stability versus change," "consensus versus conflict," and so on. Instead, most interactionists would argue that no single part of any of these dichotomies can be studied without reference to its counterpart. For example, individual social behavior only makes sense in the context of some larger community of organization and culture, so studying individual behavior (even individual self and identity) is rather pointless without reference to the individual's surrounding social world. At the same time, interactionists have difficulty with the notion that macro-level social structures (nations, states, corporations, etc.) can be meaningfully studied without any reference to the individuals, interactions, and joint acts that constitute them.

In sum, the broadest definition of symbolic interaction is that it is a sociological approach that is founded in pragmatist philosophical assumptions, especially the ones outlined in this section. Whether one uses a broad or narrow definition, however, symbolic interaction has profoundly shaped key branches of criminology.

SYMBOLIC INTERACTIONISM'S RELATIONSHIP TO CRIMINOLOGICAL THEORIES

Differential association/social learning theory and labeling theory are the two criminological theories most linked with symbolic interactionism. These theories are discussed in greater detail later. However, other key theories in criminology are also compatible with interactionism and sometimes have drawn on interactionist concepts.

Opportunity theory represents the marriage of Merton's notion of structural strain with differential association theory. Differential association theory in turn was influenced by the same pragmatist approach to sociology as social disorganization theory. Furthermore, later extensions of opportunity theory clearly link it with interactionist-friendly themes, such as situational-level interaction, definitions of situations, socialization, culture, and decision making.[6] In general, the notion of criminal opportunity is fully complementary to the interactionist depiction of human activity described earlier: that human activity is a dialectic of choice and constraint, and human agency is conditioned by the availability and attractiveness of alternative lines of action presented by situations and larger social structures.

Social control theory argues that people refrain from committing crime and deviance because of social bonds to conventional others, institutions, and norms. The four elements of these social bonds are known as stakes in conformity: attachment to conventional others, commitment to conventional lines of action, involvement in conventional activities, and belief in conventional norms. Hirschi's treatment of attachment and commitment was influenced by Goffman, and his treatment of them as stakes in conformity derives from both Goffman's and Becker's notion of commitments as "side bets."[7]

Furthermore, Gottfredson and Hirschi emphasized self-control (plus opportunities for crime) as the principle individual-level cause of crime through all stages of the life course. They also argue that self-control is a stable propensity or tendency that is largely solidified early in life, and individual differences in self-control are stable thereafter.[8] Given their dialectical thinking about stability and change and their emphasis on the malleability of the self and identity, most interactionists would likely take issue with the position that self-control (and between-individual differences in it) is necessarily solidified early in life and stable thereafter. Given that they tend to see behavior as the product of a complex interaction between self and environment, most interactionists would doubt the validity of the claim that self-control is the principle individual cause of crime.

However, nothing in symbolic interactionism conflicts, in principle, with the notion of self-control, and there is no reason why interactionism should reject the proposition that self-control is an important situational cause of criminal activity at any point in the life course. In fact, a process of self-restraint in the face of internalized social norms is inherent in interactionist conceptions of self, beginning with Mead's treatment of the "I" and the "me" (and Dewey's earlier, similar treatment). Furthermore, if the capacity for self-control is an important cause of crime, and if it is not a purely biological trait, then the capacity and desire for self-control would by definition be developed through processes of symbolic interaction and reflexivity as described by Mead and his later interpreters. An interactionist perspective might also raise the question of whether one's self-control would vary according to situational contingencies and according to particular identities the individual performs or aspires to have. Overall, a symbolic interactionist conception of self-control might have more in common with recent treatments of it in social psychology.

Differential Association/Social Learning and Neutralizations

The term *symbolic interactionism* had not yet been coined by Herbert Blumer when Edwin Sutherland first began formulating differential association theory in 1924. However, Sutherland and Blumer were products of the sociology department at University of Chicago, and both absorbed pragmatist assumptions and perspectives through the work of Robert Park, G. H. Mead, Louis Wirth, John Dewey, and others. Sutherland was especially influenced by the work of W. I. Thomas and his notion (with Dorothy Swain Thomas) of the definition of the situation. This concept, fundamental to symbolic interactionism, would become the bedrock upon which he constructed his generic theory of crime and deviance: differential association. Just as interactionists view individual definitions of situations as the most proximal cause of behavior and interaction, so to Sutherland viewed definitions of "the law" and criminal activities as the most proximal cause of crime.

Differential association/social learning theory is very much a process-oriented, life-course–oriented perspective, although Sutherland did not systematize that perspective in the way that life-course theorists do today. However, Matsueda and Heimer, for example, describe how key emphases of differential association/social learning, such as self-concepts and attitudes, are dynamic across the life course.[9] The perspective is clearly compatible with the notion that criminal propensity changes over the life course. On the other hand, the kind of differential association/social learning perspective we describe here is also consistent with the possibility of stable or enduring criminal propensity. Key themes of differential association/social learning theory, such as definitions, skills, self-concepts, commitments, and rewards, can produce both change and stability in criminal propensity over the life course.

Sutherland's theory is one of differences in association with messages and opportunities favorable to crime. It posits that deviant/criminal behavior, like other types of social behavior, is learned and that it is learned through processes of social interaction. In Sutherland's definitive statement of the theory, criminal or delinquent behavior involves the learning of techniques of committing crimes and orientations (vocabularies of motives, drives, rationalizations) and attitudes favorable to crime. Differential association/social learning theory is supported by an impressive array of quantitative, qualitative, and cross-cultural evidence.

The concept of **normative conflict** frames the theory. This is simply the notion that various groups and subgroups in society differ in terms of definitions of right and wrong and in their definitions of whether individuals are obligated to follow laws (either particular ones or laws in general). These groups can include subcultures, professions and occupational groups, ethnic groups, voluntary associations, religious groups, neighborhoods, and other collectives. This normative conflict, in turn, is played out at both the individual and group levels.

The structural, group-level manifestation of normative conflict is differential social organization. The norms, values, meanings, skills, and definitions found within these groups can be favorable to particular kinds of crime or to crime in general. In addition, cultural, subcultural, or group messages or definitions can be supportive, neutral, hostile, or mixed toward crime. Groups' organization (the social and cultural organization of various groups or settings) can potentially be favorable to crime in general, and/or specific kinds of crimes, or specific kinds of crimes in specific circumstances. Sutherland describes the effect of differential social organization this way: "The law is pressing in one direction, and other forces are pressing in the opposite direction."[10]

Later, Daniel Glaser linked "association" more closely with the symbolic interactionist perspective he acquired from Herbert Blumer at the Chicago School. Focusing on the subjective role-taking process that occurs in both social interaction and private thought, Glaser suggested that Sutherland's theory be reconceptualized as "differential identification." Defining "identification" as "the choice of another from whose perspective we view our own behavior," he expressed the essence of his revision as follows: "A person pursues criminal behavior to the extent that he identifies himself with real or imaginary persons from whose perspective his criminal behavior seems acceptable."[11]

Neutralization theory, formulated by Sykes and Matza, noted that delinquents and criminals usually have both criminal and noncriminal values, expressing the former in their lawbreaking while at the same time maintaining at least some allegiance to conventional norms. Sykes and Matza drew upon Mills's concept of "vocabularies of motive" and identified a process by which would-be deviants or criminals suspend or rationalize conduct that violates norms in which the actor otherwise believes. These vocabularies of motive allow the actor to then engage in the chosen form of deviant activity. They called this prebehavior rationalization process "neutralization" and identified five common and nonmutually exclusive techniques of neutralization: denials of responsibility (e.g., blaming their crimes on their circumstances), denials of the victim (e.g., claiming that their victim "had it coming"), denials of injury (e.g., saying that their victims have plenty and won't miss it), condemning their condemners (e.g., accusing their prosecutors of prejudice), and appeals to higher loyalties (e.g., saying that they did it for their friends or family). Accounts are similar to neutralizations but are articulated after deviant behavior to retrospectively excuse or rationalize the violation of norms to self and others. In general, accounts and neutralizations are examples of a broader concept that interactionists call "aligning actions," or rhetoric and actions aimed at repairing disrupted social situations and damaged identity.[12]

Jacobs and Topalli further extended neutralization theory in ways that broaden it and ground it on even more solid interactionist footing. These two scholars found that neutralizations and accounts do not just apply to deviance/crime. Rather, they can be used by people embedded in deviant subcultures to excuse conventional actions that would call their identity and standing in the deviant group into question. Jacobs found that undercover police had to provide accounts and neutralizations for not using drugs in front of drug dealers and other users to avoid blowing their own cover. Topalli found that "hard core" gang members often had to provide accounts and neutralizations for otherwise conventional behavior. For example, they often had to account to their gangster peers for situations where they chose not to violently retaliate against someone who wronged them.[13]

In sum, socialization favorable to crime involves learning definitions, attitudes, behaviors, skills, and vocabularies of motive favorable to given forms of deviance as well as forming relationships with deviant others and self-definitions in terms of deviant identities. Differential association/social learning theory incorporates several symbolic interactionist themes: an emphasis on interpretive meanings and definitions of situations as the proximal causes of behavior, an emphasis on the importance of interaction processes and how they are shaped by larger social contexts, and the importance of self-definitions and identities. In focusing on these themes, differential association/social learning theory would explain variation in crime across the life course primarily in terms of variation in social learning processes connected to changing social contexts, opportunities, relationships, and status passages that occur in the life course.

LABELING THEORY

Labeling theory has long been closely identified with symbolic interactionism. Its early proponents, such as Mead, Tannenbaum, Lemert, Becker, Lindesmith, Goffman, and Schur, were either proponents of interactionism or were substantially influenced by pragmatist or interactionist work. Key interactionist elements that suffuse labeling theory are an emphasis on social process; an emphasis on the importance of the construction of meaning and the interpretive definition of social objects; the importance of the dialectic between selves, identities, and situations; and the dialectic between agency and constraint.

Contemporary labeling theory's themes include the situational and larger scale definition of deviance, the role of social statuses and identities in affecting whether one is negatively labeled, the consequences of deviant labels for the careers and selves of individuals labeled as deviant, and the consequences of the creation and labeling of deviance for society. Labeling theory also predicts that negative labeling of behavior in question (either informally for deviance or formally for crime) depends on the following situational contingencies: the meaning and consequences of the act, the identity and statuses of the actor, the audience of the act, and the context of the act (for a systematic treatment of the role of status, demeanor, appearance, audience, and context in social reaction to deviance). One of labeling theory's main themes is the effect of being labeled unfavorably (e.g., as delinquent or criminal) on a person's subsequent life history. Arrests and convictions among young offenders tend to reverberate throughout the life course, producing negative consequences and cumulative disadvantages in education, employment, family life, substance abuse, and further criminal behavior throughout the life course.

Differential association and labeling theory also have a symbiotic relationship with one another. One basic contention of labeling theory is that sanctions against crime/deviance can inadvertently contribute to continuity in deviant activity and careers. This occurs because punitive sanctions can potentially restrict opportunities for conventional employment and relationships, open up illegitimate opportunities and access to deviant networks or subcultures, and lead to the development of deviant self-identities and a sense of estrangement from conventional society. In other words, labeling can open up criminal/deviant learning and performance opportunities and produce further differential association processes that entrench an individual into crime or deviance.[14]

In another example of interactionist research spanning the social learning and labeling theory traditions, Matsueda and colleagues have been extending, testing, and refining interactionist-based ideas about the role of interpersonal relationships and identity processes in producing crime and delinquency for 3 decades. In particular, Matsueda and colleagues focus on the nature and role of peer relationships and interaction in producing youth crime, as well as the role of negatively reflected appraisals from parents and other authority figures in solidifying deviant identities among juveniles and into young adulthood.[15]

INTERACTIONIST APPROACHES TO CRIMINAL CAREERS

Symbolic interactionists, particularly Howard Becker and Erving Goffman, arguably originated the concept of deviant careers. A symbolic interactionist approach suggests five distinct implications for the study of criminal careers and crime across the life course:

1. If the self, identity, and interpretive definitions are the most proximal drivers of behavior, it follows that both stable and dynamic features of the social world that shape self, identity, and definitions are central to understanding crime and desistence at any point in the life course.

2. An interactionist view fosters a dynamic and agency-oriented view of opportunities and constraints. People make decisions and act within the boundaries and constraints of situations, which are shaped by larger social structures. However, people are not passive. People seek out, aspire to, and (within limits) choose among identities and have elements of self they both prefer and devalue. People are constrained by the opportunities that their social worlds present, but they also seek out opportunities and situations they find favorable to their identities and goals.

3. The factors that may initiate a criminal career are not necessarily the same as the things that keep it going, or end it. In particular, labeling theory argues that the original causes of criminal behavior may not be the same as the eventual causes that reproduce or sustain criminal behavior.

4. There are distinct limits on the predictability of criminal careers. Criminologists can discern stable patterns and regularities and even make generalizable theoretical propositions, but individual-level criminal career predictions will always be to some degree indeterminate—not just because of methodological limitations but also because of the nature of human social activity.

5. There is no *a priori* reason to assume stability in criminality across the life course. Selves, identities, and interpretive schema are both stable and mutable. There will be as much stability in criminal propensity as is produced by the durable characteristics of individual's social relationships and life circumstances. Prior socialization is quite durable, but dramatic self-change is not uncommon.[16]

The examples of interactionist (and interactionist-friendly) research and theorizing about criminal careers described as follows draw out some or all of these implications. There are many other examples I could have chosen, but space constraints limit this chapter to these few that I believe are particularly illustrative.

Neal Shover: Aging Property Offenders

Although he does not explicitly identify his work as interactionist, Neal Shover's analyses of male ordinary property offenders exhibit many compatible features with an interactionist perspective on criminal careers. He argues that it is not the biological process of aging per se that explains exit from a career of property offending, but the socially constructed and negotiated changes in perspectives that accompany aging. Shover found four temporal contingencies or changes in definitions of oneself and significant others or patterns of events in one's life: shifts in identity, an awareness of time becoming an exhaustible resource, changes in aspirations and goals, and growing tired of the relentless effects of incarceration. Property offenders also experience an interpersonal contingency that Shover defines as an objective change in one's social relationships or networks typically involving a satisfying job or stable relationship.

As in other analyses of deviant careers, Shover found variability and unpredictability in the occurrence of, and interdependence among, these contingencies. In some cases the occurrence of separate contingencies could not be isolated, and, at times, contingencies occurred simultaneously. In many cases temporal and interpersonal contingencies:

> Interacted with or followed one another as part of a dynamic process, with one type preceding and increasing the probability of occurrence for the other(s). Imposition of a rigid temporal and causal order on this process would violate its dynamic nature, and given our present state of knowledge, would be arbitrary and premature.[17]

Shover's work on persistent property offender careers also emphasizes how these offenders choose between structurally constrained alternatives at key turning points in their lives and how these choices lead them down complicated roads of persistent and varied criminal activity, trying to go straight, reentry into crime, going straight once again, and so forth.

Lonnie Athens: Violentization and Careers of Serious Violence

Thanks to journalist Richard Rhodes's book, *Why They Kill: Discoveries of a Maverick Criminologist*, a vast academic and public audience has been introduced to Lonnie Athens's theory and research on violence and also core tenets of symbolic interactionism. Athens's theorizing and research on violent crime reflect his study under Herbert Blumer at the University of California, Berkeley, and it is steeped in the thought of Blumer and Mead. He analyzed the careers of serious violent offenders and offenders' decisions to commit homicide, rape, robbery, and/or aggravated assault.[18]

The prime movers driving Athens's theory of violent acts are situational definitions conducive to violence, self-images consistent with violent behavior, self-conversations with phantom communities that provide moral justification for violence, and corporal (physical) communities whose norms favor the use of violence in dominance disputes. Athens delineates a theory of "violentization" (or violent socialization) in which serious violent offenders are distinguished by their having passed through four progressive and contingent career phases:

1. Brutalization is a process involving three subprocesses: violent subjugation (the individual is dominated by another person's violence or threat), personal horrification (the individual witnesses the violent subjugation of another), and violent coaching (the individual is mentored in violence as a means of winning dominance disputes).
2. Defiance is a process in which the individual buys into/embraces the lessons of violent coaching and resolves to begin using violence to dominate others and achieve goals.
3. Violent dominance engagements are when the individual successfully dominates others through violence.
4. Virulency occurs when individuals attain and enjoy a reputation for violence and enjoy instilling fear in others. The culmination of virulency is "malevolency," in which the individual becomes transformed from a person who would only resort to violence to resist his or an intimate's debasement or violent subjugation to a person who relishes any opportunity to hurt others.

Steffensmeier and Ulmer: Commitment Portfolios

Steffensmeier and Ulmer integrate several core theories of criminal and deviant behavior into an interactionist-based learning/opportunity/commitment framework. For example, they argue that criminal/deviant socialization (differential association) and conventional social control both consist of a generic sociological process—the development and experience of structural, personal, and moral commitments.

Commitment processes can produce continuity in criminal behavior both by fostering a desire for criminal activities and choices and by constraining individuals from pursuing alternatives. For example, personal commitment to criminal activities develops through learning positive attitudes toward criminal activities and criminal others and adopting criminal self-definitions. Moral commitment constrains individuals from terminating criminal lines of action once one has become involved in them through internalized cultural and subcultural norms of exchange and obligation. Structural commitment constrains the termination of criminal careers and is produced by the relative availability and attractiveness of conventional and criminal opportunities, irretrievable investments in criminal activity, and social reactions from criminal associates. Both differential association into crime/deviance and conventional social control entail developing all these kinds of commitments (to either deviance or conformity, or both), and the concept of structural commitment also embeds the individual in structural contexts, large and small.

Steffensmeier and Ulmer introduce the term **commitment portfolio** to refer to an individual's "accumulated and valued interests and allegiances, as represented by his or her structural, moral, and personal commitments." They argue that stability and change in criminal careers can be explained with reference to individuals' commitment portfolios. For example, building a portfolio of conventional commitments as one gets older explains the typical desistence from crime with aging. On the other hand, some offenders do not develop such conventional commitments and/or continue to develop strong criminal commitment portfolios as they get older. These kinds of offenders would persist in crime throughout the life course (or until such time as their commitment portfolios changed to favor conventional commitments).[19]

Furthermore, there are negative consequences of labeling work through commitment processes. Commitment processes are the "intervening variables" between negative social reaction and secondary deviance, when one becomes entrenched in a deviant or criminal identity and career. That is, punitive or exclusionary social reactions to crime or deviance can rearrange structural, personal, or moral commitments to deviance versus conventional activity and entrench individuals in criminal identities and careers.

Peggy Giordano and Colleagues: An Interactionist Theory of Desistence From Crime

Finally, Giordano and colleagues have developed an explicitly symbolic interactionist theory of desistance from criminal activity. They tested it with longitudinal quantitative and qualitative data from a broadly varying sample of men and women. They find substantial support for their theory and that it outperforms alternatives. Among the benefits Giordano et al. claim for their interactionist theory of desistance are: offering a balanced emphasis on the importance of both agency and constraint in the life course, explaining why some individuals do not desist from crime even when opportunity structures are favorable for them to do so, and offering better explanations for why men's and women's criminal careers and desistance processes differ considerably. In addition, Giordano et al. expand this interactionist theory of desistence to include a focus on emotions and how what they call "emotional selves" influence long-term continuity and change in crime.

Giordano et al.'s interactionist theory of desistence from crime coined the term *hooks for change*. These hooks for change are said to be vitally necessary for individuals to choose to leave behind criminal behavior and lifestyles and embrace conventional ones. Hooks for change are opportunities, relationships, and experiences that offer those entrenched in criminal careers the incentive to adopt new, noncriminal identities and new conventional self-definitions and to foster decreased negative emotions that motivate crime and increased skill in managing one's emotions.[20]

CONCLUSION

There has been little explicit symbolic interactionist discussion of policy with regard to crime and justice. However, in general, symbolic interactionism would emphasize policies that took seriously the main themes of interactionist and interactionist-inspired research in criminology: socialization and social learning, self and identity, commitments, and labeling processes.

Symbolic interactionist–compatible efforts at preventing crime and rehabilitating offenders are similar to those suggested by differential association/social learning theory. That is, policies to prevent crime would most effectively focus on fostering communities and other social contexts that provide young people with learning opportunities for conventional norms, skills, definitions, and behaviors. Interactionist-compatible policies to prevent crime would provide people of all ages with the opportunities to build commitment portfolios that favored conventional activity over crime. These commitment portfolios would particularly include an emphasis on building conventional identities and definitions of self rather than differential identification with criminal role models.

Interactionist-inspired rehabilitation policies would foster the "hooks for change" emphasized by Giordano and colleagues' research. Giordano and colleagues make several specific recommendations for such policies in their work. In sum, symbolic interactionism favors rehabilitative policies that foster conventional social learning; the development of portfolios of conventional structural, personal, and moral commitments; and hooks for change that foster the development of conventional identities and selves. Vital to this would be encouraging offenders to internalize and embrace new, conventional generalized others or "phantom communities" to replace previous ones favorable to deviance or crime from whose standpoint the offender would judge his or her behavior.

An interactionist approach to criminal careers would support policies that take the dialectic of choice and constraint seriously. Policies that are serious about channeling individuals away from or out of criminal careers would be directed at the social structures, institutional contexts, communities, peer groups, and families that present the situational constraints and opportunities within which developmental processes take place, biographies unfold, and people make choices between conventional and criminal activity. These kinds of policies would address the availability and attractiveness of conventional and criminal learning and performance opportunities and the ways these are configured by larger scale arrangements of community, institutional structure, and stratification.

There is a distinct irony here. Symbolic interactionism is often mischaracterized as a perspective that only applies to the micro-level realm of individual minds, behavior, and interaction. However, the types of social policies it encourages are not just at the individual level; ones are also aimed at the interrelationship between larger scale social organization and situational contexts of action. Incidentally, these are exactly the type of social reform efforts in the realms of education, poverty, urban land use, and crime developed and supported by the early pragmatists.[21]

An interactionist view of criminal careers leads us away from the notion that scientific progress will bring us increasingly accurate ability to predict individuals' criminal behavior and that what we can predict, we can control. Instead, we would recognize that continuity and change are inseparable and that biology, previous behavior and experience, and constraints and opportunities may influence, but never totally determine, the contingencies and choices involved in criminal activity throughout the life course. This view and the theoretical models and research programs that derive from it provide an important complementary approach to the study of criminal careers.

Symbolic interactionism is a vital part of sociology today, and many of its key ideas and concepts are taken for granted by nearly all sociologists. This perspective has shaped sociological criminology since its inception. As long as the links between interactionism and criminology remain viable, interactionism will direct criminologists' attention to the importance of interpretive meaning, interaction processes, human agency, self, and identity and the dialectical relationships between individuals and situations and larger social contexts.

GLOSSARY

Agency—the notion that people do not act in a predetermined way, but rather people define the objects, events, and others around them and act with intention

Commitment portfolio—an individual's accumulated and valued interests and allegiances, as represented by his or her structural, moral, and personal commitments

Dialectic—two phenomena that mutually interrelate in a manner that is practically inseparable

Normative conflict—the notion that various groups and subgroups in society differ in terms of definitions of right and wrong and in their definitions of whether individuals are obligated to follow laws

Stigma—a devalued identity imposed on a person

NOTES

1. Fine, G. A. (1993). The sad demise, mysterious disappearance, and glorious triumph of symbolic interactionism. *Annual Review of Sociology, 19,* 61–87; Hall, P. (2007). Symbolic interaction. In G. Ritzer (Ed.), *Blackwell encyclopedia of sociology.* Boston, MA: Blackwell; Maines, D. R. (2001). *The faultline of consciousness: A view of interactionism in sociology.* New York: Aldine de Gruyter.
2. Athens, L. (1994). The self as a soliloquy. *The Sociological Quarterly, 35,* 521–532.
3. Blumer, H. (1969). *Symbolic interactionism: Perspective and method.* Englewood Cliffs, NJ: Prentice Hall; Glaser, B. G., & Strauss, A. L. (1971). *Status passage.* New York: Aldine de Gruyter; Strauss, A. L. (1959). *Mirrors and masks: The search for identity.* New York: The Free Press of Glencoe; Thomas, W. I., & Thomas, D. S. (1928). *The child in America: Behavior problems and programs.* New York: Knopf.
4. Perinbanayagam, R. (1986). The meaning of uncertainty and the uncertainty of meaning. *Symbolic Interaction, 9,* 105–126.
5. Becker, H. S. (1963). *Outsiders.* New York: Macmillan.
6. Cloward, R., & Ohlin, L. (1960). *Delinquency and opportunity.* Glencoe, IL: Free Press; Cullen, F. T. (1988). Were Cloward and Ohlin strain theorists? Delinquency and opportunity revisited. *Journal of Research in Crime and Delinquency, 25,* 214–241; Merton, R. K. (1997). On the evolving synthesis of differential association and anomie theory: A perspective from the sociology of science. *Criminology, 35,* 517–525; Steffensmeier, D. J., & Ulmer, J. T. (2005). *Confessions of a dying thief: Understanding criminal careers and illegal enterprise.* New Brunswick, NJ: Aldine-Transaction.
7. Becker, H. S. (1960). Notes on the concept of commitment. *American Journal of Sociology, 66,* 32–40; Goffman, E. (1961). *Encounters.* Indianapolis: Bobbs-Merrill; Goffman, E. (1963). *Stigma.* Engelwood Cliffs, NJ: Prentice Hall; Hirschi, T. (1969). *Causes of delinquency.* Berkeley, CA: University of California Press.
8. Gottfredson, M. R., & Hirschi, T. (1990). *A general theory of crime.* Stanford, CA: Stanford University Press; Hirschi, T. (2004). Self-control and crime. In R. Baumeister & K. Vohs (Eds.), *Handbook of self regulation: Research, theory, and applications* (pp. 537–552). New York: Guilford Press.
9. Matsueda, R., & Heimer, K. (1997). Developmental theories of crime. In T. Thornberry (Ed.), *Advances in criminological theory* (pp. 174–186). New Brunswick, NJ: Transaction.

10. Blumer (1969), see Note 3; Sutherland, E. (1940). White collar criminality. *American Sociological Review, 5*, 1–12, p. 12; Sutherland, E. (1947). *Principles of criminology*. Philadelphia: Lippincott.

11. Glaser, D. (1956). Criminality theories and behavioral images. *American Journal of Sociology, 61*, 433–444, p. 440.

12. Mills, C. W. (1940). Situated actions and vocabularies of motive. *American Sociological Review, 5*, 904–913; Scott, M., & Lyman, S. (1968). Accounts. *American Sociological Review, 33*, 46–62; Stokes, R., & Hewitt, J. (1976). Aligning actions. *American Sociological Review, 41*, 838–849; Sykes, G. M., & Matza, D. (1957). Techniques of neutralization: A theory of delinquency. *American Sociological Review, 22*, 664–670.

13. Jacobs, B. (1992). Undercover drug use evasion tactics: Excuses and neutralizations. *Symbolic Interaction, 15*, 435–454; Topalli, V. (2005). When being good is bad: An expansion of neutralization theory. *Criminology, 43*, 797–836.

14. Ulmer, J. T. (1994). Revisiting Stebbins: Labeling and commitment to deviance. *Sociological Quarterly, 35*, 135–157; Ulmer, J. T. (2000). Commitment, deviance, and social control. *Sociological Quarterly, 41*, 315–336.

15. Matsueda, R. (1982). Testing control theory and differential association: A causal modeling approach. *American Sociological Review, 47*, 489–504; Matsueda, R. (1988). The current state of differential association theory. *Crime & Delinquency, 34*, 277–306; Matsueda, R. (1992). Reflected appraisals, parental labeling, and delinquency: Specifying a symbolic interactionist theory. *American Journal of Sociology, 97*, 1577–1611.

16. Athens, L. H. (1995). Dramatic self change. *Sociological Quarterly, 36*, 571–586.

17. Shover, N. (1983). The later stages of ordinary property offender careers. *Social Problems, 31*, 208–218, p. 214; Shover, N. (1996). *Great pretenders: Pursuits and careers of persistent thieves*. Boulder, CO: Westview Press.

18. Athens, L. H. (1989). *The creation of dangerous violent criminals*. Chicago: University of Illinois Press; Athens, L. H. (1997). *Violent criminal acts and actors revisited*. Chicago: University of Illinois Press; Athens, L. H. (2003). Violentization in larger context. In L. H. Athens & J. T. Ulmer (Eds.), *Violent acts and violentization: Assessing, applying, and developing Lonnie Athens' theories*. Oxford, UK: Elsevier Sciences; Rhodes, R. (1999). *Why they kill: Discoveries of a maverick criminologist*. New York: Knopf.

19. Steffensmeier, & Ulmer (2005), see Note 6.

20. Giordano, P. C., Cernkovich, S. A., & Holland, D. (2003). Changes in friendship relations over the life course: Implications for desistence from crime. *Criminology, 41*, 293–328; Giordano, P. C., Cernkovich, S. A., & Rudolph, J. (2002). Gender, crime, and desistance: Toward a theory of cognitive transformation. *American Journal of Sociology, 107*, 990–1064; Giordano, P. C., Schroeder, R., & Cernkovich, S. A. (2007). Emotions and crime over the life course: A neo-Meadian perspective on criminal continuity and change. *American Journal of Sociology, 112*, 1603–1661.

21. Shalin, D. (1986). Pragmatism and social interactionism. *American Sociological Review, 51*, 9–29.

CHAPTER TWENTY

A "Good Lives" Approach to Rehabilitation

Edward Manier, Truce Ordoña, and C. Robert Cloninger

KEY TERMS

Criminogenic risks
Personality
Prison code
Rehabilitation
Triggers

INTRODUCTION

This chapter provides the reader with a glimpse of the possibilities for extending developmental and life-course theories of antisocial behavior into a comprehensive and effective theory of **rehabilitation**. In a nutshell, it argues that the "cognitive antisocial potential" of incarcerated offenders and returning former prisoners can be reduced by an instructional and interactive group therapy guided by the highly confirmed "seven factor" theory of personality—itself developmental and relevant throughout the life course.[1] This theory is unique in both distinguishing and integrating the most elementary psychobiological substrata of behavior in the human infant and child (traits of temperament) with more advanced and distinctively human capacities for reflecting on the trajectory of one's personal life course and gradually shaping a more or less mature and prosocial character.

The corresponding multistage therapy expands each client's capacities for self-awareness and self-acceptance of his or her basic personality traits and for more effective communication and cooperation with other group members. The goal is to show each offender or former prisoner how to achieve the sort of lasting and resilient happiness that flows from the mature expression of the basic virtues of independent practical reasoning and work (responsible, resourceful reliability) as well as the traits of compassion and principled generosity and helpfulness that are the keystones of healthy human social life. The final and perhaps most distinctive feature of this theory and the attendant therapeutic practice is emphasis on a form of spirituality enabling each client to grow toward the transcendent reconciliation of the tensions and conflicts inherent in human nature that, if unchecked, feed our antisocial tendencies.

LIFE-COURSE DEVELOPMENT OF ANTISOCIAL POTENTIAL

Farrington's life-course model of criminal activity distinguishes long- and short-term forms of integrated cognitive antisocial potential (ICAP).[2] Long-term levels of ICAP vary with age, peaking in adolescence. Short-term ICAP, in contrast, "varies (over the entire life course) with type of crime and the details of the situation" in which the crime is committed. A third category, "main energizing factors" or catalysts of criminal activity, includes such long-term desires and micro-social factors as those associated with materialistic values, the prevalence of antisocial role modeling in the environment, and the absence of warm and structured parenting. This combination of characteristics—some more or less stable, others likely to vary over time and from situation to situation—enables Farrington to explain both "between-individual" and "within-individual" variation in antisocial potential.

Farrington's theory distinguishes three types of factors. Biological factors include excitement seeking (low cortisol), low cognitive processing abilities (planning, problem solving), high impulsivity, and low anxiety. Early developmental, childhood, school-age factors include low school attainment, family criminality or poverty, large family size (requiring careful analysis for individual variation), poor childrearing, and disrupted families. Main energizing factors include desires for ("extrinsic goods") material goods, status among peers, and sexual satisfaction, but only antisocial strategies for attaining such goods are chosen or available, and the impact of this factor is enhanced by antisocial modeling in the environment but mitigated by warm, well structured parenting.

Farrington's explanation of criminal behavior also involves factors comprising short-term antisocial potential, including alienation; powerlessness; situational frustration; such transient states as intoxication, anger, and boredom; and encouragement by peers. Combinations of long- and short-term factors influence the assessment of criminal opportunities and/or available victims through cognitive processes involved in the calculation of subjective costs and benefits. Also influential are the available cognitive scripts or routines, available in memory, for responding to such circumstances. All these factors influence the situation- and person-specific balance of impulsivity versus rationality.

Different types of crimes, however, are likely to be precipitated by quite different reasons: property crimes by a perceived need for material gain and/or needs for heightened excitement, enjoyment, or relief from boredom and violent crimes by anger and a need to display toughness and masculinity. Antisocial behavior also varies with distinctive characteristics of the available victim. For example, experimental evidence shows that money lost in an envelope also containing minimal personal information is more likely to be taken from rich men (cash plus a bill for a payment on a yacht) than older, less advantaged women (cash plus a bill for services at a neighborhood senior center).

Persons are likely to embark on criminal careers because of three factors:

1. Increasing long-term, but thwarted, desires for material goods, status, sex, and excitement; increased physical strength and skills; and shifts in socialization (increasing influence of peers, decreasing influence of parents)
2. Increasing short-term factors such as boredom, anger, intoxication, alienation, and frustration
3. Changed decision-making processes resulting in positive shifts in the subjective utility of offending

The consequences of offending (conviction and incarceration as opposed to enhancement of wealth and status) also influence long-term integrated cognitive antisocial potential.

It is important to compare Farrington's model with an approach intended to account primarily for the between-individual risks of recidivism across populations of incarcerated felons developed by Andrews and Bonta.[3] Andrews and Bonta identify a set of risk/need factors they identify as the "central eight." Among these, the first four, the "central four," are said to be most highly correlated with and/or predictive of criminal behavior. Each "risk" is paired with a "criminogenic need," enabling the identification of a target for rehabilitation.

Andrews and Bonta argue that rehabilitative programs should be evaluated by asking how many major risk factors and how many noncriminogenic needs the program addresses. They hold that addressing noncriminogenic needs decreases the effectiveness of rehabilitative programs. The standard means of achieving this level of specificity in programming is to rely on highly standardized, manualized, cognitive-behavioral therapies. **Criminogenic risks** and needs are often directly read from public (criminal) records or from standardized psychological inventories (e.g., various versions of the Level of Service Inventory). In addition, this approach explicitly links risks for antisocial behavior with targets for change. Most often such change is sought through explicitly targeted cognitive-behavioral therapies. In contrast, ICAP is open to a wider variety of rehabilitative strategies, particularly those attending to and repairing individually specific deficits or injuries.

PSYCHOLOGICAL EFFECTS OF INCARCERATION

Theories of rehabilitation must consider the psychological and ecological effects of incarceration as well as the risks and causes of antisocial behavior. Incarceration always eliminates or sharply curtails privacy and control of such elementary features of life as dress, diet, work, recreation, and socialization. Thus, it amplifies the stigma and loss of social status associated with conviction while concurrently promoting regression to dependent and passive, rather than proactive, status. Because prisoners must be aware of the threats posed by fellow inmates, their interpersonal relations are often hypervigilant and distrusting. Standard interpersonal demeanor is cautious and even suspicious. Inmates often adopt a "tough convict" veneer, intended to keep others at a distance. Over time, this results in varying levels of social withdrawal and isolation. Prisoners disconnect from others as individuals and often retreat deep into themselves. In some individuals this starts a cascade eventuating in a sense of powerlessness resembling clinical depression.

One challenge sure to face rehabilitative programs intended for implementation behind bars is that prisoners, or significant subgroups of individuals within prison, adopt the norms of the **prison code** or culture. Candid expressions of emotion are suppressed. Failure to exploit weakness is seen as a weakness. Because signs of weakness or vulnerability are typically exploited, the persona of hypermasculinity is a frequent option, as if the use of force to dominate others were essential for identity and self-respect. Such effects linger after release and complicate the efforts of rehabilitative programs expecting normal or near-normal levels of initiative and good judgment. Ex-convicts often exhibit the sort of impairments in decision making and sociality to be expected after a relatively lengthy period in the grip of extrinsically imposed, institutional constraints controlling almost all behavior.

The adequacy of a theory of rehabilitation has to be measured against the range of issues briefly reviewed above. Visualize individuals, behind bars or in the community, convicted of a felony and incarcerated for multiple years in a state prison. In addition to their personal baggage of antisocial potential and criminogenic risks and needs, many of these individuals have been further damaged by circumstances and practices of incarceration designed to protect public safety rather than to rehabilitate the offender. These persons seek assistance in grappling with the processes of reconciliation, rehabilitation, and restoration to a productive and law-abiding life in the community.

A PROGRAM FOR DECREASING ANTISOCIAL POTENTIAL AND INCREASING CAPACITY FOR WELLBEING BEHIND BARS

Working in two distinct environments (one behind bars, one in the community), the first author began by working with small groups of convicted male felons by first attempting to rebuild notions of friendship and group loyalty through fostering intragroup use of communication skills developed for professional and paraprofessional use by William Miller, who explicitly links his approach with the "stages of change" model of recovery from substance abuse developed by Prochaska and DiClemente.[4] The underlying goal for these small groups is to develop a dynamic culture in which each person's capacity for self-scrutiny is amplified by participation in processes in which each one helps another. Given prisoner and ex-con norms against self-disclosure, this is tricky business at best.

The program focuses on teaching prisoners to weigh the pros and cons of the problem behavior and its alternatives. In contrast, the first major moves in the cognitive-behavioral therapy of the Matrix Intensive Outpatient Treatment Program for Persons with Stimulant Use Disorders involve the identification and management of **triggers**, a step well into the action, maintenance, and relapse prevention stages of the Miller-Prochaska-DiClemente approach. The key difference in the two approaches is the much greater emphasis on patient and deliberate utilization of motivational enhancement in the latter. Such enhancement is most effective in a social setting that reinforces it (the chief benefit of 12-step fellowship). This supplies the rationale for improving interpersonal conversations about motivation, values, and goals by emphasizing the techniques of intragroup communication in therapeutic settings.

PERSONALITY: TEMPERAMENT AND CHARACTER

Personality comprises the set of internal processes of self-aware learning and proactive planning by which the person adapts to changing life experiences. The maturation and integration of human personality involves growth in self-awareness and self-control across a wide range of situations. Given a definition of human personality as the dynamic organization of the psychobiological systems by which a person shapes and adapts in a unique way to a changing internal and external environment, a clear account of these psychobiological systems would seem the best place to start in mapping a strategy for the reduction of cognitive antisocial potential.[5] The basic functions of personality are perceiving, feeling, and thinking and the integration of these processes into purposeful action. These functions emanate from the three central divisions of the functional architecture of personality: temperament, character, and psyche or spirit.

The four subdimensions of temperament—novelty seeking, harm avoidance, reward dependence, and persistence—are rooted in neuromuscular and neurovisceral systems whose activities begin in utero and underlie behaviors apparent at birth. These systems are gradually shaped by the learning experiences (associative learning and behavioral conditioning) of the preverbal child. The resulting temperamental dispositions directly modify the perceptual and appetitive/avoidant processes of the young and developing organism. Temperament modifies the salience (likelihood of drawing attention and focus) of adaptively significant stimuli as well as the rates at which organisms learn to map and locate such stimuli.

Parental responses to early childhood behaviors play a significant role in succeeding or failing to shape them to conform to prosocial norms. The more or less secure attachment relationships linking parents and child are of great importance in providing the "safe base from which to explore the environment" crucial for the development of adaptive prosocial skills. The four temperament traits are closely associated with the four basic emotions of fear (harm avoidance), anger (novelty seeking), attachment (reward dependence), and ambition (persistence).

Harm avoidance involves a heritable bias toward the inhibition of behavior in response to signals of punishment or frustrating nonreward. Some children exhibit a specific form of harm avoidance ("freezing") in response to signals that are "strange" or unrecognizable. Harm avoidance is observed as fear of uncertainty, shyness, social inhibition, passive avoidance of problems/danger, rapid fatigability, and pessimistic worry in anticipation of problems even in situations that do not worry other people. Adaptive advantages of high harm avoidance are cautiousness and careful planning when hazard is likely. The disadvantages occur when hazard is unlikely but still is anticipated, which leads to maladaptive inhibition and anxiety. People who score low in harm avoidance are carefree, courageous, energetic, outgoing, and optimistic even in situations that worry most people. The advantages of low harm avoidance are confidence in the face of danger and uncertainty, leading to optimistic and energetic efforts with little or no distress. The disadvantages are related to unresponsiveness to danger or unrealistic optimism with potentially severe consequences when hazard is likely.

Novelty seeking reflects a heritable bias in the initiation or activation of appetitive approach responding to novelty, to signals of reward, active avoidance of conditioned signals of punishment, and escape from unconditioned punishment (all of which are hypothesized to covary as part of one heritable system of learning). Novelty seeking is observed as exploratory activity in response to the unfamiliar, impulsiveness or extravagance in approach to cues of reward, and active avoidance of frustration. Individuals high in novelty seeking are quick tempered, curious, easily bored, impulsive, extravagant, and disorderly. Adaptive advantages of high novelty seeking are enthusiastic exploration of new and unfamiliar stimuli, potentially leading to originality, discoveries, and reward. The disadvantages are frequent and easy boredom, impulsivity, angry outbursts, potential fickleness in relationships, and impressionism in efforts. Persons who score low in novelty seeking are slow tempered, not inquisitive, stoical, reflective, frugal, reserved, tolerant of monotony, and orderly. Their reflectiveness, resilience, systematic effort, and meticulous approach are clearly advantageous when these features are adaptively needed. The disadvantages reflect tolerance of monotony and lack of enthusiasm, potentially leading to prosaic routinization of activities. The initiation and frequency of hyperactivity, binge eating, sexual hedonism, drinking, smoking, and other substance abuse, especially stimulants, are each associated with high scores in novelty seeking.

Reward dependence reflects a heritable bias in the maintenance of behavior in response to cues of social reward. Reward dependence is characterized by sentimentality, social sensitivity, attachment, and dependence on approval by others. Individuals high in reward dependence are tender hearted, sensitive, dedicated, dependent, and sociable. One of the major adaptive advantages of high reward dependence is the sensitivity to social cues facilitating affectionate social relations and genuine care for others. The disadvantages are related to suggestibility and loss of objectivity frequently encountered with people who are excessively socially dependent. Individuals who score low in reward dependence are practical, tough minded, cold, socially insensitive, and indifferent if alone.

The advantages of low reward dependence are personal independence and objectivity not corrupted by efforts to please others. Its adaptive disadvantage is related to social withdrawal, detachment, and coldness in social attitudes.

Persistence reflects a heritable bias in the maintenance of behavior despite frustration, fatigue, and intermittent reinforcement. It is observed as industriousness, determination, ambitiousness, and perfectionism. Highly persistent people are hard working, perseverative, and ambitious overachievers who tend to intensify their effort in response to anticipated reward and perceive frustration and fatigue as a personal challenge. High persistence is an adaptive behavioral strategy when rewards are intermittent but contingencies remain stable. When the contingencies change rapidly, perseveration becomes maladaptive. Individuals low in persistence are indolent, inactive, unstable, and erratic; they tend to give up easily when faced with frustration, rarely strive for higher accomplishments, and manifest a low level of perseverance even in response to intermittent reward. Accordingly, low persistence is an adaptive strategy when reward contingencies change rapidly and may be maladaptive when rewards are infrequent but occur in the long run.

Temperament involves a relatively small set of emotions associated with one's basic needs (e.g., safety; so-called primary motives). Fear and anger are motivationally monopolistic and take over the personality by altering perception, learning, and behavior in a biased way. However, under normal circumstances, after survival needs are met, the goals of normally developing personality can change to include the integrity of both the physical and the mental self. Normal personality development also adapts to numerous social goals (e.g., education, occupation, family) and a rich spectrum of secondary (social) emotions (such as shame and compassion). These "secondary," "social," or "growth" motives are closely functionally related to character development. Specifically, basic emotions of fear, anger, and excitement are transformed into more complex secondary emotions, such as carefulness, assertiveness, and joy, through interaction with increasingly more complex internalized concepts. The secondary emotions take over as prime movers of further character development and maturation. They are not as monopolistic as the basic emotions and thus motivate development of more flexible and adaptable personality traits. Abnormal (deviant, immature) motivation derives from two or three dominant, monopolistic, elementary emotional needs associated with survival. In contrast, mature motivation develops after basic needs are met and the person is freed to experience numerous secondary motives for growth in character and in self-awareness.

Character refers to the mind—that is, the conceptual core of personality. Character involves higher cognitive functions, including abstraction, symbolic interpretation, and reasoning. It involves individual differences in self-concepts and object relations that reflect personal goals and values. In other words, character is what we make of ourselves intentionally. Character is rational and volitional. Whereas temperament involves basic emotions like fear and anger, character involves secondary emotions like purposeful moderation, empathy, and patience, and, in even more mature individuals, hope, love, and faith. As a result, character can be described as our mental self-government, which involves executive, legislative, and judicial functions. The higher functions associated with symbolic memory interact with temperament through cognitive processing of emotionally hidden sensory percepts. This temperament–character interaction leads to the development of mature, realistic internalized concepts about the self and the external world. The executive, legislative, and judicial functions of mental self-government can be measured as three distinct character traits: self-directedness, cooperativeness, and self-transcendence, respectively. High values for these character traits signal an adaptive or "mature" character, whereas low self-directedness and cooperativeness lead to better adaptation in a restricted, basically unhealthy range of situations.

Self-directedness quantifies differences in the executive competence of individuals. A highly self-directed person is self-sufficient, responsible, reliable, resourceful, goal oriented, and self-accepting. The most advantageous summary feature of self-directed individuals is that they are realistic and effective—that is, they are able to adapt their behavior in accord with individually chosen, voluntary goals. Individuals who score low in self-directedness are blaming, helpless, irresponsible, unreliable, reactive, and unable to define, set, and pursue meaningful internal goals. Such poor executive function, manifest as unrealistic expectations of behavioral results and lack of internal guidance, is rarely advantageous to the individual. It is even less likely to result in a feeling of wellbeing or resilient serenity and self-acceptance.

Cooperativeness quantifies differences in the legislative functions of individuals. Highly cooperative people are empathetic, tolerant, compassionate, supportive, and principled. These features are advantageous in teamwork and social groups, but not for individuals who must live in a solitary manner. People who score low in cooperativeness are self-absorbed, intolerant, critical, unhelpful, revengeful, and opportunistic. They primarily look out for themselves and tend to be inconsiderate of other peoples' rights or feelings.

Self-transcendence quantifies individual differences in the judicial functions of people. Self-transcendent individuals are described as judicious, insightful, spiritual, unpretentious, and humble. These traits are adaptively advantageous when people are confronted with the suffering, illness, or death, inevitable with advancing age. They may appear disadvantageous in most modern societies where idealism, modesty, and meditative search for meaning can interfere with the acquisition of wealth and power. People who score low in self-transcendence tend to be pragmatic, objective, materialistic, controlling, and pretentious. Such individuals appear to fit in well in most Western societies because of their rational objectivity and materialistic success. However, they consistently have difficulty accepting suffering, failures, personal and material losses, and death, which leads to lack of serenity and adjustment problems, particularly with advancing age, poverty, or harsh life circumstances, including insecure attachment relations. People who score low in self-transcendence live in the material world, skeptical of whatever they cannot prove objectively and use practically. In contrast, for highly self-transcendent individuals the meaning of life goes beyond material things and includes intuitive awareness of what is beautiful, true, and good, to which materialists may be insensitive.

Finally, character matures in a stepwise manner in incremental shifts from infancy through late adulthood. The timing and rate of transitions between levels of maturity are nonlinear functions of antecedent temperament configurations, systematic cultural biases, and experiences unique to each individual, which depend on individual differences in episodic memory or intuitive self-awareness that enable human beings to remember past experiences. These constructs are measured with the Temperament and Character Inventory, which is an inventory covering the seven factors of Cloninger's theory of personality. The self-report version of the Temperament and Character Inventory is a 240-item test available in both true/false and 5-point Likert formats. Its psychometric properties are presented in the Temperament and Character Inventory manual.[6]

APPLYING TEMPERAMENT AND CHARACTER TO REHABILITATION

Full assessment of personality requires consideration of a person's level of self-awareness and wellbeing, not just his or her impairments. Health and wellbeing are more than the absence of deviant traits. Wellbeing depends on a person's level of self-awareness and leads to the expression of human virtues and positive emotions that go beyond what is average in contemporary society. It is not unlikely, given the many obstacles associated with ex-convicts' reentry to society, that true rehabilitation must aim at facilitating the development of above-average character traits. There are three major stages of self-awareness along the path to wellbeing. The absence of self-awareness occurs in severe personality disorders and psychoses in which there is little or no insightful awareness of the preverbal outlook or beliefs and interpretations that automatically lead to emotional drives and actions. Lacking self-awareness, people act on their immediate likes and dislikes, which is usually described as an immature or "childlike" ego state.

The first stage of self-awareness is typical of most adults most of the time. Ordinary adult cognition involves a capacity to delay gratification to attain personal goals but remains egocentric and defensive. Ordinary adult cognition is associated with frequent distress when attachments and desires are frustrated. Hence, the average person can function well under good conditions but may frequently experience problems under stress. At this stage of self-awareness a person is able to make a choice to relax and let go of his or her negative emotions, thereby setting the stage for acceptance of reality and movement to higher stages of coherent understanding.

The second stage of self-aware consciousness is typical of adults when they operate like a "good parent." Good parents are "other centered" and capable of calmly considering the perspective and needs of their children and other people. This state is experienced when a person is able to observe his or her own subconscious thoughts and consider the thought processes of others in a similar way. Hence, the second stage is described as "meta-cognitive" awareness, or mindfulness. The ability of the mind to observe itself allows for more flexibility in action by reducing dichotomous thinking. At this stage a person is able to observe oneself and others without judging or blaming. However, in a mindful state people still experience the emotions that emerge from a dualistic perspective, so mindfulness is only moderately effective in improving wellbeing.

The third stage of self-awareness is called contemplation because it is direct perception of one's initial perspective—that is, the preverbal outlook or schemas that direct one's attention and provide the frame that organizes expectations, attitudes, and interpretation of events. Direct awareness of this outlook allows the enlarging of consciousness by accessing previously unconscious material, thereby letting go of wishful thinking and the impartial questioning of basic assumptions and core beliefs about life, such as "I am helpless," "I am unlovable," or "faith is an illusion." The third stage of self-awareness can be described as contemplative because in this state a person becomes aware of deep preverbal feelings that emerge spontaneously from a single perspective, such as hope, compassion, and reverence. Contemplation is much more powerful in transforming personality than is mindfulness, which often fails to reduce feelings of hopelessness.

The clinical utility of this property is that therapists can teach people to exercise their capacity for self-awareness, moving through each of the stages of awareness just described. Their ability to do so, and the difficulties they have, reveals the way they are able to face challenges in life. Cloninger developed an exercise, called the "silence of the mind" meditation, with explicit instructions to take people through each stage of awareness. The first phase of this meditation results in a relaxed state in the first stage of self-awareness. The second phase facilitates entry into the second stage of self-awareness, and the third phase into the third stage of self-awareness, if the person is able to do so. Each step of growth in self-awareness and wellbeing requires a person to accept the limitations of the assumptions he or she has been making about what is satisfying and valuable to begin adapting in ways that are more coherent and flexible. However, such growth requires a person to let go of what is familiar so there is a temporary increase in negative emotions before there is further growth in positive emotions and other aspects of wellbeing.

Therefore, an effective practitioner must provide appropriate conditions to facilitate this transition ("the energy of activation") with their own hope, patience, and awareness, realizing that the patient must ultimately be the one who does the work to grow in self-understanding. Each step along the pathway to wellbeing requires different procedures to facilitate growth because different aspects of the being are involved in the different steps in character development. The specific techniques for the various steps of development are described next.

Major Stages of Rehabilitation

The initial stage of rehabilitation deals with the stressors that make it difficult for the person to get into a calm enough state to develop a working alliance with the care provider. The second stage involves elevating a person's outlook on life so that the person can experience things he or she enjoys and values under relaxed conditions. This involves a spiritual awakening that has often been neglected in strictly cognitive-behavioral or psychodynamic approaches, but without which there is little capacity for fundamental change in the quality of life. The third stage of other-centered awareness involves increases in self-awareness and capacity for contemplation that elevate a person's usual thoughts, feelings, and relationships in a wide range of conditions. The fourth stage of integration of reason and love in action allows a person to be mature and happy even under conditions that were previously stressful.

A useful initial approach is to focus on healthy lifestyle choices. This provides a nonthreatening basis for evaluation of a patient's goals, values, habits, and skills (that is, their personality). Choices about diet, weight control, exercise, smoking, drinking, and ways of relaxing and managing stress are appropriate for discussion with a caregiver and do not threaten or stigmatize group participants. Goal setting and problem solving lead to the ability to admit faults and to recognize one's strengths and limitations. Respect for one's accomplishments requires acceptance of responsibility and leads to trust of others. Self-respect and respect for others usually progress hand in hand.

Neither medications nor cognitive-behavioral approaches, alone or in combination, are usually adequate to transform a person's personality in a fundamental way. Individuals who radically change their perspective on life usually attribute the change to getting a good job that provides a sense of self-respect, marrying a loving and trusted person, or experiencing a religious conversion. These kinds of life experiences change a person's initial perspective on life, which in turn transform their thoughts, feelings, and behavior. Cognitive-behavioral and psychodynamic therapies often leave patients in a tense inner struggle with themselves unless treatment provides experiences that allow a reevaluation of basic assumptions about life. Such persons cannot transcend the conflicts among their emotional drives and remain in constantly recurring struggles among parts of their personalities. Growth in self-awareness is only possible once a person is calm enough to face and accept unpleasant realities about him- or herself. As a result, therapy usually proceeds by a series of small steps that lead to progressively more mature and integrated ways of viewing and coping with the challenges of life. Figuratively speaking, a very hungry person cannot appreciate beauty and wonder while struggling to gratify basic needs for safety and food. However, it is difficult to specify the adequate conditions for the progress of any unique individual because either internal or external resources can satisfy a person's needs. In fact, internal psychological resources, rather than external material and social supports, satisfy a person's needs to an increasing degree as he or she matures.

The second stage of rehabilitation involves the awakening of the positive outlooks on life that are needed for wellbeing. The basic principles of wellbeing focus on therapeutic experiences and activities for participants designed to help them value the dignity of their life and that of others as human beings as a result of self-awareness of each person's body, mind, and psyche.

Psychiatry literally means the "healing of the psyche," and the psyche refers to a creative, self-aware consciousness that is more than analytical reasoning. Rehabilitation most consistently involves a transformation of a person's spirit or basic perspective about life. However, psychiatrists often become narrowly focused on reduction of symptoms or risk of harm and forget the importance of helping the patient maintain and expand self-respect and hope. So often the insight that psychiatry is the "healing of the psyche" has been neglected or denied as a result of the errors of materialistic and reductionistic thinking. What is meant here by an awakening and healing of the psyche is that the patient becomes directly aware that his or her basic outlook on life has an impact on thoughts, emotions, and actions. One's initial outlook on life can lead to increments in antisocial potential. Rehabilitation requires awareness of the assumptions implicit in initial perspectives about the nature of reality, about the nature of our knowledge, and about our basic moral principles. Without some degree of awareness of the consequences of this initial perspective, it is not possible to transcend the conflicts and contradictions in thoughts and emotions that arise from it. As a result, rehabilitative strategies that neglect the stage of spiritual awakening inevitably leave participants locked in an inner struggle among parts of themselves from which there is no escape.

Once there is an awakening of self-awareness at a meta-cognitive level, participants can proceed to the advanced stages of treatment. Once behavioral stability and self-awareness are achieved, it is possible to focus on the causes of the symptoms that were targets of initial intervention during the stabilization phase. The therapeutic strategies for addressing each of these sets of causes include psychoeducation, physical and other nonverbal therapies, emotional skills training, cognitive skills training, and spiritual exercises. As a result, no single form of therapy (e.g., behavioral, cognitive, interpersonal, or psychodynamic) is really comprehensive, even with addition of modules from positive psychology or mindfulness training.

In essence, the extremes of each temperament are transcended by the development of particular forms of spiritually elevated thoughts—namely self-respect reconciles the extremes of harm avoidance, self-mastery reconciles the extremes of novelty seeking, secure attachments reconcile the extremes of reward dependence, and the cultivation of virtues and transcendent meaning reconciles the extremes of persistence and the limits of the finite human intellect. Both extremes of each temperament have advantages and disadvantages, and transcendence of the underlying conflict resulting from their disadvantages allows a person to live without tension or conflict about these issues as a result of a more holistic initial perspective.

A major practical advantage of therapy modules that focus on the reconciliation of both extremes of each temperament is that heterogeneous groups of participants with widely different personality profiles can be treated together. The focus is on transcending emotional conflicts, coherent character development, and wellbeing for everyone, not particular personality subtypes. The stigma linked to status as a convicted felon is mitigated by facing the facts that everyone is imperfect but can learn to live without fear along the path to wellbeing. The design of therapies has often failed to recognize the path of development of character and wellbeing and the crucial role of spirituality in transcending and sublimating emotional conflicts.

CONCLUSION

There are striking parallels between Farrington's life-course model of antisocial potential and the coherence therapy described here. Each is distinctive in its attention to the biological roots of primary or basic emotions and the role that can be played by such emotion in the generation of impulsive, nonreflective behavior. The seven-factor model described here also spells out the role of the rational mind in generating more complex secondary emotions (purposeful moderation, empathy, patience, shame, compassion, care and caution, assertiveness, joy) necessary for mature control of more basic impulses and for the virtues necessary for development of the characteristics of self-directedness, cooperativeness, and self-transcendence. What is most distinctive about the model described here, however, is its return to the platonic emphasis on the triune nature of the human person: body, mind, and psyche.

An adequate theory of rehabilitation on this account must pay sufficient attention to the role of personal or narrative identity in ways that neither behaviorist nor radically utilitarian ("what works") approaches are likely to do. It is hard to argue with the proposition that individuals are more likely to respond to treatment if they expect to have a better life as a result of the intervention. An adequate theory of rehabilitation combines moral or political values with scientific research. It must be a hybrid theory linking accounts of the causes of crime and desistance from crime with therapeutic practices designed to protect the community from harm and to restore ex-offenders to good lives. Cloninger's theory of personality is promising. The theory of temperament, for example, has the deepest roots currently possible, with well established links to molecular neurobiology and the functional architecture of the human brain. Cloninger's views are also closely aligned with those of other leading biopsychologists, including Jerome Kagan[7] and Eric Kandel.[8] The relevance and utility of Cloninger's work for the rehabilitation of criminal offenders can be expanded by combining it with insights relating to motivational interviewing and stages of change models of recovery. The increased personal maturity resulting from his model of rehabilitation can also be mobilized to enhance the effectiveness of such otherwise disparate approaches to recovery from substance abuse and to such special problems as anger management, conflict resolution, coping with stress, and depressed and/or manic moods.

GLOSSARY

Criminogenic risks—factors that when present increase the odds of criminal behavior

Personality—set of internal processes of self-aware learning and proactive planning by which the person adapts to changing life experiences

Prison code—a set of unwritten rules that guide prisoner conduct

Rehabilitation—programs that are designed to eliminate offending behaviors among criminals

Triggers—cues associated with relapse

NOTES

1. Cloninger, C. R. (2004). *Feeling good: The science of well-being*. New York: Oxford University Press.
2. Farrington, D. P. (2005). The integrated cognitive antisocial potential (ICAP) theory. In D. P. Farrington (Ed.), *Integrated developmental and life course theories of offending: Advances in criminological theory* (Vol. 14).

New Brunswick, NJ: Transaction; Farrington, D. P. (2003). Developmental and life-course criminology: Key theoretical and empirical issues: The 2002 Sutherland Award Address. *Criminology, 41,* 221–255, p. 231.

3. Andrews, D. A., & Bonta, J. (2006). *Psychology of criminal conduct* (4th ed., pp. 67–68). Cincinnati, OH: Lexis-Nexus/Anderson.

4. DiClemente, C. C. (2003). *Addiction and change: How addictions develop and addicted people recover.* New York: The Guilford Press; Miller, W. R. *TIP #35: Enhancing motivation for change in substance abuse treatment.* Substance Abuse and Mental Health Services Administration (SAMHSA), 1999–2004. Rockville, MD: SAMHSA; Miller, W. R., & Rollnick, S. (2002). *Motivational interviewing: Preparing people for change* (2nd ed.). New York: The Guilford Press; Prochaska, J. O., & DiClemente, C. C. (1983). Stages and processes of self-change of smoking: Toward an integrative model of change. *Journal of Consulting and Clinical Psychology, 51,* 390–395.

5. Cloninger, C. R. (2002). Antisocial personality disorder: A review. In M. Gaebel (Ed.), *Psychiatric diagnosis and classification* (pp. 79–106). New York: John Wiley & Sons.

6. Cloninger, C. R. (1994). *The Temperament and Character Inventory (TCI): A guide to its development and use.* St. Louis, MO: Center for Psychobiology of Personality, Washington University.

7. Kagan, J. (2007). *What is emotion? History, measures, and meanings.* New Haven, CT: Yale University Press.

8. Kandel, E. (2005). *Psychiatry, psychoanalysis and the new biology of mind.* Washington, DC: American Psychiatric Publishing.

CHAPTER TWENTY-ONE

Never-Desisters: A Descriptive Study of the Life-Course-Persistent Offender

Matt DeLisi, Anna E. Kosloski, Alan J. Drury, Michael G. Vaughn, Kevin M. Beaver, Chad R. Trulson, and John Paul Wright

KEY TERMS

>   Chronic offenders
>   Desistance
>   Official records
>   Self-reports
>   Trajectories

INTRODUCTION

A perennial debate in criminal career research centers on the pliability of the offending careers of serious criminals over time. As noted by Sampson and Laub, "Sharply divergent portraits of the developmental course of crime characterize the current scene."[1] From one view—broadly known as the propensity perspective—serious offenders evince lifelong antisocial careers generally including conduct disorder and other problem behaviors in childhood, increasingly serious delinquency during adolescence, and versatile criminal behaviors throughout adulthood. For this categorization of offender, criminal behavior is high in rate, severe, and enduring. Over time their criminal careers continue far longer than other offenders', and although serious offenders do desist from crime, their **desistance** is delayed.[2] From another view—broadly known as the developmental perspective—there is similar recognition of the antisocial propensities and behaviors of the most serious offenders. But there is greater emphasis on the salience of normal social processes and the informal social controls that spring from them. Marriage, employment, and other investments in conventional institutions serve to mediate or moderate antisocial tendencies and hasten desistance from crime. From this perspective all offenders—even the most recalcitrant and nefarious—desist.[3]

In recent years two influential theories about the development of offending patterns over time have shaped the debate on criminal careers research. For the propensity perspective, Moffitt's life-course-persistent (LCP) offender has become a dominant prototype of an offender whose criminal career is severe and persistent. For the developmental perspective, Sampson and Laub's age-graded theory has convincingly shown that desistance characterizes all offenders and their criminal careers. Specifically, Sampson and Laub have been critical of typological approaches in criminology, particularly Moffitt's LCP prototype. The current study seeks to contribute to this debate.

LIFE-COURSE DESISTERS

Using the longest longitudinal study in criminology containing data on criminal activity from 7 to 70 years of age, Laub and Sampson produced findings that were critical of the idea of an LCP offender. They found that crime declines even for the most active offenders, and trajectories of desistance could not be prospectively identified based on typological accounts rooted in childhood and individual differences. Even when they controlled for death and incarceration—conditions that would disproportionately affect the most high-rate offenders—the results held. This means that even at the extremes of offending among offenders, who through the years have been referred to as chronic offenders, habitual criminals, career criminals, or LCP offenders, desistance happens. According to Laub and Sampson:

> It is difficult to reconcile these findings with the theoretical idea of a life-course-persistent group, which suggests that criminological terminology, if not typological theory, is in need of overhaul. We offer the concept of life-course desisters as a cornerstone for this effort, accounting for the apparent fact that all offenders desist but at time-varying points across the life course.[4]

Since the late 1990s, many studies have been published illustrating the various **trajectories** of relatively distinct groups of offenders. These groups have been referred to by a variety of names, including very low-level offenders, slow-uptake **chronic offenders**, low-level offenders, late starters, intermittent offenders, high-level chronic offenders, bell-shaped desisters, and many others.[5] Irrespective of their labels or data source, research indicates that all offenders desist from crime as they age. This retorts the strict interpretation of an offender whose criminal career is lifelong and persistent.

More recently, Moffitt defended her conceptualization of the LCP offender and criticized Sampson and Laub for making "straw man" predictions about the viability of an offender that demonstrates lifelong antisocial behavior. Moffitt notes that 84 percent of the men from the Gluecks' data were arrested between the ages of 17 and 24. Afterward, sizable proportions of offenders continued to accumulate arrests as 44 percent were arrested in their 40s, 23 percent in their 50s, and 12 percent in their 60s. In addition, the average criminal career length was nearly 26 years, which according to Moffitt is consistent with the notion of the LCP offender. In addition, Moffitt indicates that her theory never asserted that the LCP offender would be homogeneous in offending over time. In her words:

> Males who spent their youth and early adulthood on the life-course-persistent pathway can show no variation in subsequent offending during midlife and aging . . . . Such uniformity is implausible, and the taxonomic theory did not make such a prediction . . . . Life-course-persistent delinquents do not have to be arrested for illegal crimes steadily up to age 70, but they do have to maintain a constellation of antisocial attitudes, values, and proclivities that affect their behavior towards others.[6]

Recent work by Laub and Sampson and Moffitt clarified the offending patterns of the LCP offender that perhaps were not clear in the original taxonomy. First, the LCP moniker is a general label or categorization that typifies persons whose behavioral repertoires are rife with antisocial and related imprudent behaviors across the life course. Second, there is heterogeneity within the LCP group. Third, the label "life-course persistent" has been caricaturized and erroneously falsified—Moffitt did not literally mean that serious criminals would be as dangerous and recidivistic at age 70 as they were at age 20. Fourth, all offenders, including the most pathological, desist significantly over time. Fifth, research is mixed about the utility of childhood factors as predictive of later antisocial behavior. On one hand, Laub and Sampson assert that while "childhood prognoses are reasonably accurate in terms of predicting levels of crime between individuals, they do not yield distinct groupings that are valid prospectively in a straightforward test."[7] On the other hand, research indicates that childhood factors profoundly predict LCP status. For instance, maltreatment in childhood (odds ratio = 14.5), diagnosis of attention-deficit/hyperactivity disorder (odds ratio = 18.7), and childhood low IQ (odds ratio = 5.8) predicted LCP membership using the Dunedin data.[8] Thus, although some facets of the taxonomy remain controversial, studies have clarified Moffitt's hypothesis of the LCP offender, including Moffitt herself, who has questioned whether the theory needs refinement.[9]

CURRENT FOCUS

This chapter seeks to further understand the LCP offender by using two correctional samples of offenders. One group is a population of offenders delineated by a judicial district as habitual criminals who were barred from consideration for personal recognizance bonds and eligible for sentencing enhancements if applicable. Eligibility for this group was 30 career arrests—thus the study group represents a practitioner-identified group mentioned by Moffitt. The second group is a simple random sample of 500 offenders selected from the same universe of offenders during the same time period, 1995–2000. The control sample is itself a pathological offending group with an average career arrest total of six. Criminal career data span childhood until age 59. The goal of this comparison is to evaluate the viability of the LCP offender using groups other than archival Gluecks' data and selected birth cohorts.

METHODOLOGY

Data are derived from pretrial service or bond interviews with jailed criminal defendants processed at a large urban jail in the western United States between 1995 and 2000. During this time, the pretrial services unit processed nearly 26,000 arrestees and gathered information on their social and criminal history for bond setting and to alleviate jail crowding. Bond interviews were sworn legal proceedings, and all self-reported criminal history was supplemented with local records and rap sheets from the National Crime Information Center, which is the most comprehensive national computerized database of criminal defendants. The combined use of self-reported and official data bolsters the concurrent validity of these data.

Records checks isolated 500 offenders (less than 2 percent) that qualified for habitual offender status. The pretrial services supervisor, along with representatives from the local district attorney's office and chief district judge's office, chose 30 career arrests as the criterion for being a habitual offender. These offenders were ineligible for personal recognizance bonds and eligible for sentencing enhancements based on their conviction history. A comparison sample of 500 randomly selected offenders was also chosen to facilitate analyses of normative and pathological offenders.[10]

Arrests, convictions, and prison sentences for all Part I and Part II index offenses were tabulated. Offending data were collected across seven life stages beginning with: (1) childhood and adolescence (starting with the earliest arrest onset at ages 7 to 17), (2) ages 18 to 24, (3) ages 25 to 31, (4) ages 32 to 38, (5) ages 39 to 45, (6) ages 46 to 52, and (7) ages 53 to 59. The 7-year intervals were selected for consistency and measure 5 decades of offending history.

The research aim is entirely descriptive. Difference of means t-tests were conducted for criminal career parameters, including career arrests; career span; total felony convictions; total prison sentences; arrest onset; career arrests for murder, rape, robbery, kidnapping, violent, and property Part I index crimes; number of states in which the offender was arrested; juvenile confinements; juvenile felony adjudications; career noncompliance (probation and parole violations); use of aliases; and age. Two-by-two contingency tables with chi-square were used for proportion of sample with 50+ arrests and gender (Table 21-1). Difference of means t-tests were also conducted for arrests, felony convictions, and prison sentences across the seven life stages. Mean levels of offending by sample were also compared as a difference of magnitude measure (Tables 21-2, 21-3, and 21-4). Finally, arrests, prison sentences, and involvement in murder, rape, robbery, and kidnapping across seven life stages are shown graphically (Figures 21-1 through 21-6) to visually present group differences in offending.

RESULTS

As shown in **Table 21-1**, both offender groups demonstrated high-rate involvement in antisocial behavior. The random sample averaged more than six career arrests, which exceeds the traditional

TABLE 21-1 Sample Comparisons for Criminal Career Parameters and Demographics

Variable	Random Sample	LCP Sample	t-value/χ^2
Career arrests	6.40	59.76	37.37*
50+ arrests	.006	.51	331.72*
Career span	6.95	20.97	27.28*
Career felony convictions	.44	5.60	21.78*
Career prison	.28	3.48	15.98*
Arrest onset	26.32	18.64	15.47*
Career murder	.01	.12	5.91*
Career kidnapping	.02	.11	4.46*
Career rape	.05	.40	4.50*
Career robbery	.03	.81	9.56*
Career violent index	.28	3.64	16.54*
Career property index	.95	13.19	20.36*
States arrested in	1.23	2.90	12.82*
Juvenile confinement	.02	.18	4.76*
Juvenile felony adjudication	.04	.59	7.45*
Career CJS noncompliance	.27	3.86	16.47*
Career aliases	.88	7.71	16.13*
Age	33.28	39.62	9.62*
Gender	73% male	89% male	41.59*

*$p < .0001$

standard of five arrests for habitual or chronic offending.[11] Fourteen percent of the random sample were convicted of a felony over their criminal career, and less than 9 percent were sentenced to state or federal prison. On average, offenders from the random sample were first arrested at age 26, and their criminal career lasted nearly 7 years. Three of the 500 randomly selected offenders were arrested 50 or more times over their offending careers.

In contrast, the LCP sample averaged nearly 60 career arrests, and more than half of the sample had more than 50 career arrests. The average arrest onset was 18.64 years, and the average criminal career span was nearly 21 years. Nearly 20 percent were adjudicated delinquent for felony offenses, and 11 percent were committed to confinement facilities as juveniles. On every measure—spanning career arrests for murder, rape, kidnapping, and robbery; career index arrests for violent and property crimes; the number of states where offenders were arrested; juvenile involvement; noncompliance with criminal justice system sentences; and use of aliases—the LCP sample was significantly worse. These offending differences are shown graphically in **Figure 21-1**.

Arrest activity across seven life stages appears in **Table 21-2**. Arrest activity for the random and LCP samples were examined during childhood and adolescence and six 7-year age intervals through adulthood (ages 18–24, 25–31, 32–38, 39–45, 46–52, and 53–59). At all life stages LCP offenders accumulated significantly more arrests with t-values ranging from $t = 5.04$ (between ages 53 and 59) to $t = 20.04$ (between ages 25 and 31). The differences in magnitude between mean offending totals per age range are dramatic. During childhood and adolescence, LCP offenders had arrest totals that were 16 times higher than random offenders'—who again were chronic offenders themselves. The arrest differentials during adulthood were as follows: 5.4 (ages 18–24), 7.3 (ages 25–31), 8.4 (ages 32–38), 14.6 (ages 39–45), 18.1 (ages 46–52), and 14.2 (ages 53–59). LCP offenders who were

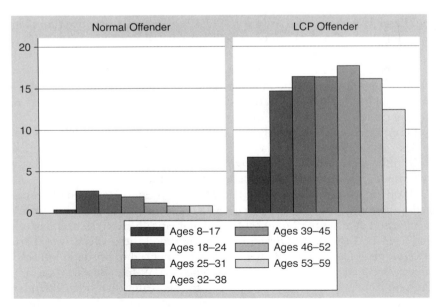

FIGURE 21-1. Total Arrests Across Seven Life Stages by Offender Group

TABLE 21-2 Difference of Means for Arrests by Life Stage and Offender Group

Age Range	Sample	n	Mean	SE	95% CI	t-value	Difference in Magnitude
8–17	Random sample	500	.42	.12	.19–.65	9.34*	16.0
	LCP sample	500	6.72	.66	5.41–8.02		
18–24	Random sample	500	2.72	.24	2.24–3.19	19.42*	5.4
	LCP sample	499	14.67	.57	13.56–15.78		
25–31	Random sample	394	2.26	.23	1.80–2.71	20.04*	7.3
	LCP sample	437	16.41	.64	15.16–17.66		
32–38	Random sample	247	1.96	.24	1.48–2.44	15.10*	8.4
	LCP sample	354	16.37	.78	14.84–17.90		
39–45	Random sample	128	1.21	.21	.80–1.62	10.72*	14.6
	LCP sample	220	17.66	1.16	15.37–19.95		
46–52	Random sample	64	.89	.18	.53–1.25	6.11*	18.1
	LCP sample	104	16.11	1.95	12.25–19.98		
53–59	Random sample	23	.87	.46	−.06–1.79	5.04*	14.2
	LCP sample	33	12.39	1.88	8.57–16.22		

Note: CI = confidence interval, SE = standard error.

*$p < .0001$

between ages 53 and 59 had arrest totals (mean = 12.39) that were nearly five times higher than the random sample during the crime-prone years between ages 18 and 24 (mean = 2.72).

Similar effects emerged for felony convictions across seven life stages for both samples of offenders. As shown in **Table 21-3**, the differences in mean levels of felony convictions are significant for all age ranges. During childhood and adolescence, LCP offenders were convicted of felonies at a level nearly 15 times higher than the random sample of offenders—again, a sample that is a chronic-offending group. The felony conviction differential hovers between 9 and 11 between ages 18 and 38 and then increases dramatically. For offenders ages 39 to 45, LCP offenders were convicted of felonies 25-fold higher than other offenders. Between ages 46 and 52 there was a 34-fold difference. LCP offenders accumulated felony convictions during ages 53 to 59 that exceeded the values of other offenders during adolescence and early adulthood.

Just as LCP offenders were convicted of felonies at high rates across the life course, so too are they sentenced to prison (**Table 21-4**). For all seven life stages, LCP offenders were sentenced to state and federal prison at levels that significantly and dramatically exceeded those of other offenders (*t*-values range from 1.93 to 10.81). During middle adulthood these effects were particularly pronounced. Between ages 39 and 45, LCP offenders were sentenced to prison at levels 43 times those of other offenders. Between ages 46 and 52, the difference was 39-fold. Between ages 53 and 59, LCP offenders were sent to prison at levels that were five times greater than other offenders during the ages 18 to 24. These differences in prison sentences are shown graphically in **Figure 21-2**.

TABLE 21-3 Difference of Means for Felony Convictions by Life Stage and Offender Group

Age Range	Sample	n	Mean	SE	95% CI	t-value	Difference in Magnitude
8–17	Random sample	500	.04	.02	.01–.08	7.45*	14.8
	LCP sample	500	.59	.07	.45–.73		
18–24	Random sample	500	.18	.03	.11–.24	14.46*	9.7
	LCP sample	498	14.67	.57	13.56–15.78		
25–31	Random sample	394	.18	.04	.10–.26	12.92*	9.3
	LCP sample	438	1.67	.10	1.47–1.87		
32–38	Random sample	247	.12	.03	.06–.19	9.31*	11.3
	LCP sample	355	1.36	.11	1.14–1.57		
39–45	Random sample	128	.05	.02	.00–.09	7.68*	24.6
	LCP sample	221	1.23	.12	1.00–1.46		
46–52	Random sample	64	.03	.02	−.01–.08	4.58*	34.0
	LCP sample	104	1.02	.17	.69–1.35		
53–59	Random sample	23	0	0	0	2.26**	—
	LCP sample	33	.67	.24	.17–1.17		

Note: CI = confidence interval, SE = standard error.

$*p < .0001, **p = .0275$

The final component of the analyses compares arrests for serious violent crimes across the life course by offender group. As shown in **Figure 21-3**, at every life stage with the exception of ages 53 to 59 when both groups had zero prevalence, LCP offenders totaled more arrests for murder on average than other offenders. At no point did offenders from the random sample total more murder arrests than the LCP group. This means that LCP offenders in their 40s and 50s were more involved in murder than other offenders in their teens and 20s. The same can be said for rape (**Figure 21-4**) and robbery (**Figure 21-5**). Offenders from the random sample between ages 18 and 24 averaged more kidnapping arrests than LCP offenders did during childhood and adolescence. Otherwise, at all life stages, LCP offenders totaled more arrests for kidnapping (**Figure 21-6**).

Consistent with Sampson and Laub's life-course desisters hypothesis, offenders in the random sample—again a high-risk, chronic-offending group—had zero prevalence for murder, rape, robbery, and kidnapping at multiple life stages. Even those in the LCP sample had zero prevalence for murder between ages 53 and 59 and kidnapping between ages 53 and 59. On the other hand, LCP offenders were arrested for rape at the highest levels between ages 53 and 59 and their arrest levels for robbery during their late 50s were comparable with robbery levels in preceding decades.

As shown in Table 21-1, LCP offenders averaged nearly 4 arrests for violent index arrests (murder, rape, robbery, aggravated assault) and more than 13 arrests for property index arrests (burglary, larceny, motor vehicle theft, arson).

TABLE 21-4 Difference of Means for Prison Sentences by Life Stage and Offender Group

Age Range	Sample	n	Mean	SE	95% CI	t-value	Difference in Magnitude
8–17	Random sample	500	.02	.01	−.00–.04	4.76*	9.0
	LCP sample	500	.18	.03	.12–.24		
18–24	Random sample	500	.11	.02	.06–.15	10.65*	9.1
	LCP sample	499	1.00	.08	.84–1.16		
25–31	Random sample	394	.13	.04	.06–.20	10.81*	8.7
	LCP sample	438	1.13	.08	.97–1.29		
32–38	Random sample	247	.08	.02	.03–.13	8.54*	12.3
	LCP sample	355	.98	.09	.81–1.15		
39–45	Random sample	128	.02	.02	−.01–.06	6.67*	43.0
	LCP sample	221	.86	.09	.68–1.05		
46–52	Random sample	64	.02	.02	−.02–.05	4.18*	39.0
	LCP sample	104	.78	.14	.50–1.06		
53–59	Random sample	23	0	0	0	1.93**	—
	LCP sample	33	.52	.22	.06–.97		

Note: CI = confidence interval, SE = standard error.

*$p < .0001$, **$p = .0591$

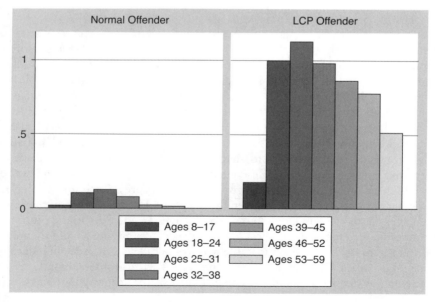

FIGURE 21-2. Prison Sentences Across Seven Life Stages by Offender Group

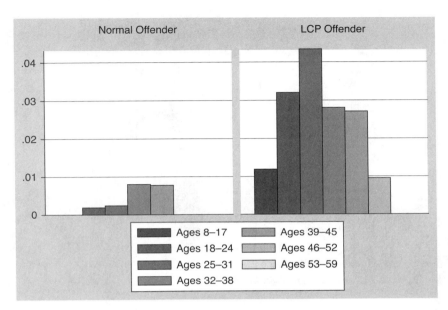

FIGURE 21-3. Murder Arrests Across Seven Life Stages by Offender Group

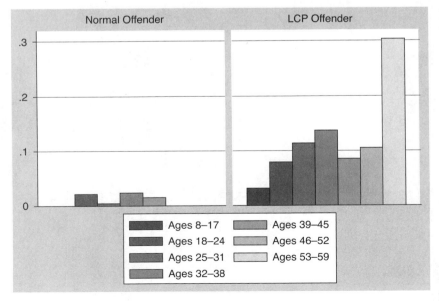

FIGURE 21-4. Rape Arrests Across Seven Life Stages by Offender Group

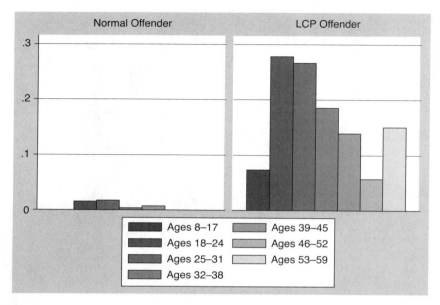

FIGURE 21-5. Robbery Arrests Across Seven Life Stages by Offender Group

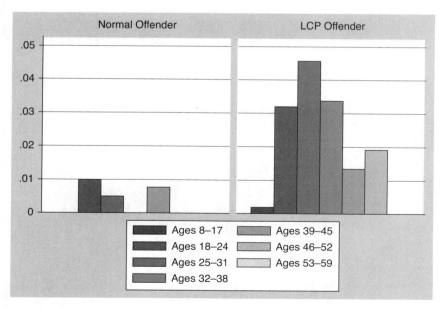

FIGURE 21-6. Kidnapping Arrests Across Seven Life Stages by Offender Group

THEORETICAL DISCUSSION

Spirited debates about the value of criminal careers research, the necessity of studying offending patterns cross-sectionally or longitudinally, and the very nature of criminal behavior are not new.[12] Empirically and conceptually, it is understood that there is important variation in criminal propensity and that various grouping of offenders by propensity—low, medium, and high—are relatively stable over time. Of course, within each group or classification of offender, there is heterogeneity such that the classifications of offenders will not behave in perfect uniformity. In Moffitt's terms, LCP offenders will always be worse than adolescence-limited offenders. In Laub and Sampson's terms, even the worst LCP offender will desist over time, particularly if followed until age 70. As noted by Laub and Sampson:

> *All offenders eventually desist. It is just that some do so later or at different rates than predicted by the conventional wisdom in criminology. Boston Billy, for example, took until his sixties to finally give up on a life in crime.*[13]

But does universal desistance refute the idea of the LCP offender?

If the Gluecks' data are hailed for their unsurpassed longevity and coverage, the current data could be hailed for their unsurpassed extremity. With an average career of six arrests, the randomly selected offenders were themselves a high-rate, chronic, or habitual sample of offenders. In this way, the randomly selected offenders were neither a community sample—which are notoriously devoid of serious offenders—nor a benign group of arrestees. Yet, their criminal careers were dwarfed by the LCP group—LCP offenders were worse in their 40s and 50s than "normal" career criminals were during their teens and early 20s. Gottfredson and Hirschi once mused that, "Mickey Mantle at 38 was better than Gene Michael at 28. Mickey Mantle at 38 was not, however, as good as Mickey Mantle at 28."[14] The spirit of Moffitt's theory arrives at the same conclusion: LCP offenders are pathologically worse than other criminals, even those whose offending careers are chronic and serious by criminological standards. Contrary to work by Uggen, which suggested that career offenders are "stylized cultural images of the hardened criminal,"[15] the current data indicate that hardened criminals continue serious offending for decades into adulthood.

In her defense of the LCP offender, Moffitt expressed concern about the reification effects that might occur if practitioners made assessments based on her LCP prototype. Indeed, the academic concern about what the criminal justice practitioner community might do with criminal career information is not new. Gottfredson and Hirschi noted, "The evidence is clear that the career criminal idea is not sufficiently substantial to command more than a small portion of the time and effort of the criminal justice practitioner or academic community."[16] The current data suggest that a practitioner-generated criterion for career criminality not only is more prudent in its assessment of bona fide LCP offending, but also more judicious in which offenders it selects for habitual offender status in terms of sentencing. The thresholds posed by the academic community are historically too low. The differences in magnitude of offending shown in Tables 21-2, 21-3, and 21-4 reinforce the sheer differences between chronic offenders as delineated with a criminological criterion and those delineated by criminal justice practitioners.

It is important to note that the current data are based on both **official records** and **self-reports**, but the face validity of self-reported criminal history from habitual criminals is generally weak. Hence, these data likely underestimate the true severity of the criminal careers of both samples of offenders. For instance, in a study of serious delinquent careers among youth in Philadelphia and

Phoenix, Mulvey et al. found that the most severe criminal history was nine prior arrests in the preceding year, and that pertained to just one offender. Half of the offenders in their samples had zero prior arrests in the prior year. Yet, more than 10 percent of the offenders in both cities self-reported more than 1000 offenses in the preceding year. By comparison, 51 percent of the LCP sample in the current study had 50 or more career arrests, and the most chronic offender had a startling 267 arrests.

CONCLUSION

The notion of a life-course desister is misleading in the sense that desistance is usually a developmental, nuanced process, not typically an abrupt switch from offender to nonoffender. Mulvey and colleagues advised:

> It is important to note that the definition of desistance as simply the absence or decline of a particular form of antisocial behavior does not allow for the possibility that other behaviors might be replacing the antisocial behavior of interest.[17]

In other words, there is a "dark figure" of desistance that is obscured by a focus on a decline in arrests. To illustrate, many LCP offenders in the current study were homeless and the bulk of their arrests were for nuisance or publicorder crimes. In this way, these transient offenders gave the appearance of being old criminals that had desisted or devolved from being serious offenders to Norman Rockwell–like hobos. Yet, the figures for involvement in crimes, such as murder, rape, robbery, and kidnapping, tell a much different story. Even if they had desisted from their peak offending levels, these LCP offenders were still formidable offenders very late in life.

Sampson and Laub suggested that there is not convincing evidence of an LCP group of offenders. We disagree. The current data show that LCP offenders can be differentiated empirically by their arrest and prison histories. They also continue to offend at extraordinarily high levels at a time when the age–crime curve would suggest otherwise. In conclusion, we caution criminologists not to treat offending careers as abstractions that are fodder for sophisticated statistical study. Crime touches lives in important and devastating ways. And the LCP offender is truly devastating.

GLOSSARY

Chronic offenders—a small group of offenders who engage in crime over a prolonged period of time and account for the majority of all serious offenses

Desistance—a developmental, nuanced process, where an offender moves from being an offender to a nonoffender

Official records—statistics and data about crimes (e.g., arrests) that is collected from official sources, such as the Federal Bureau of Investigation

Self-reports—information about criminal involvement that is collected from individuals

Trajectories—groups of relatively homogenous offenders who share similar developmental pathways to criminal involvement

NOTES

1. Sampson, R. J., & Laub, J. H. (2003). Life-course desisters? Trajectories of crime among delinquent boys followed to age 70. *Criminology, 41*, 555–592, p. 555.

2. Gottfredson, M. R. (2005). Offender classifications and treatment effects in developmental criminology: A propensity/event consideration. *Annals of the American Academy of Political and Social Science, 602*, 46–56; Gottfredson, M. R., & Hirschi, T. (1990). *A general theory of crime.* Stanford, CA: Stanford University Press; Moffitt, T. E. (1993). Adolescence-limited and life-course-persistent antisocial behavior: A developmental taxonomy. *Psychological Review, 100*, 674–701; Wilson, J. Q., & Herrnstein, R. J. (1985). *Crime and human nature: The definitive study of the causes of crime.* New York: Simon and Schuster.

3. Laub, J. H., & Sampson, R. J. (2003). *Shared beginnings, divergent lives: Delinquent boys to age 70.* Cambridge, MA: Harvard University Press; Sampson, R. J., & Laub, J. H. (1993). *Crime in the making: Pathways and turning points through life.* Cambridge, MA: Harvard University Press; Sampson, R. J., & Laub, J. H. (2005). When prediction fails: From crime-prone boys to heterogeneity in adulthood. *Annals of the American Academy of Political and Social Science, 602*, 73–79.

4. Laub, & Sampson (2003), p. 588, see Note 3.

5. Blokland, A. A. J., Nagin, D. S., & Nieuwbeerta, P. (2005). Life span offending trajectories of a Dutch conviction cohort. *Criminology, 43*, 919–954; Blokland, A. A. J., & Nieuwbeerta, P. (2005). The effects of life circumstances on longitudinal trajectories of offending. *Criminology, 43*, 1203–1240; Brame, R., Bushway, S. D., & Paternoster, R. (2003). Examining the prevalence of criminal desistance. *Criminology, 41*, 423–448; Bushway, S. D., Thornberry, T. P., & Krohn, M. D. (2003). Desistance as a developmental process: A comparison of static and dynamic approaches. *Journal of Quantitative Criminology, 19*, 129–152.

6. Moffitt, T. E. (2007). A review of research on the taxonomy of life-course-persistent versus adolescence-limited antisocial behavior. In D. J. Flannery, A. T. Vazsonyi, & I. D. Waldman (Eds.), *The Cambridge handbook of violent behavior and aggression* (pp. 49–76). New York: Cambridge University Press, pp. 66–67.

7. Laub, & Sampson (2003), p. 585 see Note 3.

8. Odgers, C. L. Caspi, A., Broadbent, J. M., Dickson, N., Hancox, R. J., Harrington, H., et al. (2007). Prediction of differential adult health burden by conduct problem subtypes in males. *Archives of General Psychiatry, 64*, 476–484.

9. Moffitt, T. E. (2003). Life-course-persistent and adolescence-limited antisocial behavior: A 10-year research review. In B. B. Lahey, T. E. Moffitt, & A. Caspi (Eds.), *Causes of conduct disorder and juvenile delinquency* (pp. 49–75). New York: The Guilford Press; Moffitt, T. E. (2006). A review of research on the taxonomy of life-course-persistent versus adolescence-limited antisocial behavior. In F. T. Cullen, J. P. Wright, & K. R. Blevins (Eds.), *Taking stock: The status of criminological theory, advances in criminological theory* (Vol. 15, pp. 277–312). New Brunswick, NJ: Transaction.

10. DeLisi, M. (2001). Extreme career criminals. *American Journal of Criminal Justice, 25*, 239–252; DeLisi, M. (2003). The imprisoned non-violent drug offender: Specialized martyr or versatile career criminal? *American Journal of Criminal Justice, 27*, 167–182.

11. Wolfgang, M. E., Figlio, R. M., & Sellin, T. (1972). *Delinquency in a birth cohort.* Chicago: University of Chicago Press.

12. Blumstein, A., Cohen, J., & Farrington, D. P. (1988a). Criminal career research: Its value for criminology. *Criminology, 26*, 1–35; Blumstein, A., Cohen, J., & Farrington, D. P. (1988b). Longitudinal and criminal career research: Further clarifications. *Criminology, 26*, 57–74; Gottfredson, M. R., & Hirschi, T. (1986). The true value of lambda would appear to be zero: An essay on career criminals, criminal careers, selective incapacitation, cohort studies, and related topics. *Criminology, 24*, 213–234; Gottfredson, M. R., & Hirschi, T. (1987). The methodological adequacy of longitudinal research on crime. *Criminology, 25*, 581–614; Gottfredson, M. R., & Hirschi, T. (1988). Science, public policy, and the career paradigm. *Criminology, 26*, 37–55; Nagin, D. S., & Paternoster, R. (2000). Population heterogeneity and state dependence: State of evidence and directions for future research. *Journal of Quantitative Criminology, 16*, 117–144; Robins, L. N. (2005).

Explaining when arrests end for serious juvenile offenders: Comments on the Sampson and Laub study. *Annals of the American Academy of Political and Social Science, 602,* 57–72; Ulmer, J. T. (2008). Book review of Career criminals in society. *Criminal Justice Review, 33,* 103–104.

13. Laub, & Sampson (2003), pp. 176–177, see Note 3.

14. Gottfredson, & Hirschi (1986), p. 223, see Note 12.

15. Uggen, C. (2000). Work as a turning point in the life course of criminals: A duration model of age, employment, and recidivism. *American Sociological Review, 67,* 529–546, p. 542.

16. Gottfredson, & Hirschi, (1986), p. 231 see Note 12.

17. Mulvey, E. P., Steinberg, L., Fagan, J., Cauffman, E., Piquero, A. R., Chassin, L., et al. (2004). Theory and research on desistance from antisocial activity among serious adolescent offenders. *Youth Violence and Juvenile Justice, 2,* 213–236, p. 220.

CHAPTER TWENTY-TWO

Evolutionary Psychological Perspectives on Men's Partner-Directed Violence

Farnaz Kaighobadi and Todd K. Shackelford

KEY TERMS

Cuckoldry
Evolutionary psychology
Intimate partner violence
Jealousy
Paternity uncertainty

INTRODUCTION

Approximately one in five reported incidents of nonfatal violence against women are perpetrated by an intimate partner. This amounts to nearly 600,000 reported incidents of nonfatal violence against women by an intimate partner in just a single year in the United States. Partly in response to the tragically high incidence of female-directed violence in intimate relationships and the devastating negative physical and psychological consequences of these behaviors, a large body of literature has investigated the predictors of men's violence against intimate partners. Previous research has identified several proximate predictors of men's partner-directed violence, such as family history of aggression and cultural influences, as well as ultimate or evolutionary predictors of such costly behaviors, notably male sexual jealousy as a solution to the adaptive problem of **paternity uncertainty**.[1]

Increasingly over the past several decades, social scientists have recognized the value of an evolutionary perspective for guiding their research. Evolutionary psychological theories have been applied successfully to the investigation of diverse human behaviors, including altruism and cooperation, conflict and violence, decision making, interpersonal relationships, and mating.[2] **Evolutionary psychology** is concerned with identifying and describing the design and function of psychological adaptations that evolved to solve the specific problems our ancestors faced recurrently over human evolutionary history. These evolved mechanisms are information-processing devices that motivate behavior in response to particular environmental inputs. An evolutionary psychological perspective can guide research on **intimate partner violence**, notably research on the evolved mechanisms that motivate these behaviors. This chapter reviews different forms of men's partner-directed violence, including insults, sexual coercion, physical violence, and homicide, from an evolutionary psychological perspective.

PATERNITY UNCERTAINTY AND MALE SEXUAL JEALOUSY _____

Over human evolutionary history, men and women have faced the adaptive problems of maintaining relationships and retaining intimate partners. **Jealousy** is an emotion that motivates behaviors that deter mate-poaching rivals and prevents partner infidelity or outright desertion from the relationship. Men and women do not differ in the frequency or intensity with which they experience jealousy. However, men and women respond differently to two different types of partner infidelity: emotional infidelity and sexual infidelity. Men are more distressed about a partner's sexual infidelity than about her emotional infidelity, whereas the opposite pattern is found for women. This sex difference has been documented in more than a dozen empirical studies using various methods, including forced-choice self-report assessments, physiological assessments, experimental methods, and archival and cross-cultural data.[3]

The sex difference in the experience of jealousy may be attributable to sex-specific adaptive problems humans faced over our evolutionary history. It is hypothesized that ancestral women faced the recurrent problems of paternal investment and acquisition and retention of resources with which to raise offspring. A partner's emotional infidelity might have predicted his current or future investment of resources in another woman's children. Ancestral men, in contrast, faced the adaptive problem of paternity uncertainty. Female sexual infidelity and subsequent **cuckoldry**—her regular partner's unwitting investment in offspring to whom he is not genetically related—carried substantial reproductive costs for ancestral men. The reproductive costs of cuckoldry, including the loss of time, energy, resources, and alternative mating opportunities, are potentially so great that men have evolved to be sensitive to and to experience more distress about a partner's sexual infidelity. Men also may have evolved mechanisms that assess the risk of partner sexual infidelity and mechanisms that motivate the performance of anticuckoldry tactics. To assess the likelihood or risk of partner sexual infidelity, these information-processing mechanisms may use cues such as greater time spent apart from the partner, the presence of potential rivals, and a partner's attractiveness, or her "mate value" as a short-term or long-term partner.[4] The behavioral output of male sexual jealousy varies from subtle, nonviolent mate retention behaviors to outright physical violence.

MALE SEXUAL JEALOUSY AND MATE RETENTION BEHAVIORS _____

One class of behavioral output of sexual jealousy is men's mate retention behaviors, which are designed to prevent a partner's infidelity or outright relationship defection, or to thwart attempts to encroach on the relationship by rivals. Buss developed a taxonomy of mate retention behaviors organized into five categories: acts of direct guarding function to keep a partner under surveillance; acts of intersexual negative inducements, which include threats to punish a partner's infidelity; acts of positive inducements, which include expressions of affection and care; acts of public signals of possession, which include acts intended to signal possession of a partner to potential rivals; and acts of intrasexual negative inducements, which include acts intended to threaten potential rivals and thereby deter them from encroaching on the relationship. As the risk of female infidelity increases, men perform more frequent mate retention behaviors. For example, Buss and Shackelford found that men mated to younger, more attractive partners (cues to reproductive value or expected future reproduction) and men who perceive greater probability of partner infidelity guard their partners more intensely. Also, men perform more frequent mate retention behaviors when they are mated to women who possess qualities that predict her infidelity, including her personality characteristics such as surgency and openness to experience. Because time spent physically apart from a partner increases the risk of a partner's infidelity, men who have spent a greater proportion of time apart from their partners also report engaging in more frequent mate retention behaviors. In addition, McKibbin and colleagues identified positive relationships between men's accusations of their partner's infidelity and the frequency with which they performed several categories of mate retention.[5]

A more physically damaging output of male sexual jealousy is partner-directed violence. Male sexual jealousy is one of the most frequently cited causes of men's partner-directed violence, both physical and sexual.[6] The frequency with which men perform nonviolent mate retention behaviors predicts the frequency with which they inflict physical violence against their partners, because both classes of behavior are hypothesized to be outputs of sexual jealousy. In three studies, Shackelford et al. investigated the relationship between men's nonviolent mate retention behaviors and men's partner-directed violence. Based on men's self-reports, women's partner reports, and cross-spouse reports, men's use of emotional manipulation as a mate retention tactic—marked by the performance of acts such as, "I told my partner I would 'die' if she ever left me"—predicted female-directed violence. Men's monopolization of their partner's time and men's sexual inducements also predicted men's physical violence against their partners.[7] Because suspicions of partner infidelity predict men's mate retention behaviors and because men's mate retention behaviors predict men's partner-directed violence, Kaighobadi, Starratt, Popp, and Shackelford hypothesized that men's suspicions of their partner's infidelity is linked directly to partner-directed violence. In two studies, using men's self-reports and women's partner reports, Kaighobadi et al. found that men's accusations of their partner's infidelity predict men's partner-directed violence. Moreover, this relationship is mediated by nonviolent mate retention behaviors. Kaighobadi et al. hypothesized that men perform nonviolent and violent mate retention behaviors in a temporal hierarchical fashion. Less severe, less costly behaviors might be deployed first, followed by more severe behaviors such that the hierarchy of events leading to female-directed violence is initiated with men's suspicions of partner infidelity followed by nonviolent mate retention behaviors and ending in acts of physical violence.[8]

Puente and Cohen conducted a series of studies investigating third-party observer perceptions of male sexual jealousy and female-directed violence. They found that observers were more accepting of violent behavior in scenarios in which intimate partner violence followed the male aggressor's experience of sexual jealousy. When the male aggressor was described as sexually jealous, observers were less inclined to convict him of a crime and assumed that the female victim also would be less inclined to file charges against her partner. Puente and Cohen also found that jealous abusers were perceived by observers as more romantically in love with their partner than nonjealous abusers. It therefore appears that people readily and intuitively render judgments and reach conclusions based on an expected relationship between men's jealousy and men's partner-directed violence. A third-party observer's ready inference of a relationship between male sexual jealousy and men's partner-directed violence is consistent with the operation of an implicit mechanism underlying these judgments.[9]

RISK OF SPERM COMPETITION AND SEXUAL COERCION

Sexual coercion or rape of an intimate partner also is a hypothesized manifestation of male sexual jealousy, which may be a response to perceived risk of sperm competition. Between 10 and 26 percent of women report being raped by their husband. In a sample of young adults in a committed, sexual relationship, Goetz and Shackelford found that 7.3 percent of men admitted to at least one incidence of raping their current partner, and that 9.1 percent of women reported having experienced at least one incidence of rape by their current partner.[10]

Many studies have investigated men's sexual coercion in an intimate relationship. A number of hypotheses have been formulated to test the proximate or immediate predictors of men's sexual coercion of an intimate partner and also the ultimate or evolutionary predictors of men's sexual coercion of an intimate partner. Several scholars have argued that men's sexual coercion of their partner is motivated by a desire to dominate and control their partners.[11] For example, several studies have found that men who are physically abusive of their partners also are more likely to be sexually coercive toward their partners than are men who are not physically abusive.[12] Shackelford and Goetz also found a positive relationship between men's nonviolent controlling behavior and men's sexual coercion of their partners. Gage and Hutchinson found that women's experience of sexual coercion by their partner is predicted by their partner's jealousy and nonviolent controlling behavior but is not predicted by differences in social power between the partners.

The desire to dominate and control a partner may explain some portion of the individual differences in men's sexually coercive behaviors, but the proponents of the domination and control hypothesis argue that men as a group are motivated to exert "patriarchal terrorism" or "patriarchal power" over all women through sexual coercion of their own partners.[13] We are unaware of research that has empirically tested these hypotheses. It should be noted, however, that hypotheses that propose coordinated male–male cooperation to dominate and control women are not consistent with substantial theoretical and empirical work that highlights the frequency and intensity of male–male competition (rather than cooperation) for attracting women as intimate partners.[14]

Sexual coercion also is hypothesized to function as an anticuckoldry tactic.[15] It has been hypothesized that by forcing their partners to have sex, men who are suspicious of their partner's infidelity introduce their own sperm into their partner's reproductive tract and thereby decrease the risk of cuckoldry. This sperm competition hypothesis for partner rape has been applied to nonhumans (notably, several avian species) to account for observations of partner rape immediately after female extra-pair copulations. Rape of an intimate partner in humans also often follows accusations of female sexual infidelity.

Thornhill and Thornhill argued that women who resist or avoid copulating with their partners might thereby be signaling to their partners a recent sexual infidelity; hence, forced copulation might function to decrease men's paternity uncertainty. The fact that rape of a woman is more likely to occur during or after a breakup (when men's concerns about her infidelities are greatest) may provide preliminary support for this hypothesis. A number of studies have documented a positive relationship between men's sexual jealousy and men's sexual coercion of their partners. For example, men who sexually coerced their wives are more sexually jealous than men who did not. Previous research has found a direct positive relationship between men's suspicions and accusations of partner infidelity and men's sexual coercion of their partners.[16]

In two studies securing data from men's self-reports and women's partner-reports, Goetz and Shackelford found that men's sexual coercion correlated positively with women's past and future likelihood of engaging in sexual infidelity. They also found that men who perform more mate retention behaviors also are more likely to perform sexually coercive behaviors against their partners, as reported by men and by their partners. The domination and control hypothesis and the sperm competition hypothesis reflect different levels of analysis. The domination and control hypothesis offers a proximate explanation of partner sexual coercion, including social or cultural causes of behavior. The sperm competition hypothesis offers an ultimate explanation of partner sexual coercion and addresses how adaptations that produce such costly behaviors could have evolved. Goetz and Shackelford do not argue that all sexually coercive behaviors are produced by evolved mechanisms that motivate anticuckoldry behaviors. Instead, they attempted to explain the increased likelihood of sexual coercion in the context of risk of female infidelity. It may be that some instances of sexual coercion are the result of, for example, an antisocial man's motivation to control, dominate, or humiliate his partner. Future research might investigate the interaction of individual differences in men's perpetration of sexual coercion and evolutionarily relevant contexts such as the risk of female infidelity and sperm competition to predict men's sexual coercion and rape of their partners.

INTIMATE PARTNER HOMICIDE

According to the U.S. Department of Justice, between 1976 and 2005, 30 percent of female homicide victims were killed by an intimate partner, making it the largest class of victim–offender relationship. In sharp contrast, just 5 percent of all male homicide victims were killed by an intimate partner. In most categories of intimate partner homicide, men far outnumber women as the perpetrator and women far outnumber men as the victim.[17] Several hypotheses have been advanced to explain the occurrence and frequency of intimate partner homicide, and here we review briefly these hypotheses, with reference to intimate partner homicide of women by men. According to one hypothesis, sheer proximity might account for the frequency of intimate partner homicide. Because intimate partners frequently engage in interactions with one another, partner killing might occur as an extreme manifestation of interpersonal conflict. According to a second hypothesis, the "killing-as-a-byproduct" hypothesis, because killing a partner carries substantial and severe costs that might not have consistently produced sufficient benefits over human evolutionary history, it is unlikely to be the product of specialized adaptations. The costs associated with killing a partner include the risk of retaliation by the victim's kin and the local community; the loss of time, energy, and resources a man invested in maintaining the relationship; and the loss of maternal investment in any shared offspring. Instead, partner killing might be a byproduct of other male psychological adaptations, including adaptations specialized to motivate nonlethal punishment of a partner's suspected or actual infidelity and to control her sexual interactions. Wilson and Daly argue that men's "sexual proprietariness"—which they define as a combination of sexual jealousy and men's "presumptions of entitlement" and motivation to control their partner's sexual behavior—is a key cause of 80 percent of spousal homicides. Other factors also affect rates of violence and homicide of female partners, including the age of the woman and the presence of a woman's children from previous relationships.[18]

Proponents of a third hypothesis for intimate partner homicide argue that the byproduct hypothesis cannot explain the large incidence of apparently premeditated partner homicides. Premeditated homicides include hiring someone to kill the partner, aiming at and shooting a partner, or deliberately poisoning a partner. Buss and Duntley argue that partner killing by men might be the designed outcome of a specialized adaptation in the context of suspected or actual female infidelity (and, as a consequence, paternity uncertainty). This "evolved homicide module" hypothesis does not imply that discovery of female sexual infidelity will always or even frequently lead to partner killing by men, but instead that the relevant evolved mechanisms are likely to be activated with suspicions of female partner infidelity and may very occasionally result in partner killing.[19]

CONCLUSION

Evolutionary psychologists are interested in identifying the ultimate (or distal) explanations for a trait or behavior. Intimate partner violence and homicide are costly behaviors, for both the victim and the perpetrator. It is useful to consider an evolutionary perspective to investigate properly the design features and evolved function of the psychological mechanisms that motivate these behaviors. An evolutionary psychological perspective can guide identification of the contexts that trigger the relevant information-processing mechanisms and motivate the subsequent behaviors. Many instances of intimate partner violence and homicide co-occur with and may be triggered by men's suspicions or knowledge of their female partner's sexual infidelity. However, this does not mean that individual men never perpetrate violence for other reasons, such as to gain control and dominance over their partners.

Moreover, previous research identified several proximate correlates of female-directed violence, such as a family history of aggression and acceptance of local cultural norms. Previous research also investigated men's personality traits as predictors of men's partner-directed violence. For example, Hellmuth and McNulty documented that neurotic men are more likely to inflict violence on their partners, especially if the men suffer from chronic stress.[20] Situational factors, including the characteristics of the perpetrator, the victim, and the circumstances in which the violence occurs, also have been considered in research investigating intimate partner violence and homicide. For example, the perpetrator's age and the victim's age, the perpetrator's mental health, and the availability of weapons have been investigated as situational factors predicting intimate partner violence.[21]

In conclusion, it is important to investigate both the proximate and ultimate causes of intimate partner violence. The relevant evolved psychological mechanisms interact with stable dispositions and situational factors to produce manifest behavior. Future research might benefit by taking an evolutionary perspective to build models of intimate partner violence that include both stable dispositions such as personality traits and environmental factors such as a family history of aggression. To achieve a fuller understanding of intimate partner violence and homicide, researchers must include a careful consideration of the evolved psychological mechanisms that motivate these costly behaviors.

GLOSSARY

Cuckoldry—males raising children who they believe are biologically their own but who are really the biological offspring of another mate

Evolutionary psychology—a perspective that is concerned with identifying and describing the design and function of psychological adaptations that evolved to solve the specific problems our ancestors faced recurrently over human evolutionary history

Intimate partner violence—violence that occurs between mates

Jealousy—an emotion that generates feelings of envy and that has been linked to behaviors that deter mate-poaching rivals and prevents partner infidelity or outright desertion from the relationship

Paternity uncertainty—refers to the fact that males are never 100 percent certain that the child their mate bears is really theirs

NOTES

1. Archer, J. (2006). Cross-cultural differences in physical aggression between partners: A social-role analysis. *Personality and Social Psychology Review, 10,* 133–153; Busby, D. M., Holman, T. B., & Walker, E. (2008). Pathways to relationship aggression between adult partners. *Family Relations, 57,* 72–83; Gage, A. J., & Hutchinson, P. L. (2006). Power, control, and intimate partner sexual violence in Haiti. *Archives of Sexual Behavior, 35,* 11–24; Goetz, A. T., Shackelford, T. K., Romero, G. A., Kaighobadi, F., & Miner, E. J. (2008). Punishment, proprietariness, and paternity: Men's violence against women from an evolutionary perspective. *Aggression and Violent Behavior, 13,* 481–489; Riggs, D. S., & O'Leary, K. D. (1996). Aggression between heterosexual dating partners: An examination of a causal model of courtship aggression. *Journal of Interpersonal Violence, 11,* 519–540; Shackelford, T. K., Goetz, A. T., Buss, D. M., Euler, H. A., & Hoier, S. (2005). When we hurt the ones we love: Predicting violence against women from men's mate retention tactics. *Personal Relationships, 12,* 447–463.

2. Ackerman, J. M., & Kenrick, D. T. (2008). The costs of benefits: Help-refusals highlight key trade-offs of social life. *Personality and Social Psychology Review, 12,* 118–140; Buss, D. M. (2000). *The dangerous passion.* New York: Free Press; Cosmides, L., & Tooby, J. (1992). Cognitive adaptations for social exchange. In J. H. Barkow, L. Cosmides, & J. Tooby (Eds.), *The adapted mind: Evolutionary psychology and the generation of culture* (pp. 163–228). New York: Oxford University Press; Daly, M., & Wilson, M. (1988). *Homicide.* Hawthorne, NY: Aldine de Gruyter; Gigerenzer, G. (2008). Why heuristics work. *Perspectives on Psychological Science, 3,* 20–29; Kruger, D. J. (2003). Evolution and altruism: Combining psychological mediators with naturally selected tendencies. *Evolution and Human Behavior, 24,* 118–125; Shackelford, T. K., Schmitt, D. P., & Buss, D. M. (2005). Universal dimensions of human mate preference. *Personality and Individual Differences, 39,* 447–458; Wilson, M., & Daly, M. (1998). Lethal and nonlethal violence against wives and the evolutionary psychology of male sexual proprietariness. In R. E. Dobash & R. P. Dobash (Eds.), *Rethinking violence against women* (pp. 199–230). Thousand Oaks, CA: Sage.

3. Betzig, L. (1989). Causes of conjugal dissolution: A cross-cultural study. *Current Anthropology, 30,* 654–676; Buss, D. M., Larsen, R. J., Westen, D., & Semmelroth, J. (1992). Sex differences in jealousy: Evolution, physiology and psychology. *Psychological Science, 3,* 251–255; Daly, M., Wilson, M., & Weghorst, J. (1982). Male sexual jealousy. *Ethology and Sociobiology, 3,* 11–27; Shackelford, T. K., LeBlanc, G. J., & Drass, E. (2000). Emotional reactions to infidelity. *Cognition and Emotion, 14,* 643–659; Schützwohl, A. (2005). Sex differences in jealousy: The processing of cues to infidelity. *Evolution and Human Behavior, 26,* 288–299; Schützwohl, A. (2008). The disengagement of attentive resources from task-irrelevant cues to sexual and emotional infidelity. *Personality and Individual Differences, 44,* 633–644; Symons, D. (1979). *The evolution of human sexuality.* New York: Oxford University Press; Thomson, J. W., Patel, S., Platek, S. M., & Shackelford, T. K. (2007). Sex differences in implicit association and attentional demands for information about infidelity. *Evolutionary Psychology, 5,* 569–583.

4. Goetz, A. T., & Shackelford, T. K. (2006). Sexual coercion and forced in-pair copulation as sperm competition tactics in humans. *Human Nature, 17,* 265–282; Peters, J., Shackelford, T. K., & Buss, D. M. (2002). Understanding domestic violence against women: Using evolutionary psychology to extend the feminist functional analysis. *Violence and Victims, 17,* 255–264; Schmitt, D. P., & Buss, D. M. (2001). Human mate poaching: Tactics and temptations for infiltrating existing mateships. *Journal of Personality and Social Psychology, 80,* 894–917; Shackelford, T. K., & Buss, D. M. (1997). Cues to infidelity. Personality and *Social Psychology Bulletin, 23,* 1034–1045; Shackelford, T. K., Goetz, A. T., McKibbin, W. F., & Starratt, V. G. (2007). Absence makes the adaptations grow fonder: Proportion of time apart from partner, male sexual psychology, and sperm competition in humans (*Homo sapiens*). *Journal of Comparative Psychology, 121,* 214–220; Trivers, R. L. (1972). Parental investment and sexual selection. In B. Campbell (Ed.), *Sexual selection and the descent of man 1871–1971* (pp. 136–179). Chicago: Aldine; Wilson, M., & Daly, M. (1993). An evolutionary psychological perspective on male sexual proprietariness and violence against wives. *Violence and Victims, 8,* 271–294.

5. Buss, D. M. (1988). From vigilance to violence: Tactics of mate retention in American undergraduates. *Ethology and Sociobiology, 9,* 291–317; Goetz, A. T., Shackelford, T. K., Weekes-Shackelford, V. A., Euler, H. A., Hoier, S., Schmitt, D. P., & LaMunyon, C. W. (2005). Mate retention, semen displacement, and human sperm competition: A preliminary investigation of tactics to prevent and correct female infidelity. *Personality and Individual Differences, 38,* 749–763; McKibbin, W. F., Goetz, A. T., Shackelford, T. K., Schipper, L. D., Starratt, V.G., & Stewart-Williams, S. (2007). Why do men insult their intimate partners? *Personality and Individual Differences, 43,* 231–241.

6. Dobash, R. E., & Dobash, R. P. (1979). *Violence against wives.* New York: Free Press; Dutton, D. G. (1998). *The abusive personality.* New York: Guilford Press; Frieze, I. H. (1983). Investigating the causes and consequences of marital rape. *Signs: Journal of Women in Culture and Society, 8,* 532–553; Russell, D. E. H. (1982). *Rape in marriage.* New York: Macmillan; Walker, L. E. (1979). *The battered woman.* New York: Harper & Row.

7. Shackelford, T. K., Pound, N., & Goetz, A. T. (2005). Psychological and physiological adaptations to sperm competition in humans. *Review of General Psychology, 9,* 228–248.

8. Kaighobadi, F., Starratt, V. G., Shackelford, T. K., & Popp, D. (2008). Male mate retention mediates the relationship between female sexual infidelity and female-directed violence. *Personality and Individual Differences, 44,* 1422–1431.

9. Puente, S., & Cohen, D. (2003). Jealousy and the meaning (or nonmeaning) of violence. *Personality and Social Psychology Bulletin, 29,* 449–460.

10. Finkelhor, D., & Yllo, K. (1985). *License to rape: Sexual abuse of wives.* New York: Holt, Rinehart, & Winston; Goetz, A. T., & Shackelford, T. K. (2009). Sexual coercion in intimate relationships: A comparative analysis of the effects of women's infidelity and men's dominance and control. *Archives of Sexual Behavior, 38,* 226–234; Hadi, A. (2000). Prevalence and correlates of the risk of marital sexual violence in Bangladesh. *Journal of Interpersonal Violence, 15,* 787–805; Painter, K., & Farrington, D. P. (1999). Wife rape in Great Britain. In R. Muraskin (Ed.), *Women and justice: Development of international policy* (pp.135–164). New York: Gordon and Breach; Watts, C., Keogh, E., Ndlovu, M., & Kwaramba, R. (1998). Withholding of sex and forced sex: Dimensions of violence against Zimbabwean women. *Reproductive Health Matters, 6,* 57–65.

11. Basile, K. C. (2002). Prevalence of wife rape and other intimate partner sexual coercion in a nationally representative sample of women. *Violence and Victims, 17,* 511–524; Bergen, R. K. (1996). *Wife rape: Understanding the response of survivors and service providers.* Thousand Oaks, CA: Sage; Gelles, R. (1977). Power, sex and violence: The case of marital rape. *Family Coordinator, 26,* 339–347; Meyer, S., Vivian, D., & O'Leary, K. D. (1998). Men's sexual aggression in marriage: Couple's reports. *Violence Against Women, 4,* 415–435.

12. Apt, C., & Hurlbert, D. F. (1993). The sexuality of women in physically abusive marriages: Comparative study. *Journal of Family Violence, 8,* 57–69; DeMaris, A. (1997). Elevated sexual activity in violent marriages: Hypersexuality or sexual extortion? *Journal of Sex Research, 34,* 361–373; Donnelly, D. A. (1993). Sexually inactive marriages. *Journal of Sex Research, 30,* 171–179; Koziol-McLain, J., Coates, C. J., & Lowenstein, S. R. (2001). Predictive validity of a screen for partner violence against women. *American Journal of Preventative Medicine, 21,* 93–100; Shackelford, T. K., & Goetz, A. T. (2004). Men's sexual coercion in intimate relationships: Development and initial validation of the Sexual Coercion in Intimate Relationships Scale. *Violence and Victims, 19,* 541–556.

13. Brownmiller, S. (1975). *Against our will: Men, women, and rape.* New York: Simon & Schuster; Johnson, M. P. (1995). Patriarchal terrorism and common couple violence: Two forms of violence against women. *Journal of Marriage and the Family, 57,* 283–294.

14. Bleske, A. L., & Shackelford, T. K. (2001). Poaching, promiscuity, and deceit: Combating mating rivalry in same sex friendships. *Personal Relationships, 8,* 407–424; Schmitt, D. P., & Buss, D. M. (1996). Strategic self-promotion and competitor perogation: Sex and context effects on the perceived effectiveness of mate attraction tactics. *Journal of Personality and Social Psychology, 70,* 1185–1204.

15. Lalumière, M. L., Harris, G. T., Quinsey, V. L., & Rice, M. E. (2005). *The causes of rape: Understanding individual differences in male propensity for sexual aggression.* Washington, DC: APA Press; Thornhill, R., & Thornhill, N. W. (1992). The evolutionary psychology of men's coercive sexuality. *Behavioral and Brain Sciences, 15,* 363–421;

Wilson, M., & Daly, M. (1992). The man who mistook his wife for a chattel. In J. H. Barkow, L. Cosmides, & J. Tooby (Eds.), *The adapted mind* (pp. 289–322). New York: Oxford University Press.

16. Starratt, V. G., Goetz, A. T., Shackelford, T. K., Stewart-Williams, S. (2008). Men's partner-directed insults and sexual coercion in intimate relationships. *Journal of Family Violence, 23,* 315–323.

17. Daly, & Wilson (1988), see Note 2; Dobash, & Dobash (1979), see Note 6.

18. Daly, M., Wiseman, K. A., & Wilson, M. I. (1997). Women with children sired by previous partners incur excess risk of uxoricide. *Homicide Studies, 1,* 61–71; Shackelford, T. K., Buss, D. M., & Weekes-Shackelford, V. A. (2003). Wife-killings committed in the context of a "lovers triangle." *Basic and Applied Social Psychology, 25,* 127–133; Wilson, M., Daly, M., & Daniele, A. (1995). Familicide: The killing of spouse and children. *Aggressive Behavior, 21,* 275–291.

19. Buss, D. M. (2005). *The murderer next door.* New York: Penguin Press; Buss, D. M., & Duntley, J. D. (1998). Evolved homicide modules. Paper presented at the Annual Meeting of the Human Behavior and Evolution Society, Davis, California, July 10; Buss, D. M., & Duntley, J. D. (2003). Homicide: An evolutionary perspective and implications for public policy. In N. Dress (Ed.), *Violence and public policy* (pp. 115–128). Westport, CT: Greenwood.

20. Hellmuth, J. C., & McNulty, J. K. (2008). Neuroticism, marital violence, and the moderating role of stress and behavioral skills. *Journal of Personality and Social Psychology, 95,* 166–180.

21. Dutton, D. G., & Kerry, G. (1999). *Modus operandi and personality disorder in incarcerated spousal killers. International Journal of Law and Psychiatry, 22,* 287–299; Goetz, A. T., & Shackelford, T. K., Starratt, V. G., & McKibbin, W. F. (2008). Intimate partner violence. In J. D. Duntley & T. K. Shackelford (Eds.), *Evolutionary forensic psychology* (pp. 65–78). New York: Oxford University Press; Paulozzi, L. J., Saltzman, L. E., Thompson, M. P., & Holmgreen, P. (2001). Surveillance for homicide among intimate partners: United States, 1981–1998. In *MMWR: CDC Surveillance Summaries* (50, SS-3). Atlanta, GA: U.S. Department of Health and Human Services; Wilkinson, D. L., & Hamerschlag, S. J. (2005). Situational determinants in intimate partner violence. *Aggression and Violent Behavior, 10,* 333–361.

Index

A

Abnormal hormones
  antisocial behavior and, 13–15
  biosocial interactions, 14–15
  environmental toxins, 14
  hormonal imbalances, 13–14
  neurotransmitter dysfunction, 14
Abrahamse, A., 203
Abuse
  brain and, 64–65
  drug, 54
  of infants, 36
  Michigan Alcoholism Screen Test,
    Drug Abuse Screening Test, 172
  of mothers, 36
Adaptability, 84
Addiction
  Addiction Severity Index, 39
  alcohol, 55
  drug, 55
Addiction Severity Index, 39
ADHD. See Attention deficit
  hyperactivity disorder
Adolescence. See also Adolescents
  brain at, 66–68
  offending during, 117
  strains, 119–120
  substance use careers and, 172
Adolescence-limited (AL) offenders,
  117, 136, 237–238, 244–245
  GST explanation of, 119–121
Adolescent-limited pathway, 60, 66
Adolescents, 60, 61
  behavior, 66–68
  hormone surge, 66–67
  youths of incarcerated parents, 37, 39
Affect, 161
Affectionate bonds, 61, 62. See also
  Loving touch
African Americans, 36, 154. See also
  Blacks
  arrest rate for crimes, 134
  infant mortality rates, 52
  juvenile justice system, 133
  life-course-persistent offenders, 136
  and Whites, 138

Age–crime curve, 66
Agency, 272
  human, 274
Agents, 274
Aggression. See also Aggressive
  behavior; General aggression
  model (GAM)
  antisocial behavior and, 150–151
  defined, 150
  physical, 150
  relational, 150
  theories, 158–159
  verbal, 150
  violence and, 64, 150
Aggressive behavior, media violence
  and, 149–163
Agnew, R., 117, 137
Akers, R. L., 187–188, 191, 197–199
AL offenders. See Adolescent-limited
  offenders
Alcohol, 171–172
  addiction, 55
  consumption, 52
  dose–response relationship, 50
  FAS, 50–51
  FASD, 51
  genetic factors and, 51
  during pregnancy, 50–52
  as prenatal insult, 50–52
Alleles, concept of plasticity, 78
Allostasis, 65
Allostatic load, stress-inducing
  events, 65
Altschuler, D. M., 208, 210
Amen, D. G., 9
American Pragmatism, 273
American Psychiatric Association,
  38, 168
American Society of Criminology,
  28, 167
Amino acid production, genes code
  for, 87
aMRI. See Anatomical magnetic
  resonance imaging
Amygdala, 31, 63, 64
Anatomical magnetic resonance
  imaging (aMRI), 9

Anderson, C. A., 149, 154, 159
Andrews, D. A., 287
Androgens, 30–31
Anger, leading to crime, 118
Anglin, M. D., 173
Anoxia, 31
ANS. See Autonomic nervous system
Anticuckoldry tactics, 312, 314
Antisocial behavior, 3, 4, 74, 85, 87, 89
  abnormal hormones,
    neurotransmitters, and toxins,
    13–15
  aggression and, 150–151
  biosocial model of, 15–17
  brain deficits, 8–12
  and career criminality, 223–224
  causes of, 249
  children engage in, 188
  conclusion, 17–18
  defined, 4
  developmental neurobiology and, 60
  family concentration of, 76
  gangs and, 215–226
  intergenerational transmission of, 76
  measures of, 78, 241
  multidisciplinary approach, 3
  neuropsychological research, 12–13
  obstetrical factors, 6–8
  overview, 3–4
  persistent, 13
  propensity-based approach to, 218
  psychophysiological research, 4–6
  risk factors, 3, 74
  substance use careers and, 167–174
Antisocial homophily, 216, 221–223
Antisocial personality disorder (APD),
  9–10, 12
Antisocial potential (AP), 240–243
  decreasing, 288
  life-course development, 286–287
Antisocial tendencies, 31
Antisocial youths, 222
AP. See Antisocial potential
APD. See Antisocial personality
  disorder
Appraisal, 161–162
Approach–avoidance imbalance, 67

Arizona Arrestee Drug Abuse Monitoring (ADAM) program, 222
Arousal, present internal state, 161–163
Arseneault, L., 6
Association, 277. *See also* Cognitive neoassociation theory; Differential association theory
Athens, Lonnie, 272, 280
Attention deficit hyperactivity disorder (ADHD), 220
Attitudes, 162
Atypical EEG frontal asymmetry, 5
Autonomic nervous system (ANS), 64–65
Autonomic underarousal, 3

B

Baker, J. H., 78, 103
Baler, R. D., 170
Barker, E. D., 179
Barnes, J. C., 76, 80, 83
Baron, S. W., 117, 123, 125–128
Bateman, A. L., 266
Bates, J. E., 222
Beadman, M., 223
Beaver, K. M., 76, 78–79, 99, 220, 255, 297
Becker, H., 272, 278
Behavior, 29, 168–169, 262. *See also* Aggressive behavior; Antisocial behavior
  adolescent, 67
  contextual/situational influences on, 263–265
  criminal, 54, 250
  gang, 206
  legal/psychological definitions, 267
  mate retention behaviors, 312–313
  pools of, 266
Behavioral consistency, 262–263
Behavioral disorders, disruptive, 224
Behavioral genetics, 29, 85–87, 169
  introduction of, 100
  risk and protective factors, intersection of, 101–103
Belsky, J., 78
Bennetto, H., 11
Benson, J. S., 217
Berman, M. E., 14
Bernard, T. J., 138
Between-individual stability, 234
Bihrle, S., 11

Biological criminology, 87–89
Biological determinism, 69
Biological research
  biosocial effects, 32–33
  central nervous system factors, 31
  early physiological factors, 32–33
  genetic/cytogenetic factors, 29–31
  head trauma, 31–32
  hormones, 33–34
  international studies, 35
  neurotransmitters, 33–34
  prenatal/perinatal predictors, 29–35
Biological risk factors, 3–4, 15
Biosocial criminology, 83
  behavior genetics, 85–87
  biological criminology, 87–89
  evolutionary psychology, 84–85
  informing public policy, 90–91
  molecular genetics, 87
  neurocriminology, 89
  policy implications of contemporary, 92–95
Biosocial effects, physiological factors and, 32–33
Biosocial interactions, 4
  abnormal hormones, neurotransmitters, and toxins, 14–15
  brain deficits, 11–12
  neuropsychological research, 13
  obstetrical factors, 7–8
  psychophysiological research, 6
Biosocial life-course theory. *See also* Developmental and life-course theory
  substance use careers and, 169–173
Biosocial policies *vs.* criminological policies, 93, 95
Birth complications, 7
Birth defects, 53
Birth weight reduction, and smoking, 49
Bishop, D., 140, 143
Blacks, 194–197, 254
Blalock, H. M., 141
Blood lead levels, 53–54
Blood–brain barrier, 47
Blumer, H., 272, 277
Blunted arousal, 65
Bolden, C. L., 217
Bonta, J., 287
Bouchard, M., 217
Bouffard, J., 256
Boutwell, B. B., 73, 76, 79
Braga, A. A., 225–226

Brain, 59–60. *See also* Developmental neurobiology; Gray matter; *specific parts of brain*
  abuse and, 64–65
  at adolescence, 66–68
  damage, 31
  early environmental influences on wiring, 62–63
  early postnatal development, 60–61
  environmental circumstances of, 59
  experience-expected developmental input, 68
  and exposure to violence, 63–64
  growth, 62
  imaging, 4
  key systems, 59
  neglect and, 64–65
  species-typical structures, 61
  stress and, 64–65
  systems, 59–60
  violence and, 63–64
Brain deficits, 3, 8–9
  biosocial interactions, 11–12
  brain lesions, 10–11
  structural/functional abnormalities, 9–10
Brain lesions, 10–11
Breastfeeding
  IQ and, 63
  oxytocin release, 62
  positive developmental effects, 62–63
Brennan, P. A., 6–8, 13
Bridges, G., 142
Brody, G. H., 137
Broken family structure, 32
Brook, A. S., 14
Brutalization, 280
Buchsbaum, M., 11
Bullying, 179
Bureau of Justice Statistics, 167
Burgess, E. W., 189
Burns, J. M., 14
Buss, D. M., 312, 315
Bynum, T. S., 221, 223

C

Cadmium, 14
Cadoret, R. J., 77
Calcium, 53
  lead and, 53–54
  in neurotransmission, 53
Caldwell, L., 208, 210
Calling card. *See* Signatures

Cambridge Study in Delinquent Development, 235–236, 240–241
Career criminality, antisocial/gang behavior and, 223–224
Carnagey, N. L., 154, 159
Caspi, A., 15, 77, 171
Caucasian, infant mortality rates, 52
Caudill, J. W., 224
CD. *See* Conduct disorder
Centers for Disease Control and Prevention (CDC), 51, 53, 93
Central nervous system factors, 31–32
Chang, J., 36
Chapple, C. L., 255
Character, 285
   applying to rehabilitation, 291–294
   defined, 290
   maturation, 288
Chen, Y. F., 137
Chicago School, 273, 277
Child abuse, 123
Child development, 190
Child maltreatment, stress responses, 65
Children, 37–38, 49, 78–79. *See also* Adolescence; Adolescents; Infants
   born to drinking mothers, 50–52
   born to smoking mothers, 48–50
   child development, 190
   childhood, 171–172, 179–183
   criminal behavior in, 74–75
   with dietary insufficiencies, 93
   interventions effectiveness, 157
   longitudinal study of, 193
   neighborhood influences on, 187, 192
   physical punishment of, 150
   protection/nurturing of young, 59
   from violent homes, 123
Christenson, P. G., 155
Chronic offenders, 234, 298
Chu, R., 222
Cigarette smoking, 32. *See also* Tobacco
   secondhand smoke, 54
Cincinnati, children blood lead levels in, 54
CJA. See Criminal Justice Abstracts
Cleveland, H. H., 79, 99
Cloninger, C. R., 285, 295
Cloninger, R., 16
Cloward, R. A., 221
Cocaine, 52
Coccaro, E. F., 14
Codina, E., 221
Cognitive ability, 32, 161

Cognitive neoassociation theory, 158–159
Cognitive scripts, 263
Cohen, D., 313
Cohen, D. B., 75
Cohen, R., 219
Collective socialization model, 190
Colvin, C. R., 264
Commitment portfolios, 280–281
Communities. *See also* Neighborhoods
   definition of, 46
   inner-city, 54
Comorbidity, 178
Competition
   competition model, 190
   male–male, 314
   sperm competition, 313–315
Concentric zone model, 189
Conduct disorder (CD), 220–221
Conflict Tactics Scale 2, 38
Conflict theory, 140
Contact stimuli, 62
Contagion model. *See* Epidemic model
Contemplation, 292
Continuity, 206
Convit, A., 5
Co-offending, 244
Cooley, Charles Horton, 272
Cooperativeness, 291
Cortical gray matter, 50
Cortisol, 13–15, 34, 65
"Crack baby," 52
Craig, W. M., 221
Crick, N. R., 222
Crime, 37, 202, 216, 224–225. *See also* Homicide; Serial crime; Violence; Violent crime
   age–crime curve, 66
   and delinquency, 188–189
   gender differences in, 251
   homelessness increases, 125
   ICAP explaining commission, 243–244
   key mechanism for understanding, 170–171
   racial/ethnic differences in, 251–252
   strain with, 120–121
   types, 286
Criminal behavior, 54, 76, 250
   in children, 74–75
Criminal careers, 234
   aging property offenders, 279–280
   commitment portfolios, 280–281
   desistance from, 281
   implications, 278–279

   parameters and demographics, 300
   reasons for, 286
   symbolic interactionism and, 278–281
   violentization and, 280
Criminal coping
   homelessness, 125
   strains, 118–119, 122
   unemployment lead to, 126
   victimization, 128
Criminal Justice Abstracts (CJA), 28, 31, 35, 37, 40
Criminal justice system, 224–225
Criminal offending, 136
Criminality
   biological factors of, 29–32
   career. *See* Career criminality
   environmental factors of, 35–36
Criminogenic risk factors, genetic and environmental influences on, 108–109
Criminogenic risks, 287
Criminological literature
   prenatal and perinatal factors in, 27–29
   research in, 36
Criminological policies, biosocial policies *vs.*, 93, 95
Criminological research, 85
Criminological theories, 275–277. *See also specific theories*
Cross-sectional correlational studies, 202–203
   defined, 151
   film/television, 153
   gangs, 202
   music, 155
   video games, 156
Cross-sectional survey studies, 183
Cuckoldry, 312, 314
"Cuddle chemical," 62
Cullen, F. T., 68, 94, 253
Culprits, list of possible, 74
Culture, 275
Curran, D. A., 142
Curry, G. D., 204, 209, 217
CYP1A1 gene, 49
Cytogenetic factors of criminality, 29–31
Czobor, P., 5

D

Dabbs, J. M., 14
Daly, M., 85, 315
Danger Assessment Instrument, 39
Daniels, D., 76

Decision making, 238, 275
Decker, S. H., 205, 208–210, 217–218
Defiance, 280
Definition of situation, 273
DeFries-Fulker (DF) equation
    analysis, 107–108
Delinquency, 54, 188–189
    neighborhoods and, 190–192
    in parent–child relationships, 100
    race differences in, 136
    SSSL and, 190–192
Delinquent peers, 104–105, 188,
    195–197
DeLisi, M., 77–78, 99, 215, 220,
    255, 297
Deoxyribonucleic acid (DNA), 46
Department of Justice, U.S., 203, 315
Depression, risk factors, 105
Desensitization to violence, 163
Desistance, 206–210, 217–219
    conclusion, 308
    from criminal careers, 281
    introduction, 297
    LCP and, 298–299
Developmental and life-course theory
    (DLC), 206–207
    conclusion, 245–246
    ICAP and, 240–241
    individual differences vs. individual
        change, 235–236
    introduction, 233–235
    policy implications, 244–245
    protective/promotive factors,
        236–237
    theories, 237–240
Developmental neurobiology.
    See also Brain
    abuse/neglect/stress and, 64–65
    antisocial behavior and, 60
    brain at adolescence, 66–68
    brain/violence and, 63–64
    conclusion, 68–69
    early environmental influences on
        brain wiring, 62–63
    early postnatal brain development,
        60–61
    introduction, 59–60
    neural readiness, 61
Developmental trajectories
    discussion, 183–184
    female victimization, 180–182
    identification, 180
    male victimization, 180
    sample/statistical analyses, 180

teachers' knowledge of, 182–183
    to violence exposure, 177–184
Dewey, J., 272, 276
Diagnostic and Statistical Manual of
    Mental Disorders, Fourth Edition
    Text Revision (DSM-IV-TR), 38,
    168, 170
Dialectic, 274
DiClemente, C. C., 288
Dietary interventions, 38
Differential association theory, 271,
    275–277
Differential identification, 277
Differential offending, 133
    theoretical explanations for, 136–139
Differential selection theories, 140
    macro-level theories, 141–142
    micro-level theories, 142–143
Differential social location, 191
Differential social organization,
    191, 276
Direct assessment of whole blood, 54
Disadvantaged neighborhoods, 187
Disney, E. R., 51
Disproportionate minority
    confinement/contact (DMC),
    143–144
Disruptive behavioral disorders, 224
Dizygotic (DZ) twins, 86, 101
DLC. See Developmental and life-
    course theory
DMC. See Disproportionate minority
    confinement/contact
DNA. See Deoxyribonucleic acid
Dodge, K. A., 222
Dolan, M. C., 10
Domination and control
    hypothesis, 314
D'Onofrio, B. M., 51
Dopamine, 30, 170
    dopaminergic system, 48
    receptor gene, 77
    receptor polymorphism, 30
Dose–responsive relationship, 50
Douglas, J. E., 265
DRD2, 77, 78
    and DRD4, 30
Drug Abuse Screening Test, 172
Drugs, 160
    addiction, 55, 168
    homelessness, 125
    maternal abuse, 54
    physical abuse associated with, 124
    as prenatal insult, 52–53

Drury, A. J., 297
DSM-IV-TR. See Diagnostic and
    Statistical Manual of Mental
    Disorders, Fourth Edition Text
    Revision
Dual hazards, 15
Dual pathway model, 66
Dual taxonomy, 178
Dualism, 274–275
Duntley, J. D., 315

E

Earls, F., 137
Early physiological factors, 32–33
Eckel, L., 65
Economic strains, street youth, 127
Edelman, G., 61
EEG. See Electroencephalogram
EF. See Executive functioning
Egan, V., 223
Egley, A., 203, 209, 216
Eitle, D., 137
Electroencephalogram (EEG), 4, 17
Elliott, D., 135
Ellis, L., 29
Embryo, 47–48, 52. See also Prenatal
    insults; Womb
Emotional abuse, 123
Emotional neglect, 123
Empirical literature, 253
Enhancement model, 203, 216
Environmental effects, on cumulative
    risk and protective factor index,
    109–110
Environmental influence on human
    behavior, 88–89
Environmental research, 35–36
    abuse of mothers, 36
    infant maltreatment, 35–36
    socioeconomic status/poverty,
        35–36
Environmental toxins, 14
Epidemic model, 190
Epilepsy, 33
Eron, Leonard, 263
ERP responses. See Event-related
    potential responses
Esbensen, F. A., 210, 218, 221
Ethnic diversity, 189
Ethnic selection bias, 139–140
Eubanks, J., 154
Event-related potential (ERP)
    responses, 4–5, 17
Evolutionary psychology, 84–85, 311

Evolutionary theory, 84
Evolved homicide module, 315
Excessive slow-wave EEG, 5
Excitation transfer theory, 159
Excitatory neurotransmitters, 67
Executive functioning (EF), 12–13
Expectation schemata, 162
Experience dependent, 61, 64
Experience expected, 61, 63, 68
Experimental studies
    defined, 151
    film/television, 152
    music, 154–155
    video games, 155–156
External risk, and protective factors,
    102–103
Exterogestation, 62
Eysenck, H. J., 15

F

Facilitation model, 203, 216
Failure to achieve positively valued
    goals, 118
Family, 39, 74, 76
    broken family structure, 32
    concentration of antisocial
        behavior, 76
    conflicts, 100
    environment, 11–12, 75
    risk, 104
    role of, 77
Family Stress Checklist, 39–40
Farrington, D. P., 6, 73–74, 220, 233,
    236–237, 245, 286
FAS. *See* Fetal alcohol syndrome
FASD. *See* Fetal alcohol spectrum
    disorders
Fast, D. K., 7
FBI. *See* Federal Bureau of
    Investigation
Fear, 289, 290
    lack of, 5
Federal Bureau of Investigation (FBI),
    202–203
Felony convictions, 302–303
Female sexual infidelity, 312, 314–315
Female victimization trajectories,
    180–182
Ferracuti, F., 138
Fetal alcohol spectrum disorders
    (FASD), 51
Fetal alcohol syndrome (FAS), 7,
    50–51
Fetal development, 46–47

Fetus, prenatal insults mechanisms to,
    47–54
Feyerherm, W., 140
Film/television
    cross-sectional studies, 153
    experimental studies, 152
    longitudinal studies, 153–154
    media violence, 151–154
Fine motor abilities, 53
Flannery, D. J., 177
fMRI. *See* Functional magnetic
    resonance imaging
Foley, D. L., 15
Forrest. W., 254
Fox, K., 222
Fragile Families and Child Wellbeing
    Study, 79
Frisell, T., 73
Functional magnetic resonance
    imaging (fMRI), 9, 11, 64, 67, 169
Funder, D. C., 264
Furr, R. M., 264

G

Gaarder, E., 142
Gage, P., 10–11, 314
Gagnon, C., 221
Galvanic skin response, 34
GAM. *See* General aggression model
Gang behavior and career criminality,
    223–224
Gang concept, critiques of
    low commitment and disloyalty,
        217–218
    low organization, 217
    nonceremonial onset and
        desistance, 218–219
Gang concept, reformulation of
    antisocial/gang behavior and career
        criminality, 223–224
    criminal justice system
        involvement, 224–225
    strained relations, peer rejection,
        and antisocial homophily,
        221–223
    temperamental and neurocognitive
        deficits, 219–221
Gang enhancement laws, 205
Gang members
    behavior, 225
    chronic involvement of, 225
    noncompliance of, 224
    prison, 224
Gang organization, 217

Gang-reduction programs, 225
Gangs
    and antisocial behavior, 215–226
    behavior, 206
    ceremonial aspects of, 217
    conclusion, 210–211, 225–226
    continuity, 206
    defined, 202
    desistance, 206, 217–219
    DLC, 206–207
    gang enhancement laws, 205
    introduction, 201, 215–216
    life course and, 202–206
    longitudinal studies, 204, 209
    National Youth Gang Center,
        203–204, 209
    onset, 206
    prison, 205–206
    violence and, 202–206
    youth, social deficits of, 220
Gao, Y., 3
Gavazzi, S., 221
GxE. *See* Gene–environment
    interaction
Gender differences, 251
Gender gap in offending, 85
Gene–environment correlation (rGE),
    78–79
    types of, 103
Gene–environment interaction (GxE),
    49, 77–78, 87
General aggression model (GAM),
    151
    person inputs, 159–160
    present internal state, 160–163
    situational inputs, 160
General strain theory (GST), 117, 137
    adolescence-limited offending,
        119–121
    life-course-persistent offending,
        121–122
    overview of, 118–119
    strain types, 126–127
A General Theory of Crime
    (Gottfredson and Hirschi), 170,
    202, 249
Genes, 75–76
    code for amino acid production, 87
    plasticity. *See* Plasticity genes
Genetic determinism, 69
Genetic effects
    on cumulative risk and protective
        factor index, 109–110
    heritability estimates, 101

Genetic factors, 29–31
  risk and protective factors, 102, 111
Genetic plasticity, 78
Genetic polymorphisms, 29–30, 87
Genetic/cytogenetic factors, 29–31
Gentile, D. A., 155
Gibbs, J. J., 253
Gibson, C. L., 187
Giever, D. M., 253
Giordano, P., 281
Glaser, B., 272
Glaser, D., 277
Glenn, A., 3
Glueck, E., 28, 238, 307
Glueck, S., 28, 307
Goetz, A. T., 314–315
Goffman, E., 272, 275, 278
Goldbaum, S., 179
Goldstein, D., 65
Goldstein, P. J., 173
Good lives approach, 285–295
Gordon, R. A., 220
Gottfredson, M. R., 136, 170, 202–203,
  218, 223, 249–257, 275, 307
Gould, S. J., 74
Goyer, P. F., 9
Grafman, J., 10
Grasmick, H. G., 256
Grasmick scale, 256
Gray matter, 31–32, 50
  reduced, 54
Grella, C. E., 173
Gross motor abilities, 53
Groves, W. B., 189
GST. See General strain theory
GSTT1 gene, 49
Gunnison, E., 36

H

Hagan, J., 124
Harm avoidance, 289
Harrell, E., 204
Hartnagel, T. F., 126
Hawkins, D. F., 138
Hawkins, J. D., 244
Hay, C., 254
He, N., 221
Head trauma, 31–32
"Healing of the psyche," 294
Heart rate activity/reactivity, 4–5
Heilbrun, K., 172
Heimer, K., 276
Hematoencephalic barrier. See blood–
  brain barrier
Hemming, J. H., 6

Heritability component, 86
Heritability estimates, 101
  of antisocial behavior, 102
  of external measures, 103
Heritable, 169
Higgins, G. E., 249, 254–256
Hindelang, M., 134
Hippocampus, 64
Hirono, N., 9
Hirschi, T., 136, 170, 202–203, 218,
  249–257, 275, 307
Hispanics, 137–138, 194–197, 208
  illicit drug use in, 52
HIV/AIDS, 55
Hoffman, V., 173
Hokkaido Study of Environments and
  Children's Health, 49
Home visitation programs, 37–38
Homelessness, street youth, 124–125
Homicide, 138, 218, 223, 225, 265–267
  evolved homicide module, 315
  gang, 202, 203
  intimate partner, 315
  premeditated, 315
Homo sapiens, 62
Hooks for change, 281–282
Hormones, 33–34, 88. See also
  specific hormones
  abnormal, 13–15
  biological research, 33–34
  imbalances, 13–14
  surge in adolescents, 66–67
Howell, J. C., 203–204, 216, 223
HPA axis. See Hypothalamic-
  pituitary-adrenal axis
Hser, Y. I., 173
5-HTTLPR, 30
Huebner, B. M., 223
Huesmann, L. R., 263
Huff, C. R., 223
Human agency, 274
Human behavior, environmental
  influence on, 88–89
Hutchinson, P. L., 314
Hyperactivity, 60
Hypercortisolism, 65
Hypermasculinity, 287
Hypoglycemia, 33
Hypothalamic-pituitary-adrenal
  (HPA) axis, 64–65
Hypothalamus, 62, 64–65

I

ICAP. See Integrated cognitive
  antisocial potential theory

Identification, 277
Identity, spoiled, 272
Illicit drug, use by pregnant women,
  52–53
Inattentiveness, 60
Incarcerated parents, youths of, 37
Incarceration. See also Prisons
  psychological effects of, 287–288
  youths of incarcerated parents, 37
Individual differences vs. individual
  change, 235–236
Individual differentiation, 262–263
Infants. See also Breastfeeding;
  Obstetrical factors; Prenatal
  insults; Prenatal/perinatal
  predictors
  abuse, 36
  "crack baby," 52
  mortality, 52
  mother–infant contact, 62
  SIDS, 52
Infidelity, 312–315
Informal social control, 191
Inhibition, 256
Inhibitory neurotransmitters, 67
Inhumane policies an inevitable
  byproduct of biosocial research,
  91–92
Institutional resource model, 190
Integrated cognitive antisocial
  potential theory (ICAP), 286
  chart, 240
  DLC and, 240–241
  explaining crime commission,
  243–244
  long-term risk factors in, 241–243
Intelligence quotient (IQ), 32, 49,
  51, 239
  breastfeeding and, 63
  lead and, 53
  low, 63
Interactional theory, 239
Interactionist theory of desistence, 281
Interactionist-inspired rehabilitation
  policies, 282
Intergenerational transmission of
  criminality, 74
Internal risk, behavioral genetics
  with, 102
International studies on perinatal
  factors, 35
Interpretation, 273
Interventions, 37
  home visitation programs, 37–38
  media violence research, 157–158
  nutrition, 38

prenatal/perinatal predictors, 37–39
risk assessment instruments, 38–39
youths of incarcerated parents, 37
Intimate partner homicide, 315
evolved homicide module, 315
killing-as-a-byproduct
hypothesis, 315
proximity hypothesis, 315
Intimate partner violence (IPV), 36,
311, 313, 316
Intoxicants, psychoactive, 167
IPV. *See* Intimate partner violence
IQ. *See* Intelligence quotient

J

Jacobs, B., 277
Jaffee, S. R., 78
James, W., 272
Jang, S. J., 137
Jasinski, J. L., 36
Jealousy
defined, 312
male, 312
male sexual, 312–313
partner-directed violence and, 313
Jencks, C., 190
JJDP Act. *See* Juvenile Justice and
Delinquency Prevention Act
Johnson, B. R., 137
Jonson, C. L., 94
Josephson, W. L., 158
Juhasz, C., 9
Justice system, criminal, 224–225
Juvenile Justice and Delinquency
Prevention Act (JJDP Act),
143–144
Juvenile justice system
overrepresentation of minority
youth in, 133, 135
policy implications, 143–144
race/ethnic selection bias, 139–140
Juvenile offending, 137

K

Kagan, J., 295
Kaighobadi, F., 311, 313
Kandel, E. S., 6–7
Kaplan, C. D., 221
Katz, C. M., 222
Kaufman, J. M., 137
Keenan, K., 220
Kendler, K. S., 78, 103
Kennedy, D. M., 225
Kerr, M., 224
Killing-as-a-byproduct hypothesis, 315

Kinship pairs, 101
analysis of, 110–111
Kishi, R., 49
Klein, M. W., 202, 205, 208–209
k-line x-ray fluorescence
spectroscopy, 54
Knifing off, 239
Koob, G. E., 168
Kopin, I., 65
Kort-Butler, L. A., 223
Kosloski, A. E., 297
Krohn, M. D., 204, 222, 239–240, 245
Kruesi M. J., 4

L

Laakso, M. P., 10
Labeling theory, 271, 275
symbolic interactionism and, 278
themes, 278
Lahey, B. B., 220
Language, 61
Latent trait, 168
Laub, J. H., 60, 141, 238–239,
244–245, 297–299
Lauritsen, J. L., 208–210
Law, 205, 276
LCP. *See* Life-course-persistent
offenders
Lead, 93
calcium and, 53–54
in children, 48
elevated blood levels, 54
IQ and, 53
poisoning, 54
as prenatal insult, 53–54
Lee, G., 192
Lefkowitz, M. M., 263
Leiber, M. J., 133, 140
Lemert, E., 278
Lewis, D. O., 11, 13, 31
Liberation hypothesis, 142–143
Lidz, V., 172
Life course, 135, 140. *See also*
Biosocial life-course theory;
Developmental and life-course
theory
development of AP, 286–287
gangs and, 202–206
violence and, 202–206
Life events, 235
Life-course-persistent offenders
(LCP), 136, 237–238, 244
current studies, 297
desistance and, 298–299
study methodology, 299–300

study results, 300–306
theoretical discussion, 307–308
Life-course-persistent pathway, 60
Life-course-persistent offending, 117
GST explanation of, 121–122
Lindesmith, A., 278
Linking, 261
Lipsky, S., 36
Lizotte, A. J., 222
Loeber, R., 137, 220, 237
Lombroso, C., 27–28
Loney, B., 65
Longitudinal studies
of children, 193
defined, 151
film/television, 153–154
gangs, 204, 209
PHDCN consists of, 188
Risk Longitudinal Twin
Study, 49
video games, 156–157
Love, 62, 64. *See also* Affectionate bonds
Loving touch, 64
Low constraint
criminal coping, 119
individuals with, 121–122
Low impulse control, 60
Lower intelligence, 12
Lynam, D. R., 12

M

MacKenzie, D. L., 254
Macro-level theories of differential
selection, 141–142
Magnetic resonance imaging (MRI),
49–50
Mahoney, M., 249
Major, A. K., 203
Male sexual jealousy, 312–313
Male victimization trajectories, 180
Male–male competition, 314
Malevolency, 280
Manganese, 14
Manier, E., 285
Mantle, M., 307
MAO-A. *See* Monoamine oxidase A
MAO-A polymorphism, 29–30
Marcum, C. D., 256
Marijuana, 52
Martin, J., 253
Marx's theory of socialism, 91
Mata, A. D., 177
Mataró, M., 11
Mate selection, 59
Mate retention behaviors, 312–313

Maternal smoking during pregnancy, 49–50
Matrix Intensive Outpatient Treatment Program for Persons with Stimulant Use Disorders, 288
Matsueda, R., 276, 278
Maturity gap, 238
Matza, D., 277
Maughan, B., 49
Maxson, C. L., 202, 204, 208, 218
Mayer, S., 190
Mazur, A., 15
McCartan, L. M., 36
McCarthy, J., 124
McGarrell, E., 221
McGloin, J. M., 218
McKay, H. D., 187, 189
McKenzie, R. D., 189
McKibbin, W. F., 313
Mead, G. H., 272, 276, 278
Meaning, 273
Media violence
    aggression theories and, 158–159
    aggressive behavior and, 149–163
    conclusion, 163
    film/television, 151–154
    intervention research, 157–158
    introduction, 149–150
    music, 154–155
    research, 151–158
    video games, 155–157
Meditation, 292
Mednick Perinatal Complications Scale, 38
Mednick, S. A., 6–8, 15, 38
Melde, C., 218, 221
Merton, R., 275
Mesolimbic reward system. See Reward pathways
Meta-analysis, 5
Michael, G., 307
Michelson, N. M., 6
Michigan Alcoholism Screen Test, 172
Micro-level theories of differential selection, 142–143
Miller, W. B., 220
Mills, C. W., 277
Minnesota Twin Family Study, 51
Minor physical anomalies, 6–8
Minority youth, 141
Mirrors and Masks (Strauss), 272
Mitchell, O., 254
Moffitt, T. E., 12–13, 15, 60, 66, 76, 136–137, 178, 237–238, 244–245, 298, 307

Moffitt's developmental theory, 136
Molecular genetic data, 77
Molecular genetics, 87
Monoamine oxidase A (MAO-A), 77, 112
    polymorphism, 29–30
Monozygotic (MZ) twins, 86, 101
Montagu, A., 62
Moral commitment, 280–281
Morris, R. G., 15, 80
Mothers
    abuse of, 36
    mother–infant contact, 62
MRI. See Magnetic resonance imaging
Multifactor approach, 28
Mulvey, E. P., 308
Munn, C., 265
Music
    cross-sectional correlational studies, 155
    experimental studies, 154–155
    media violence, 154–155
Mutations, 30
MZ twins. See Monozygotic twins

N

Nascent gang members, 223
National Commission on the Causes and Prevention of Violence, 149
National Council on Crime and Delinquency Research Center West, 37
National Crime Victimization Survey (NCVS), 134, 204
National Gang Threat Assessment, 224
National Incident-Based Reporting System (NIBRS), 134
National Institute of Justice, 206
National Longitudinal Study of Adolescent Health, 224
National Youth Gang Center, 203–204, 209
National Youth Survey, 135
NATO Advanced Study Institute on the Biosocial Bases of Violence, 35
Natural selection, 61–62, 66, 68–69
NCVS. See National Crime Victimization Survey
Negative emotionality, 60
Negative emotions
    development of, 124
    homelessness, 125
    individuals with, 121–122
    leading to crime, 118

Negative person-environmental interactions, 60
Negative stimuli, presentation of, 118
Neglect, 64–65
Neighborhood structural elements, 190
Neighborhoods. See also Communities; Project on Human Development in Chicago Neighborhoods
    delinquency and, 190–192
    disadvantaged, 187
    SSSL and, 190–192
Neonaticide, 35
Neural Darwinism, 61
Neural networks, 60–61
Neural readiness, 61
Neurocriminology, 89
Neurons, frequently-activated, 61
Neuropsychological deficits, 239
    risk factors, 105
Neuropsychological impairments, 3, 12
    biosocial interactions, 11–12
    EF, 12–13
    lower intelligence, 12
Neuroscience, 169
    literature, 63
Neurotoxins, 47–48
Neurotransmission, 53
Neurotransmitters, 30, 33–34
    biological research, 33–34
    defined, 34
    dysfunction, 14
Neutralization theory, 277
Neutralizations, 276–277
NHVP. See Nurse Home Visitation Program
NIBRS. See National Incident-Based Reporting System
Nicotine exposure, 7–8
Nonshared environmental, 86
    effects, 101, 111
Non-White juveniles, 135
Norepinephrine, 34
Normative conflict, 276
Novelty seeking, 170
Nurse Home Visitation Program (NHVP), 93–94
Nutrition, 33
    and dietary interventions, 38
    interventions, 38
    of pregnant mothers, 33

O

Objective strains, 118
Obstetrical factors, 3
  alcohol exposure, 7
  biosocial interactions, 7–8
  birth complications, 7
  minor physical anomalies, 7–8
  pregnancy complications, 32
  prenatal nicotine exposure, 7
Obstetrics/posture conflict, 62
Oddball tasks, 5
Offender group
  arrests by life stage and, 301–302
  felony convictions by life stage and, 302–303
  kidnapping arrests across seven life stages by, 303, 306
  murder arrests across seven life stages by, 303, 305
  prison sentences across seven life stages by, 302, 304
  prison sentences by life stage and, 302, 304
  rape arrests across seven life stages by, 303, 305
  robbery arrests across seven life stages by, 303, 306
Offenders. *See also* Adolescence-limited offenders; Life-course-persistent offenders
  aging property offenders, 279–280
  chronic, 298
  profiling, 261
  victim relationship, 315
Offending, 233–234
  co-offending, 244
  differential, 133
  Hispanic, 137–138
  race differences in, 134–135
Offenses, 117
  adolescence-limited offending, 119–121
  among street youth. *See* Street youth
  life-course-persistent offending, 121–122
Office of Juvenile Justice and Delinquency Prevention, 206
Official records, 307
Ohlin, L. E., 221
O'Leary, M., 65
O'Neill, M. L., 172
Onset, 206
Opportunity theory, 275

OPRM1 gene, 95
Ordoña, T., 285
Orphanage-reared children, 63
Orphans, 63
Ortiz, J., 88
Oxytocin, 62–63. *See also* "Cuddle chemical"

P

PACT. *See* Parents and Children Together
Parent(s), 37, 74, 250–251
  biological, 63
  child attachment, 79
  and child behavior, 74
  child relationships, 79, 94, 100
  emotional bonds with, 66
Parental management, 250
Parent–child relationships, 79, 94, 100
Parenting effects, 75
Parents and Children Together (PACT), 37
Park, R., 189, 273, 276
Partner-directed violence. *See also* Intimate partner violence
  conclusion, 316
  intimate partner homicide, 315
  introduction, 311
  male jealousy and, 312
  male sexual jealousy and, 312–313
  mate retention behaviors and, 312–313
  paternity uncertainty, 312
  sexual coercion, 313–315
  sperm competition and, 313–315
Pasamanick, B., 7
Paternity uncertainty, 311–312
Patriarchal power, 314
Patriarchal terrorism, 314
Patterson, G., 250
PCL:SV. *See* Psychopathy Checklist Revised: Screening Version
PCL:YV. *See* Psychopathy Checklist: Youth Version
Peabody Picture Vocabulary Test (PPVT), 105
PeaceBuilders, 180
Peck, J. H., 133
Peeples, F., 137
Peer rejection, 221–223
Peers
  bonding, 60
  in self-control theory, 252–253
Pennington, B. F., 11

Perceptual schemata, 162
Perron, B. E., 167
Perry, B., 59
Persistence, 290. *See also* Life-course-persistent offenders
  life-course-persistent pathway, 60
Person inputs, 159–160
Personality, 169
  APD, 12–14
  rehabilitation and, 288–291
Peskin, M., 3
PET. *See* Positron emission tomography
Peterson, D., 210
Pettit, G., 222
PFC. *See* Prefrontal cortex
Phantom community, 272
PHDCN. *See* Project on Human Development in Chicago Neighborhoods
Phenotype, 84
Physical abuse, 124
Physical aggression, 150
  video game exposure on, 157
Physical punishment of children, 150
Physical victimization, 128
Physiological arousal, 161
Pine, D. S., 8
Piquero, A. R., 35, 254–256
Pittsburgh Youth Study, 236–237
Pizarro, J. M., 218
Plasticity genes, 78
Pleasanton Children's Center, 37
Plomin, R., 76
Pogrebin, M. B., 205
Poles, T. B., 187
Polymorphism, 20, 29–30
  dopamine receptor, 30
  MAO-A, 29
Pope, C., 134–135, 140
Popp, D., 313
Porteus Maze test, 13
Positively valued goals, failure to achieve, 118
Positively valued stimuli, removal of, 118
Positron emission tomography (PET), 9, 11
Postnatal brain development, 60–61
Poverty, stresses associated with, 122
PPVT. *See* Peabody Picture Vocabulary Test
Pratt, T. C., 253, 255
Pre-, peri-, and postnatal care (PPPC), 93

Preformationism, 60
Prefrontal cortex (PFC), 61, 63–64, 67
Pregnancy
  alcohol during, 50–52
  smoking during, 48–50
Premeditated homicides, 315
Prenatal insults. *See also* Embryo
  alcohol as, 50–52
  drugs, 52–53
  introduction, 45–47
  lead as, 53–54
  mechanisms of, 47–54
  neurotoxins and, 47–48
  policy consequences, 54–55
  soft/hard effects of, 48
  tobacco as, 48–50
Prenatal nicotine/alcohol exposure, 7
Prenatal/perinatal predictors
  biological research, 29–35
  conclusion, 40
  environmental research, 35–36
  history of research, 27–29
  interventions, 37–39
  introduction, 27
  legal issues, 39
Preschool intellectual enrichment programs, 244
Present internal state, 160
  affect, 161
  arousal, 161–163
  cognitions, 161
Presentation of negative stimuli, 118
Presumptions of entitlement, 315
Prison
  codes, 287
  gang members, 224
  gangs, 205–206
Process, 273–274
Prochaska, J. O., 288
Project on Human Development in Chicago Neighborhoods (PHDCN), 188
  current focus, 192
  discussion, 197–199
  methods, 192–194
  results, 195–197
Promotive factors, 236–237
Propensity perspective, 297
Propensity-based approach to antisocial behavior, 218
Property victimization, 128
Protection/nurturing of young, 59
Protective factors, 177, 236–237
  academic achievement, 106

criminal and delinquent involvement, 99
  etiology of, 100–103
  genetic and environmental influences on, 109
  index, 107
Proximate mechanisms of mother–infant bonds, 62
Proximity hypothesis, 315
Psychiatry, 294
Psychoactive intoxicants, 167
Psychological themes
  in serial crime, 266–267
  signatures *vs.*, 265–266
Psychopaths, 9–10, 17
Psychopathy, 168, 170, 172
Psychopathy Checklist Revised: Screening Version (PCL:SV), 221
Psychopathy Checklist: Youth Version (PCL:YV), 172
Psychophysiological impairments, 4
  atypical EEG frontal asymmetry, 5
  biosocial interactions, 6
  ERP responses, 5
  excessive slow-wave EEG, 5
  heart rate activity/reactivity, 4–5
  reduced skin conductance, 4–5
PsycINFO database, 28
Puberty, 66
Puente, S., 313
Pyrooz, D. C., 217

Q
Quantitative behavioral genetics, 86
Quinsey, V. L., 14

R
Race, 194, 251–252
Race differences
  in delinquency, 136
  in offending, 134–135
Race/ethnic selection bias, 139–140
Racial/ethnic differences, 251–252
Raine, A., 3–12, 14, 16, 88–89, 171
Raine's biosocial model of violence, 4
Randa, R., 223
Rape. *See* Sexual coercion
Rasanan, P., 8
Raudenbush, S. W., 137
Rebellon, C. J., 137
Recent Exposure to Violence Scale, 179
Reduced skin conductance, 4–5
Reflexivity, 272
Regression model, 195–197

Rehabilitation
  character applied to, 291–294
  conclusion, 295
  good lives approach, 285–295
  personality and, 288–291
  programs, 94–95
  stages of, 293–294
  temperament applied to, 291–294
Relational aggression, 150
Relative deprivation models, 190
Relative stability, 234
Religiosity, 106–107
Removal of positively valued stimuli, 118
Reproduction, 59
Reproductive fitness, 84
Research on media violence effects, 151–158
Residential instability, 189
Restak, R., 68
Resting heart rate, 88
Revenge, 205
Reward dependence, 289–290
Reward pathways, 170–171
rGE. *See* Gene–environment correlation
Rhodes, R., 280
Richmond Youth Survey, 256
Ricketts, M. L., 254
Risk assessment instruments, 38–39
Risk factor index, 105–106
  genetic and environmental effects on, 109–110
Risk factors, 178
  antisocial behavior, 3, 74
  criminal and delinquent involvement, 99
  criminogenic risks, 287
  etiology of, 100–103
  long term, in ICAP, 241–243
Risk Longitudinal Twin Study, 49
Roberts, D. F., 155
Rochester Youth Development Study, 239
Rodriguez, N., 141–142
Roper v. Simmons, 90

S
SAAF. *See* Strong African American Families Program
Salfati, C. G., 261, 266
Sampson, R. J., 60, 137–138, 141, 189, 238–239, 244–245, 297–299, 308
Schemata, 162
Schneider, F., 9

Schools. *See also* Chicago School
  protective factors, 106
  in self-control theory, 252
Schug, R. A., 3
Schur, E., 278
Screen violence, 149
Script theory, 159
Secondhand smoke, 54
Selection model, 203, 216
Selective evolutionary pressures, 62
Self, 272
Self-awareness, 292
Self-control, 249–250
  development of, 252
  variance in, 86
Self-control theory, 170, 250, 275
  conclusion, 257
  gender differences, 251
  introduction, 249
  literature review, 253–256
  peers in, 252–253
  racial/ethnic differences, 251–252
  schools in, 252
Self-reports, 135, 307
Self-restraint, 275
Self-transcendence, 291
Serial crime
  behavioral consistency, 262–263
  contextual/situational influences
    on, 263–265
  defined, 267
  individual differentiation, 262–263
  introduction, 261
  psychological themes in, 266–267
  signatures, 265–266
Serious behavioral disorders, 220
Serotonergic system, 48
Serotonin, 14–15, 30, 34
  serotonergic system, 48
Serotonin gene, 95
Severity of Violence Against Women
  Scale, 38
Sex-specific adaptive problems, 311
Sexual coercion, 313–315
Sexual proprietariness, 315
Shackelford, T. K., 311, 313–315
Shaffer, J. N., 222
Shared environment, 86
Shared environmental effects, 101
Shaw, C. R., 187, 189
Sheldon, W., 27–28
Shoal, G., 13
Shoplifting, 234
Short, J. F., Jr., 219–220
Shover, N., 279–280

Side bets, 275
SIDS. *See* Sudden infant death
  syndrome
Signatures
  defined, 265
  psychological themes *vs.*, 265–266
  in serial crime, 265–266
Silberg, J. L., 49
"Silence of the mind," 292
Silva, P. A., 12
Simons, R. L., 137
Single photon emission computed
  tomography (SPECT), 9
Situational factors, 263
Situational influence on aggression, 160
Situational inputs, 160
Situational-level interaction, 275
Skin conductivity, 34
Skinner, B. F., 263
Smoking during pregnancy, 48–50
Snares, 238
Snyder, H., 134–135
Social control theory, 275
Social disorganization theory, 45,
  137, 188
  in SSSL, 188–189
Social forces, 54
Social information–processing
  theory, 222
Social learning, 188
Social learning theory, 158, 190–191,
  275–277
Social pathologies, 68
Social push, 6, 16
Social risk factors, 3–4, 8
Social structure social learning (SSSL)
  child development, 190
  discussion, 197–199
  introduction, 187–188
  neighborhoods/delinquency and,
    190–192
  social disorganization theory in,
    188–189
Social support systems, protective
  factors, 106
Socialism, Marx's theory of, 91
Socialization, 275
Societal violence, 149
Socioeconomic status/poverty, 35–36
Socioemotional deprivation, 63
Sociological Abstracts, 28
Sociological criminologists, 68
Sociology, 272
Spandrel, 74
  criminology, 75

Species-typical brain structures, 61
SPECT. *See* Single photon emission
  computed tomography
Spergel, I. A., 225
Sperm competition, 313–315
Spindler, A., 217
Spoiled identity, 272
Spontaneous abortion, 53
Spousal homicides, 315
SSSL. *See* Social structure social
  learning
SSSM. *See* Standard social science
  method
Stages of change model, 288
Stakes in conformity, 275
Standard social science method
  (SSSM), 85
Starratt, F., 313
Starzyk, K. B., 14
Status Passage (Glaser), 272
Steen, S., 142
Steffensmeier, D. J., 143, 280–281
Sterzer, P., 9
Stewart, E. A., 137
Stigma, 272, 294
Stoddard, J., 11
Stone, G., 272
Stouthamer-Loeber, M., 220
Strain proliferation, 122
Strains
  adolescence, 119–120
  with crime, 120–121
  of homelessness, 125
  leading to crime, 118
  street youth, 123–124
  types of, 137
  and unemployment, 126
Strauss, A., 272
Street code, 139
Street youth
  economic strains, subjective, 127
  homelessness, 124–125
  offending patterns among, 122
  strains, 123–124
  unemployment, 125–127
  victimization, 128–129
Stress, 64–65
  brain and, 64–65
  defined, 64
  response, 64–65
Stretesky, P. B., 205
Strodtbeck, F. L., 219–220
Strong African American Families
  Program (SAAF), 95
Structural commitment, 281

Structural strain, 275
Structural/functional abnormalities in
    brain imaging, 9–10
Studies. *See also* Biological research;
    Cross-sectional correlational
    studies; Environmental
    research; Experimental studies;
    Longitudinal studies; Media
    violence
    Cambridge Study in Delinquent
        Development, 235–236, 240–241
    Hokkaido Study of Environments
        and Children's Health, 49
    LCP, 298–299
    Minnesota Twin Family Study, 51
    NATO Advanced Study Institute on
        the Biosocial Bases of Violence, 35
    Pittsburgh Youth Study, 236–237
    Risk Longitudinal Twin Study, 49
    Rochester Youth Development
        Study, 239
Subcultural theories, 138
Subjective strains, 118
Substance abuse, 167
Substance careers and biosocial
    life-course theory, 169–173
Substance dependence, 168
Substance use careers
    adolescence and, 172
    antisocial behavior and, 167–174
    biosocial life-course theory and,
        169–173
    childhood and, 171–172
    conclusion, 173–174
    introduction, 167–169
    prenatal/early development
        factors, 171
    reward pathways and, 170–171
    vulnerability to, 169–170
Sudden infant death syndrome
    (SIDS), 52
Sutherland, E., 28, 276
Sykes, G. M., 277
Symbolic interactionism
    conclusion, 282–283
    criminal careers and, 278–281
    criminological theories and, 275–277
    defined, 272
    introduction, 271–272
    labeling theory and, 278
    propositions, 273
Synapses, 61
Synaptogenesis, 60–61
Systematic desensitization, 163
Systemic reductionism, 28

T

Tactile stimulation and
    breastfeeding, 63
Tannenbaum, F., 278
Taylor, T. J., 210, 221
Teachers, 182–183
Television violence, 149
Temperament, 285
    applying to rehabilitation, 291–294
    defined, 290
    extremes of, 294
Temperament and Character
    Inventory, 291
Teplin, L. A., 167
Teratogens, 46, 47
Terracciano, A., 170
Terrie Moffitt's developmental theory
    of criminal behavior, 60
Testosterone, 13, 88
    antisocial behavior, 14
    levels in youth, 33
    linked to chronic offending, 30
    pubertal surge of, 66
Tewksbury, R., 254
Thaxton, S., 137
Thomas Theorem, 273
Thomas, W. I., 273
Thornberry, T. P., 202, 204, 206–207,
    209, 222, 239–240, 245
Thornhill, N. W., 314
Thornhill, R., 314
Thrasher, F., 215, 225
Thumb, basic rule of, 32
Thyroxin, 15
Tibbetts, S. G., 27, 35
Tita, G., 203
Tittle, C. R., 142
Tobacco. *See also* Cigarette smoking;
    Nicotine exposure
    compounds, 50
    during pregnancy, 48–50
    as prenatal insults, 48–50
    secondhand smoke, 54
Toch, H., 263
Topalli, V., 277
Tough convict, 287
Toxins, 13–14
    environmental, 13
    eradication, 93
    neurotoxins, 47–48
Traditional criminologists, 69
Traits, 263
Trajectories, 298
Tremblay, R. E., 221
Triggers, 288

Trulson, C. R., 297
Tsai, H., 49
Turner, M. G., 33, 254
Turner, R. J., 137
Type of group model, 203
Type of person model, 203

U

Ulmer, J. T., 271, 280–281
Unemployment, 235–236
    street youth, 125–127
Uniform Crime Reports (UCR), 134
United States, illicit drug use in, 52
Uterogestation, 62

V

Valdez, A., 221
van Dulmen, M. H. M., 177
Van Winkle, B., 205, 208, 210
Vandalism, 238
Varano, S. P., 223
Variable-centered techniques, 178
Vaughn, M. G., 99, 167, 174, 220, 297
Venables, P. H., 4–6
Verbal aggression, 150
Vicarious violence, 64
Victimization, 204, 208–210
    data, 134–135
    female victimization trajectories,
        180–182
    male victimization trajectories, 180
    street youth, 128–129
Victims, 134, 262
    offender relationship, 315
Video games
    cross-sectional studies, 156
    experimental studies, 155–156
    longitudinal studies, 156–157
    media violence, 155–157
Vigil, J. D., 205, 208
Violence. *See also* Intimate partner
    violence; Media violence;
    Partner-directed violence;
    Violent crime
    aggression and, 64, 150
    brain and, 63–64
    defined, 150
    desensitization to, 163
    evolutionary purpose, 63
    exposure to, 63–64, 177–184
    gangs and, 202–206
    life course and, 202–206
    National Commission on the
        Causes and Prevention of
        Violence, 149

Recent Exposure to Violence
Scale, 179
Severity of Violence Against
Women Scale, 38
systematic desensitization to, 163
television, 149
vicarious, 64
youth, 225
Violence exposure
conclusion, 184
correlates/antecedents of, 177–179
defined, 177
developmental trajectories, 177–184
group-based modeling of middle
childhood, 179–183
introduction, 177
Recent Exposure to Violence
Scale, 179
Violent crime, 8, 73–74
Violentization, 280
Virulency, 280
Vitaro, F., 221
Vocabularies of motive, 277
Volavka, J., 5
Volkow, N. D., 9, 170

W

Wadsworth, M. E. J., 4
Wakschlag, L. S., 220
Walder, L. O., 263
Walsh, A., 59, 99
Wang, X., 49
Ward, J. T., 222
Webb, V. J., 222
Weisburd, D. L., 225
Wellbeing threats, 59, 64
Werner, E. E., 8
Whites, 31–32, 194–197, 254
African Americans and, 138
arrest rate for crimes, 134
juvenile offenders, 135
*Why They Kill: Discoveries of a Maverick
Criminologist* (Rhodes), 280
Wiebe, R., 99
Wilcox, P., 223
Williams, M., 5
Wilson, M. I., 85, 315
Wilson, W. J., 138
Winfree, L. T., Jr., 221
Wirth, L., 276

Woermann, F. G., 10
Wolfe, S. E., 256
Wolfgang, M. E., 138
Womb, 46–47, 62. *See also* Embryo
fetal development in, 46–47
Women, 35–36, 62. *See also* Homicide;
Intimate partner; Intimate
partner violence; Mothers;
Partner-directed violence; Sexual
coercion; Womb
Women's Experience with Battering
Scale, 38
Wright, J. P., 45, 59, 76, 99, 220,
255, 297

Y

Yablonsky, L., 208
Yang, Y., 3, 10
Youth violence, 225
Yun, I., 59

Z

Zatz, M., 142